SYMPOSIA OF THE ZOOLOGICAL SOCIETY OF LONDON NUMBER 69

Miniature Vertebrates

The Implications of Small Body Size

SYMPOSIA OF THE ZOOLOGICAL SOCIETY OF LONDON NUMBER 69

Miniature Vertebrates

The Implications of Small Body Size

The Proceedings of a Symposium

held at the Zoological Society of London

on 11th and 12th November 1994

Edited by P. J. MILLER

School of Biological Sciences,
University of Bristol

Published for THE ZOOLOGICAL SOCIETY OF LONDON

by CLARENDON PRESS . OXFORD

1996

Oxford University Press, Walton Street Oxford OX2 6DP

Oxford New York
Athens Auckland Bangkok Bombay
Calcutta Cape Town Dar es Salaam Delhi
Florence Hong Kong Istanbul Karachi
Kuala Lumpur Madras Madrid Melbourne
Mexico City Nairobi Paris Singapore
Taipei Tokyo Toronto

and associated companies in
Berlin Ibadan

Oxford is a trade mark of Oxford University Press

Published in the United States
by Oxford University Press Inc., New York

A catalogue record for this book is available from the British Library

Library of Congress Cataloging in Publication Data

Miniature Vertebrates : the implications of small body size : the proceedings of a symposium held at the Zoological Society of London on 11th and 12th November 1994 / edited by P.J. Miller. – 1st ed.
(Symposia of the Zoological Society of London : no. 69)
Includes bibliographical references and index.

1. Vertebrates–Size–Congresses. 2. Vertebrates–Ecology–Congresses.
I. Miller, P. J. (Peter James) II. Zoological Society of London. III. Series.
QL1.Z733 no. 69 [QL739]
591 s–dc20 [596'.04] 95–52659

ISBN 0 19 857787 7 (Hbk)

Typeset by Hewer Text Composition Services, Edinburgh

Printed in Great Britain by Biddles Ltd, Guildford, Surrey

Preface

The present volume results from a Zoological Society of London symposium on 'miniature vertebrates: the implications of small size', held on 11 and 12 November 1994. It is probably the first book to be dedicated to the biology of small body size across the entire vertebrate spectrum from fishes to mammals, with problems and advantages being considered from the major aspects of functional allometry and ecological role in the natural world.

The size of objects in relation to our own physique is a primary factor in human perception and assessment of our surroundings. Depending on their dimensions, other vertebrates may prompt in our subconscious one of three possible responses: feeding, associating, or fleeing. Human interest in the vertebrates considered in the present volume might derive essentially from the first of these instincts. With *Homo sapiens* essentially a strolling omnivorous primate, gathering small food items, perhaps it is not surprising that most of the vertebrates regarded as small for the present academic survey also fall into the category of having a girth less than the diameter of the average human mouth and a mass no greater than that of a typical brown rat. However, given this as a practical guide, inclusion as a miniature vertebrate can also follow a more flexible definition of miniature as 'representation on a small scale', in either absolute or relative terms. The recognition of finer size categories within the overall situation of being small and the possibility of defining 'miniature' on the basis of radical genomic, developmental or ecological shifts are ongoing lines of research. Some of the conventional vertebrate classes such as the amphibians, reptiles, and birds (with notable extant exceptions like crocodiles and ostriches) comprise species which are virtually all small by the present concept. Within these classes, attention in the symposium has tended to concentrate on the very smallest forms. In fish and mammals, discussion has included relatively small representatives but not exclusively the tiniest; in these groups, despite many familiar large representatives, the majority of species are also appreciably smaller than humans.

Our biological knowledge of vertebrates at the lower end of the size range is inevitably polarized into a great deal for some species and very little, at best, for the majority. A limited number of small vertebrates, such as the house mouse and the three-spined stickleback, have been intensively investigated, because their physiological resilience and small size permit the maintenance of large numbers under laboratory conditions and often a rapid succession of generations gives the opportunity to follow genetic change. Abundant and accessible populations of other species in the wild have also yielded much information. However, many other small vertebrates are more difficult to find and investigate, especially in contrast to popular larger species, simply because a small size lends itself to concealment or precise habitat requirements are difficult to reproduce under experimentation. For such species, for example caecilian amphibians, most information is limited to systematics and morphology.

Scientific study of the phenomenon of being small (miniaturism) and the evolutionary processes (miniaturization) behind this condition can be seen to address two main topics. These are essentially functional and ecological: the implications of small size for fitness—physical or Darwinian—with respect to functional limitations and ecological opportunity. The very smallest species raise intriguing questions of how the basic complexity of a vertebrate body can operate and have evolutionary success within a framework less than that of the largest beetle.

For the physiology of scaling effects on small vertebrate function, the symposium provides an overview of potential constraints, followed by more detailed discussion of interface problems in water, competence in cephalization, and, for homeotherms, the functional complications of seasonality on land, taking to the air and locating an aerial food resource in the dark. The topics of function and ecological interaction are linked by the developmental genetics of size reduction and the role of size in species diversification which concluded the first day of the symposium.

Scaling down has permitted great extension of the limits of adaptive radiation among the vertebrates, opening up new resources in what becomes a vastly expanded world. Whatever functional burdens may be identified, above limiting sizes within the classes, small vertebrates can achieve impressive ecological success in terms of abundance and exploitation of specialized ecological niches only available for the tiny – scuttling over sea fans, living in a termite colony, hunting spiders under the snow, or rearing their young in bromeliad reservoirs.

In the second part of the symposium, discussion of the ecology of small vertebrates is taken further with a consideration of whether functional allometries might be expected to exert a strong influence on the temporal pattern of energy expenditure over life in an ecosystem, but a survey of life-history features in mammals shows a range in adaptive response little constrained by size. Other contributions deal with the occurrence and ecology of small species within the teleost fishes, the amphibians, and the reptilian sauropsids, culminating in the hummingbirds, exemplifying a radiation in adaptation and co-evolution linked with small body size. Apart from the many species which are small in absolute terms, size reduction in phylogeny has occurred in a number of situations to produce forms which are relatively dwarfed by comparison to ancestral stock. This phenomenon is discussed here in the context of 'miniature' deer, elephants, and mammoths.

Small vertebrates interact with mankind in a number of ways, usually complex and economically advantageous or deleterious. The tropical fish trade may be compared with the depredations of rats and mice. Another relationship is an emotional one and the psychology of associations with small vertebrates for the pleasure of companionship, rather than intellectual satisfaction, is a suitable conclusion to the present volume.

Big game and large objects attract more human attention than the small, but diminutive size obviously has selective advantage in a variety of natural situations.

Other than in global catastrophe, the expanded world of small vertebrates offers many more niches for survival and evolutionary persistence than those in which African elephants or Indian tigers might find comparable security. After all, miniature mammals came through when Cretaceous giants failed.

Finally, my thanks go to all contributors and participants in the symposium and to the Zoological Society of London, for supporting this review of miniature vertebrates. In particular, I must thank Ms Unity McDonnell for all her organizational and editorial work.

Bristol
September 1995

P. J. M.

CONTENTS

Function

Biophysical problems of small size in vertebrates

R. McN. ALEXANDER

Interface areas in small fish

I. J. HARRISON

Miniaturization in tetrapods: consequences for skull morphology

OLIVIER RIEPPEL

Energetics and the evolution of body size in small terrestrial mammals

J. R. SPEAKMAN

Biomechanical constraints on size in flying vertebrates

JEREMY M. V. RAYNER

Does echolocation constrain the evolution of body size in bats?

GARETH JONES

The genetics and physiology of size reduction in mice

IAN M. HASTINGS

Speciation in small mammals

J. B. SEARLE

Ecology

Miniature mammals: life-history strategies and macroevolution

ANDY PURVIS & PAUL H. HARVEY

The functional ecology of small fish: some opportunities and consequences

P. J. MILLER

Small size in amphibians—its ecological and evolutionary implications

B. T. CLARKE

Ecology of small reptile-grade sauropsids

R. A. AVERY

Body size and the ecology of hummingbirds

PETER A. COTTON

Ecology of very small terrestrial mammals

SARA CHURCHFIELD

Dwarfing in island elephants and deer: processes in relation to time of isolation

A. M. LISTER

Contributors

ALEXANDER, R. MCN., Department of Biology, University of Leeds, Leeds LS2 9JT, UK.

AVERY, R. A., School of Biological Sciences, University of Bristol, Woodland Road, Bristol BS8 1UG, UK.

CHURCHFIELD, SARA, Division of Life Sciences, King's College London, Campden Hill Road, London W8 7AH, UK.

CLARKE, B. T., Department of Zoology, The Natural History Museum, Cromwell Road, London SW7 5BD, UK.

COTTON, PETER A., Edward Grey Institute, Department of Zoology, University of Oxford, South Parks Road, Oxford OX1 3PS, UK.

HARRISON, I. J., The National Fishing Heritage Centre, Alexandra Dock, Great Grimsby, South Humberside DN13 1UZ, UK.

HARVEY, PAUL H., Department of Zoology, University of Oxford, South Parks Road, Oxford OX1 3PS, UK.

HASTINGS, IAN M., Institute of Cell, Animal & Population Biology, University of Edinburgh, West Mains Road, Edinburgh EH9 3JT, UK.

JONES, GARETH, School of Biological Sciences, University of Bristol, Woodland Road, Bristol BS8 1UG, UK.

LISTER, A. M., Department of Biology, University College London, Gower Street, London WC1E 6BT, UK.

MCBRIDE, ANNE, Anthrozoology Institute, University of Southampton, Bassett Crescent East, Southampton SO16 7PX, UK.

MILLER, P. J., School of Biological Sciences, University of Bristol, Woodland Road, Bristol BS8 1UG, UK.

PURVIS, ANDY, Department of Zoology, University of Oxford, South Parks Road, Oxford OX1 3PS, UK; *present address* Department of Biology, Imperial College, Silwood Park, Ascot SL5 7PY, UK.

RAYNER, JEREMY M. V., School of Biological Sciences, University of Bristol, Woodland Road, Bristol BS8 1UG, UK.

RIEPPEL, OLIVIER, Department of Geology, The Field Museum, Roosevelt Road at Lake Shore Drive, Chicago, IL 60605–2496, USA.

SEARLE, J. B., Department of Biology, University of York, PO Box 373, York YO1 5YW, UK.

SPEAKMAN, J. R., Department of Zoology, University of Aberdeen, Aberdeen AB9 2TN, UK.

Organizer of symposium

DR P. J. MILLER, School of Biological Sciences, University of Bristol, Bristol BS8 1UG, UK.

Chairmen of sessions

PROFESSOR R. MCN. ALEXANDER, FRS, Department of Biology, University of Leeds, Leeds LS2 9JT, UK.

DR P. H. HARVEY, FRS, Department of Zoology, University of Oxford, Oxford OX1 3PS, UK.

DR P. J. MILLER, School of Biological Sciences, University of Bristol, Bristol BS8 1UG, UK.

PROFESSOR J. M. V. RAYNER, School of Biological Sciences, University of Bristol, Bristol BS8 1UG, UK.

Function

Symp. zool. Soc. Lond. (1996) No. 69: 3–14

Biophysical problems of small size in vertebrates

R. McN. ALEXANDER

Department of Biology
University of Leeds
Leeds LS2 9JT, UK

Synopsis

Geometrically similar animals of different sizes have areas proportional to length squared and volumes to length cubed. Metabolic rates of similar animals of different sizes tend to be proportional to (body mass)$^{0.75}$. Many important consequences follow from these simple scaling rules. Small endotherms require very high metabolic rates in cold conditions, small ectotherms heat up and cool down quickly, small animals in dry environments lose water rapidly by evaporation both through the skin and in the breath, maximum rates of oxygen uptake in lungs or gills of small animals may be smaller multiples of basal metabolic rate than for larger animals, and small birds and bats fly slowly.

Many of these consequences of small size involve processes occurring faster in smaller animals. Whether this presents a problem or not may depend on comparisons in which account is taken of the higher rates of physiological processes generally in small animals.

In other cases, problems may arise because animals of all sizes have similar-sized cells. Smaller vertebrates have room for fewer cells in their brains, which presumably have less processing power. Their muscles contain fewer fibres, so cannot be so precisely controlled.

Problems depending on the wavelength of light make small eyes less acute than large ones can be.

Introduction

In evolution, as in manufacturing industry, miniaturization presents problems. Many of these are biophysical in nature, relating, for example to exchange of heat or water with the environment or to the optics of eyes. In some cases, the problems arise from geometric scaling rules: animals of similar shape but different sizes have surface areas proportional to the squares of their lengths, but volumes proportional to length cubed. In other cases the problems are due to the fixed sizes of body components, either cells or molecules: if cells cannot be made smaller than some lower limit of size, fewer cells can be fitted into a small animal than into

ZOOLOGICAL SYMPOSIUM No. 69
ISBN 0–19–857787–7

a big one. In yet other cases, the problem relates to the wavelength of solar radiation. This paper reviews these various kinds of problem. Subsequent papers in this symposium will show how they have affected the evolution of small vertebrates.

Endotherms

One of the most familiar problems of small size concerns maintenance of an elevated body temperature. The smallest vertebrate endotherms are shrews and hummingbirds of approximately 2 g body mass (Brooke & Birkhead 1991; Macdonald 1984; Cotton, this volume, pp. 239–258; Churchfield, this volume, pp. 259–276). It is widely agreed that smaller mammals and birds would require excessive metabolic rates to maintain the body temperatures of 37–40°C which are usual for their classes.

The simplistic argument found in elementary textbooks is that the rate of metabolic heat production should be proportional to the volume of the body and the rate of heat loss to the surface area. Smaller animals have larger ratios of area to volume, so may find it difficult to produce heat fast enough for temperature maintenance. That argument may lead to the right conclusion, but is misleadingly simple.

First, basal metabolic rates (M) of animals are not simply proportional to body volume (or mass), but to a power of body mass (m), so that

$$M = am^b \quad (1)$$

where a and b are constants. The factor a varies between taxa (for example, it is approximately 10 times lower for reptiles at 39°C than for mammals), but exponent b is remarkably uniform, approximately 0.75 in each vertebrate class (Calder 1984). Attempts have been made to explain this exponent (for example, by Spatz 1991) but none known to me explains successfully why it is 0.75 for ectotherms as well as endotherms and for protozoans as well as metazoans. For the present we can only accept it as an empirical scaling rule.

Secondly, the rate of heat loss (H) from an animal is not simply proportional to surface area (A). If the temperatures of the animal's body and of the environment are T_{body} and T_{env}, respectively, then

$$H = (kA/R)(T_{body} - T_{env}) \quad (2)$$

where k is the volumetric heat capacity of air and R is the resistance to heat loss (see Campbell 1977). In the case of a bird or mammal, the lost heat is first conducted through an insulating layer (feathers, fur, or blubber), then radiated or convected from the outer body surface. The overall resistance to heat loss is thus made up of an array of three resistances, a conductive resistance (R_{cond}) plus radiative and convective resistances (R_{rad} and R_{conv}) in parallel (Fig. 1a). By analogy with calculations of electrical resistance,

$$R = R_{cond} + 1/[(1/R_{rad}) + (1/R_{conv})] \tag{3}$$

R_{cond} is proportional to insulation thickness, which for geometrically similar animals is proportional to diameter or to $m^{0.33}$. However, R_{rad} is proportional to m^0 and R_{conv} approximately to the square root of diameter or to $m^{0.17}$ (Campbell 1977). For large animals, R_{cond} is much larger than the other resistances and the rate of heat loss depends on it. For small animals, however, R_{cond} is the smallest of the resistances and R depends mainly on the smaller of R_{rad} and R_{conv}, that is on R_{conv}. Thus, for large animals, R approaches proportionality to $m^{0.33}$ and for small animals to $m^{0.17}$. Thus if surface area A is proportional to $m^{0.67}$, the rate of heat loss (equation 2) should approach proportionality to $m^{0.33}$ for large animals and to $m^{0.50}$ for small ones.

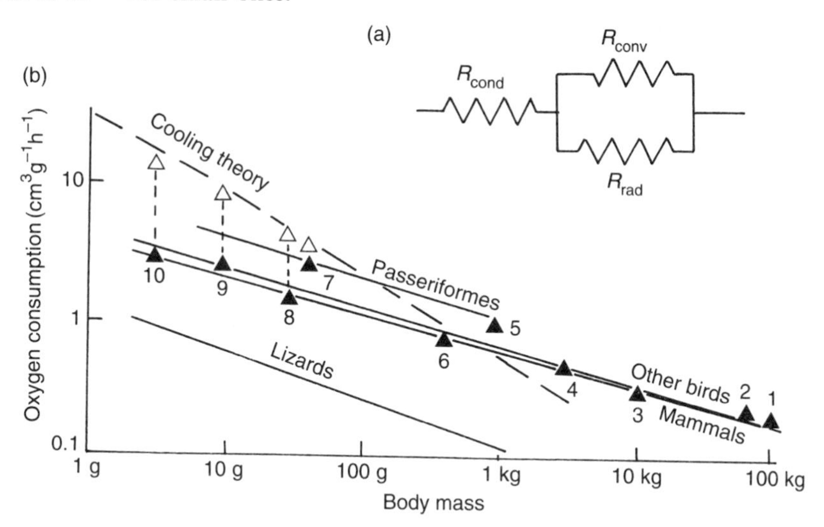

Fig. 1 (a) A diagram showing how the conductive, convective, and radiative resistances to heat loss from an endotherm interact. (b) A graph of metabolic rate against body mass for vertebrate animals. Continuous lines and filled points show basal metabolic rates (or in the case of lizards, resting metabolic rates at 37°C). Hollow points show metabolic rates of some of the species at 10°C. The broken line shows metabolic rates calculated theoretically (using equations 2 and 3) as needed to keep an endotherm's body temperature at 37°C, in an environment at 10°C. Points 5 and 7 refer to passerine birds, 1, 3, and 10 to non-passerine birds, and the remainder to mammals. From Alexander (1981).

Thus the metabolic rate needed to balance heat loss in a cool environment is proportional to body mass to some power between 0.33 and 0.50, but basal metabolic rates are proportional to $(\text{mass})^{0.75}$. The smaller the animal, the bigger the multiple of basal metabolism which must be used to maintain body temperature in a given environment (Fig. 1b). Required metabolic rates might have to be impracticably high for vertebrate endotherms of less than 1 g. However, it should be noted that many insects of less than 1 g function as endotherms when flying, maintaining mammal-like body temperatures (Heinrich 1993).

Small endotherms such as hummingbirds and shrews save energy when inactive by reducing their metabolic rate and allowing body temperature to fall. They can warm up again faster than larger mammals could do, for the following reason. Consider a torpid animal, at equilibrium with its environment. If it suddenly increases its metabolic rate, its temperature will start to rise at a rate proportional to (metabolic rate)/(body mass). If metabolic rate is proportional to $m^{0.75}$, the rate of rise of temperature will be proportional to $m^{-0.25}$.

Ectotherms

Now consider an ectotherm such as a lizard, warming in a warm environment. It gains heat at a rate $-H$ (equation 2) and its temperature T_{body} will rise at a rate given by

$$\begin{aligned} \mathrm{d}T_{body}/\mathrm{d}t &= -H/mC \\ &= (kA/mCR)(T_{env} - T_{body}) \end{aligned} \tag{4}$$

where C is the specific heat capacity of the body. This is exponential warming with a time constant (mCR/kA). If R is proportional to mass to some power between 0.17 and 0.33 (see above) and A is proportional to $m^{0.67}$, heating (and cooling) time constants should be proportional to mass to some power between 0.50 and 0.67. Alexander (1989: fig. 7.4) shows that this is approximately so for lizards.

Lizards often bask in the sun, to heat themselves by solar radiation. To describe this we need a more complicated version of equation (4):

$$\mathrm{d}T_{body}/\mathrm{d}t = (A/mC)[S + (k/R)(T_{env} - T_{body})] \tag{5}$$

where S is the solar flux density averaged over the whole body surface. If initially $T_{body} = T_{env}$, the rate of increase of temperature is As/mC, which is larger for small animals, because A is proportional to $m^{0.67}$ while S and C should be the same for animals of all sizes. The lizard will reach equilibrium ($\mathrm{d}T_{body}/\mathrm{d}t = 0$) when

$$T_{body} = T_{env} + SR/k \tag{6}$$

For small animals without feathers or fur, the conductive resistance to heat loss can be ignored and R can be expected to approximate to R_{conv}, proportional to $m^{0.17}$ (see above). Thus, small ectotherms should reach lower temperatures in the sun than large ones, despite heating faster. In some circumstances this may be an advantage because high temperature may be a danger. Lizards living in hot environments cannot stay out in the sun for long, but have to retire frequently to cool burrows (Bradshaw 1986).

Water loss

Terrestrial animals lose water by evaporation from the surface of the skin. The rate E can be calculated from the equation

$$E = (A/R_{vap})\ (\rho_{sat} - \rho_{atm}) \tag{7}$$

(see Campbell 1977). Here A is the surface area of the body, R_{vap} is the resistance to water vapour loss, ρ_{sat} is the saturation vapour density of air at body temperature and ρ_{atm} is the vapour density in the atmosphere. Water vapour diffuses through the skin, and then is convected away from the body surface. Thus, R_{vap} is the sum of two resistances in series: the skin resistance which can be expected to be proportional to skin thickness or to $m^{0.33}$ for geometrically similar animals and a resistance to convection which, as in the case of heat convection, should be proportional to $m^{0.17}$. By an argument like the one for heat loss of endotherms (above) skin resistance can be expected to dominate for large animals and convective resistance for small ones. Thus, if A is proportional to $m^{0.67}$ we can expect the rate of water loss per unit body mass (E/m) to be proportional to some power of body mass between $m^{-0.50}$ and $m^{-0.67}$: in a dry atmosphere, small animals will desiccate much faster than large ones. Measured values of (E/m) are approximately proportional to $m^{-0.5}$ for lizards of 0.3 g to 1 kg from all but the driest environments (Mautz 1982).

That argument refers to water loss from the skin. Water is also lost by evaporation from lungs. If the air breathed out is water-saturated, different-sized animals at the same temperature will lose water at rates proportional to the volume of air breathed and so to the volume of oxygen used. If metabolic rate is proportional to $m^{0.75}$, the rate of water loss by evaporation from the lungs, per unit body mass, should be proportional to $m^{-0.25}$, reinforcing the conclusion that small animals should dry up faster than large ones.

Amphibians and reptiles generally lose water faster from the skin than from the lungs at low temperatures, though the reverse was true of a desert lizard with highly waterproof skin, at 40°C (Bentley & Schmidt-Nielsen 1966). Birds and mammals, with faster metabolism and higher body temperatures, might be expected to lose water by evaporation mainly from the lungs, but some birds lose more from the skin (Skadhauge 1981). Measured rates of water loss per unit body mass, for birds and mammals, seem to be very roughly proportional to $m^{-0.25}$ (Calder 1984).

Note that the above refers to water loss in the absence of heat stress, i.e. when the animal is neither sweating nor panting.

Respiration

In both the lungs and gills, oxygen enters the bloodstream by diffusion. The rate of diffusion Q depends on the diffusion constant D, the area of respiratory surface A_{resp}, the thickness d of the barrier across which diffusion occurs, and the partial pressures of oxygen in the medium and the blood, P_{med} and P_{blood}, so that

$$Q = (DA_{resp}/d)\ (P_{med} - P_{blood}) \tag{8}$$

(Schmidt-Nielsen 1990). If animals of different sizes had geometrically similar respiratory organs, A_{resp} would be proportional to $m^{0.67}$ and d to $m^{0.33}$, so Q would be proportional to $m^{0.33}$. Since the metabolic rate M is proportional to $m^{0.75}$, small animals would be in an advantageous position: the ratio of the rate at which they need to take up oxygen to the maximum rate at which they could take it up (M/Q) would be smaller for them than for large animals.

However, the respiratory organs of small animals cannot be geometrically similar to those of large ones. Consider the lungs first. The diffusion distance d is the sum of the thickness of the tissue barrier between air and blood and some fraction of the diameter of the blood capillaries. Both these depend on cell dimensions, since blood capillaries must be large enough for blood corpuscles to pass through. Animals of different sizes have similar-sized cells, as discussed in a later section. More specifically, mammals ranging from shrews to whales have red blood corpuscles of approximately 7 μm diameter (Altman & Dittmer 1961) and mammals ranging from shrews to cows have a tissue barrier 0.3–0.6 μm thick between air and blood in the lungs (Gehr *et al.* 1981). Thus, diffusion distance d is approximately the same for mammals of all sizes. Also, the pressure required to inflate an alveolus of radius r is $2\gamma/r$, where γ is the surface tension of the surfactant film on the inner surface (see Alexander 1983). Small animals cannot have very small alveoli because, if they had, they could not inflate their lungs. The alveoli of mammals vary in size, but without any apparent relationship to body mass (Gehr *et al.* 1981). If different-sized mammals have alveoli of equal size and lungs the same proportion of body volume, A_{resp} must be proportional to m. Hence, with d constant, equation (8) predicts that Q will be proportional to m, making M/Q (the ratio of metabolic rate to the rate at which oxygen can be taken up) proportional to $m^{-0.25}$. Excessively small mammals could not take up oxygen fast enough.

A set of gills can be thought of as an array of parallel tubes, with tubes carrying water interspersed among tubes carrying blood. If the oxygen capacity of the blood is constant, the volumetric flow rates of water and blood can be expected to be in the same ratio, in fishes of different sizes. The diameter of the capillaries is approximately the same in fish of all sizes so if the ratio of the pressure needed to drive the water, to the pressure needed to drive the blood, is to be the same in all sizes of fishes, they should all have water channels of the same diameter. (For Poiseuille's equation, which relates pressure to flow rate in tubes, see Alexander

(1983).) This is observed very nearly to be the case: the water-channel width is proportional to $m^{0.09}$ and $m^{0.02}$ in ontogenic series of two species of teleost (Hughes 1977).

If the water channels have the same width in fish of different sizes, the diffusion distance d (equation 8) is constant and (if the gills occupy a constant fraction of body volume) the gas exchange area A_{resp} is proportional to m. This leads, as in the discussion of lungs, to the conclusion that the maximum rate of gas uptake Q is proportional to m and the ratio of metabolic rate to Q is proportional to $m^{-0.25}$. Excessively small fishes, like excessively small mammals, might not be able to take up oxygen fast enough.

There is more extensive discussion than this of the scaling of gills and lungs in Perry (1992).

Osmotic regulation

The osmotic concentration of the blood in teleost fishes is substantially lower than that of sea water but also substantially higher than that of fresh water. Consequently, water tends to move osmotically out of marine fishes and it tends to move into freshwater ones. The skin is highly waterproof but the gills much less so: it is difficult to see how they could be made impermeable to water while remaining highly permeable to oxygen. Consequently, much more osmotic movement of water occurs through the gills than through the skin (Schmidt-Nielsen 1990).

If the gills of different-sized fishes are adapted to take up oxygen at rates corresponding to metabolic rate (that is, proportional to $m^{0.75}$) then the rate at which water leaves or enters the gills osmotically (diffusing through the same barrier as the oxygen) should also be proportional to $m^{0.75}$. Consequently, the rate at which energy has to be used for osmoregulation can be expected to be the same fraction of metabolic rate, in different-sized fishes. Small size presents no obvious problem in osmoregulation.

Measured flux rates of water into teleost fishes are proportional to body mass to powers between 0.83 and 0.94, which are rather higher than predicted by the above simple argument (Evans 1969).

Digestion

Geometrically similar animals would have gut walls with areas (through which foodstuffs could be absorbed) proportional to (body mass)$^{0.67}$. This does not quite match the proportionality of metabolic rate to (mass)$^{0.75}$, so might be expected to lead to problems for large animals. That would not concern us in this paper, but in any case the problem may not arise if guts have fractal surfaces (Pennycuick 1992).

There is a problem, however, for small herbivores that depend on microbial fermentation. No known vertebrate has enzymes capable of breaking down cellulose, but microbes in the guts of most mammalian herbivores (and of a few members of other vertebrate classes, as noted by Foley *et al.* (1992)) break down cellulose and some other indigestible materials to fatty acids which the animal can absorb. The metabolic rates of these herbivores, as of other vertebrates, are about proportional to $(\text{body mass})^{0.75}$: small herbivores metabolize faster, relative to body mass, than large ones. However, there is no reason to expect the rate of fermentation, per unit volume of gut, to be different in herbivores of different sizes. This suggests that it will be more difficult for small herbivores than for large ones to satisfy their energy requirements.

Some plant foods (notably soft fruits, grains, and storage roots) contain large proportions of digestible foodstuffs such as sugars, starch, fat, and protein. However, mature leaves (both of grasses and of trees) contain large proportions of cellulose and other indigestible materials and very little that can be digested by vertebrate enzymes. It seems unlikely that a very small herbivorous mammal could do well on a diet of mature leaves. Few mammals below 15 kg body mass feed unselectively on grasses or other leaves and very few below 0.7 kg include much leaf in their diets (Cork 1994). The most notable exceptions are voles of less than 0.1 kg, with highly specialized guts (Hume 1994).

Flight

Small birds generally fly more slowly than large ones (Norberg & Rayner 1991). This is to be expected because optimum flight speeds are proportional to the square root of wing loading. The wing loading of an aircraft is the craft's weight divided by the wing area, so for geometrically similar structures made of the same materials it is proportional to $(\text{body mass})/(\text{body mass})^{0.67} = (\text{body mass})^{0.33}$. Hence speeds should be proportional to $(\text{body mass})^{0.17}$. The optimum speeds in question are the speed at which flight needs least power and the rather higher speed at which the work required to travel unit distance is least (Norberg 1990).

A small bird or bat could acquire high wing loading and fly fast, if it evolved very small wings. However, if this were the case, its body would be very large compared to the wings, parasite drag (the drag on the body) would be relatively high, and flight performance would be poor.

Low flying speed may be a problem for animals flying in windy conditions. For example, slope soaring (commonly used by sea birds) depends on being able to glide faster than the wind (Alexander 1982). Accordingly, all slope-soaring birds are moderate to large in size.

Brains and eyes

The problems discussed in this section depend on the uniformity of the sizes of cells and of the wavelengths of solar radiation. Animals of all sizes have cells of the order of 10 μm diameter. Cells cannot be smaller than some minimum size because the nucleus must be large enough to contain the DNA of the genome (see Roth,

Rottluff, & Linke 1988) and the cytoplasm must contain the organelles needed for its metabolism. They probably cannot be larger than some maximum size because of the difficulty of controlling parts of the cell that are excessively far from the nucleus. This seems to be confirmed by the presence of multiple nuclei in exceptionally large cells such as giant amoebae (*Pelomyxa*).

Because different-sized animals are built from similar-sized cells, smaller animals have fewer cells. In some organs this may matter little, but there are others whose performance is critically dependent on the number of constituent cells. In particular, a brain built of a small number of cells cannot be expected to perform functions as complex as those achieved by a brain built from more cells. A shrew cannot accommodate a brain with as many cells as that of a whale and cannot be expected to match the whale's brain power.

The sizes of brains presumably tend to be optimized by evolution. A large brain confers the benefit of being able to perform complex operations, but must also exact costs. One of these must be the cost of metabolism, which may be quite large: for example, though the human brain is only 2% of body mass, it accounts for 16% of our basal metabolic rate (Schmidt-Nielsen 1990).

The optimization may work like this. Consider an animal of mass m with a brain of mass Bm. The net advantage G given by a brain of this size is the difference between a benefit $a(Bm)^b$ reflecting the complexity of the tasks it can perform and a cost $c.Bm.m^{-0.25} = cBm^{0.75}$ equal to its metabolic rate (assuming that the metabolic rate per unit mass of brain tissue, like that of the whole animal, is proportional to $m^{-0.25}$), so that

$$G = a(Bm)^b - cBm^{0.75} \quad (9)$$

where a, b and c are constants. If $b < 1$, G has a maximum value for a particular value of B, at which dG/dB is zero. Measurements of brain masses of similar vertebrates of different sizes show that relative brain mass B tends to be proportional to some power of body mass between -0.2 and -0.5. If this is the solution to the optimization problem represented by equation (9), the exponent b lies between 0.5 and 0.7. However, it is by no means clear that the optimization problem has been correctly formulated. The function of this paragraph is simply to show that an optimization argument could explain the scaling of brain size.

Small animals must have small organs as well as small brains. Consequently, they have fewer receptor cells and the quantity of information available to the brain is limited, as well as the brain's processing power. The performance of eyes depends less on the diameters of receptor cell bodies than on those of their rod and cone processes. For optical reasons (waveguide effects) these would not work well if their diameters were less than approximately 1 μm (Land 1980). Consequently, receptors cannot be centred less than approximately 1 μm apart and images less than this distance apart on the retina cannot be resolved.

A similar limit is set by a different optical effect. Owing to diffraction, the image on the retina of a point source of light is not a simple point of light, but a pattern of concentric bright and dark rings. This makes it impossible to distinguish images less than about λF apart, where λ is the wavelength of light and F is the F-number, the ratio of the focal length to the aperture (Land 1980). Animals are dependent for sight on solar radiation, whose energy peaks at a wavelength of approximately 0.5 μm (green light). It would be difficult to design an eye with an F-number less than approximately 2. Consequently, images less than approximately 1 μm apart on the retina cannot be distinguished and eyes must be large if they are to see much detail. Eyes must tend to evolve to optimal size, balancing the benefits of the finer detail that a larger eye can see against the cost of large size, whatever that may be. Smaller animals have relatively larger eyes. For example, Davis (1962) found that the masses of the eyes of Felidae were proportional to (body mass)$^{0.43}$.

Small size may also affect the control of muscles. Motor units in the medial gastrocnemius muscle of cats consist of an average of approximately 500 muscle fibres (Burke & Tsairis 1973). Compare a 2 kg cat with a 2 g shrew. The shrew is 1/1000 of the mass of the cat, so its muscle fibres can be expected to be only 1/10 as long. If that is true, if the fibres have the same diameter as in the cat and if the number of motor units is the same in both animals, the shrew will have an average of only five fibres in each motor unit. A still smaller mammal might have to make do with fewer motor units, because it had even fewer muscle fibres.

Vertebrates control the forces exerted by their muscles by all-or-none recruitment of different numbers of motor units. If there were fewer motor units, control would be less fine. Unlike vertebrates, arthropods achieve fine control of muscle composed of only a few fibres by graded activation of fibres (as Aidley 1989). Tiny muscles consisting of only a few fibres can be controlled precisely in arthropods, but not in vertebrates.

Discussion

In many cases, the conclusions of this paper are that processes occur faster in smaller animals: for example, that smaller animals heat up faster in the sun and dehydrate faster in dry conditions. In some cases this may present no problem because the tempo of life is faster for smaller animals: they have higher heartbeat frequencies and stride frequencies and shorter gestation periods and lifespans (Calder 1984). The times taken by these and other physiological processes generally scale approximately in proportion to (body mass)$^{0.25}$. In some cases it may be appropriate to adjust the rates we have been discussing to this physiological time scale, before comparing them. In other cases, the significant times are independent of body size—for example, the hot part of the day, during which dehydration may be a danger, lasts as long for small animals as for large ones. In such cases we should compare absolute rates of processes.

Many of the processes we have been considering occur faster in smaller animals, even when the physiological time scale is the basis of comparison. This is true if the rate is related to body mass with an exponent more negative than -0.25, or if the time constant scales with mass with an exponent greater than 0.25. The metabolic rates per unit mass required by endotherms, the rates of heating and cooling of ectotherms, and rates of water loss through the skin all satisfy this criterion, so seem clearly problematic for small animals. However, rates of respiratory water loss are predicted to be proportional to (body mass)$^{-0.25}$. Whether or not this presents a problem for small animals will depend on whether the important period, which the animal must survive, is the hot part of the day (an absolute time period) or the duration of a bout of activity (possibly a physiological time period that is shorter for small animals).

Further discussion of the consequences of difference in size between animals can be found in Schmidt-Nielsen (1984) and McMahon & Bonner (1983).

References

Aidley, D. J. (1989). *The physiology of excitable cells.* (3rd edn). Cambridge University Press, Cambridge.

Alexander, R. McN. (1981). *The chordates.* (2nd edn). Cambridge University Press, Cambridge.

Alexander, R. McN. (1982). *Locomotion of animals.* Blackie, Glasgow.

Alexander, R. McN. (1983). *Animal mechanics.* Blackwell, Oxford.

Alexander, R. McN. (1989). *Dynamics of dinosaurs and other extinct giants.* Columbia University Press, New York.

Altman, P. L. & Dittmer, D. S. (1961). *Blood and other body fluids.* Federation of American Societies for Experimental Biology, Washington.

Bentley, P. J. & Schmidt-Nielsen, K. (1966). Cutaneous water loss in reptiles. *Science* **151**: 1547–1549.

Bradshaw, S. D. (1986). *Ecophysiology of desert reptiles.* Academic Press, Sydney.

Brooke, M. & Birkhead, T. (Eds) (1991). *The Cambridge encyclopedia of ornithology.* Cambridge University Press, Cambridge.

Burke, R. E. & Tsairis, P. (1973). Anatomy and innervation ratios in motor units of cat gastrocnemius. *J. Physiol.* **234**: 749–765.

Calder, W. A. (1984). *Size, function and life history.* Harvard University Press, Cambridge, MA.

Campbell, G. S. (1977). *An introduction to environmental biophysics.* Springer, New York.

Cork, S. J. (1994). Digestive constraints on dietary scope in small and moderately-sized mammals: how much do we really understand? In *The digestive system in mammals: food, form and function*: 337–369. (Eds Chivers, D. J. & Langer, P.). Cambridge University Press, Cambridge.

Davis, D. D. (1962). Allometric relationships in lions vs domestic cats. *Evolution* **16**: 505–514.

Evans, D. H. (1969). Studies on the permeability to water of selected marine, freshwater and euryhaline teleosts. *J. exp. Biol.* **50**: 689–703.

Foley, W. J., Bouskila, A., Shkolnik, A., & Choshniak, I. (1992). Microbial digestion in the herbivorous lizard *Uromastyx aegyptius* (Agamidae). *J. Zool., Lond.* **226**: 387–398.

Gehr, P., Mwangi, D. K., Ammann, A., Maloiy, G. M. O., Taylor, C. R., & Weibel, E. R. (1981). Design of the mammalian respiratory system. V. Scaling morphometric pulmonary diffusing capacity to body mass: wild and domestic mammals. *Respirat. Physiol.* **44**: 61–86.

Heinrich, B. (1993). *The hot-blooded insects: strategies and mechanisms of thermoregulation*. Harvard University Press, Cambridge, MA.
Hughes, G. M. (1977). Dimensions and the respiration of lower vertebrates. In *Scale effects in animal locomotion*: 57–81. (Ed. Pedley, T. J.). Academic Press, London.
Hume, I. D. (1994). Gut morphology, body size and digestive performance in rodents. In *The digestive system in mammals: food, form and function*: 315–323. (Eds Chivers, D. J. & Langer, P.). Cambridge University Press, Cambridge.
Land, M. F. (1980). Optics and vision in invertebrates. In *Handbook of sensory physiology* 7 (**6B**): 471–592. (Ed. Antrum, H.). Springer Verlag, Berlin.
Macdonald, D. (1984). *The encyclopaedia of mammals*. Allen & Unwin, London.
McMahon, T. A. & Bonner, J. T. (1983). *On size and life*. Scientific American Library, New York.
Mautz, W. J. (1982). Patterns of evaporative water loss. In *Biology of the Reptilia* **12C**: 443–481. (Eds Gans, C. & Pough, F. H.). Academic Press, London.
Norberg, U. M. (1990). *Vertebrate flight*. Springer, Berlin.
Norberg, U. M. & Rayner, J. M. V. (1991). Flight. In *The Cambridge encyclopaedia of ornithology*: 53–66. (Eds Brooke, M. & Birkhead, T.). Cambridge University Press, Cambridge.
Pennycuick, C. J. (1992). *Newton rules biology: a physical approach to biological problems*. Oxford University Press, Oxford.
Perry, S. F. (1992). Morphometry of vertebrate gills and lungs: a critical review. In *Oxygen transport in biological systems*: 57–77. (Eds Egginton, S. & Ross, H. F.). Cambridge University Press, Cambridge.
Roth, G., Rottluff, B., & Linke, R. (1988). Miniaturization, genome size and the origin of functional constraints in the visual system of salamanders. *Naturwissenschaften* **75**: 297–304.
Schmidt-Nielsen, K. (1984). *Scaling. Why is animal size so important?* Cambridge University Press, Cambridge.
Schmidt-Nielsen, K. (1990). *Animal physiology: adaptation and environment*. (4th edn). Cambridge University Press, Cambridge.
Skadhauge, E. (1981). *Osmoregulation in birds*. Springer, Berlin.
Spatz, H. C. (1991). Circulation, metabolic rate, and body size in mammals. *J. comp. Physiol. (B)* **161**: 231–236.

Symp. zool. Soc. Lond. (1996) No. 69: 15–45

Interface areas in small fish

I. J. HARRISON

The National Fishing Heritage Centre
Alexandra Dock, Great Grimsby
South Humberside DN31 1UZ, UK

Synopsis

Morphometric relationships between selected body tissues and body mass are analysed for small gobies (Teleostei: Gobiodei). These species show large body surface areas relative to mass, which might be advantageous for facultative cutaneous respiration but potentially disadvantageous for osmoregulation and locomotion. Large body surface areas per unit mass and correlated drag forces could account for 'transient' modes of swimming in small fish. Some species show large mass-specific development of gill area and, in this respect, are thought to be paedomorphic. Gill development in miniature teleosts must be constrained ultimately by pharyngobranchial space and a compromise must be reached between gills and skin in respiratory and osmoregulatory roles. Miniature gobies show approximately isometric development of gut length, without loops, perhaps because of limited abdominal volume. The compensatory development of mucosal rugosity is discussed, together with other possible consequences for diet and assimilation. In the kidneys, small gobiids possess relatively large glomeruli; glomerular diameter increases approximately isometrically with body mass throughout ontogeny. In some species glomerular frequency and total glomerular volume apparently show a diphasic relationship with body mass. Morphometric glomerular development is compared with species ecotope and physiological control is discussed. Miniaturization could be constrained by the large metabolic costs associated with renal osmoregulatory control. The morphological and possible physiological characters of diminutive teleosts indicate that they are not necessarily less 'specialized' than larger fishes.

Introduction

Miniaturization is important for macroevolutionary phenomena such as the origin of morphological novelty (Roth, Rottluff, Grunwald, Hanken, & Linke 1990). Small body size has been described either as an 'escape from specialization' (when governed by paedomorphosis; L. G. Marshall & Corruccini 1978; Balon 1981) or, conversely, as a route towards specialization (Hutchinson 1959; Miller 1979). Irrespective of these hypotheses, the organism must still be constructed within anatomical and physiological parameters that determine its functional competence (Schmidt-Nielsen 1984; Roth *et al.* 1990). The relationship between body size and structure becomes particularly important under extreme miniaturization, when special changes in the structure and function of some organs may become necessary.

ZOOLOGICAL SYMPOSIUM No. 69
ISBN 0-19-857787-7

Phyletic reduction in body size is widespread within the Teleostei. Extreme dwarfism is common and minimum body size seems to be constrained by the need to reproduce (Miller 1979; this volume, pp. 175–199). Small body size also limits the space available for other organ systems and affects the ratios of body interface areas to body weight (Huxley 1943; Alexander, this volume, pp. 3–14).

The present study analyses variation in the allometric relationships between organ size or area (Y) and body size (X), using the power function $Y = aX^b$ (Schmidt-Nielsen 1984). Parameters studied include body surface area, gill area, gut surface area and total renal glomerular volume. Gobiid fishes were chosen for examination because the group is diverse, with minute representatives from various marine and freshwater environments (Nelson 1994). The species examined were as follows (with source and maximum standard length for the species): marine *Aphia minuta* (Plymouth, 58 mm), *Crystallogobius linearis* (Plymouth, 47 mm), and *Pomatoschistus minutus* (Plymouth, 95 mm); brackish *Pomatoschistus microps* (Teignmouth, 64 mm) and *Gobiopterus chuno* (Bangkok, 24 mm); and freshwater *Brachygobius doriae* (aquarist source, 35 mm) and *Sicydium plumieri* (Puerto Rico, post-larvae) (size data from Miller 1984, 1986; Harrison 1987).

Methodology

Changes in body form can be more accurately described against body mass than body length (Zihler 1982; Harrison 1987). Methodology and morphometric techniques are detailed by Harrison (1987).

Body surface area

Schmidt-Nielsen (1984) outlined the difficulties in defining and measuring the true body surface area of animals. The present study concerns only the external surface area, which is governed by the geometric form of the fish. Exclusion of bilateral fin areas in this study does not significantly affect the allometric exponents for the relationships between body surface area and body mass (Harrison 1987). Body surface areas of the present fish were estimated by multiplying body length by the average circumference of the transversely sectioned body (Webb & Skadsen 1979; Oikawa & Itazawa 1985).

Gill morphometry

Techniques of gill morphometry were reviewed by Hughes (1984a) and their application by Harrison (1987).

Alimentary tract morphometry

Expansion of gut surface area is achieved by extension of the alimentary canal in length, diameter, and through developing mucosal folds (Siankowa 1966). Hence, gut length and mucosal circumference at selected points along this length were measured. Gut mucosal surface area is estimated as the product of gut length and mean circumference of the mucosal surface (Siankowa 1966; Hofer 1988).

Renal glomerular morphometry

Total glomerular filtration surface area is dependent on the size and frequency of the glomeruli. Glomerular frequency (*GF*) can be estimated from histological sections (Nash 1931; Te Winkel 1935; Oikari 1977; Colville, Richards & Dobbie 1983; Eastman & DeVries 1986). The mean diameter of each glomerulus was measured from the section which most closely cut through its centre. Subsequently, the mean glomerular diameter was calculated for the whole kidney. Milton (1971) and Oikari (1977) took glomerular surface area as the critical parameter to be analysed. This assumes the glomerulus to be a simple ball, with water filtration occurring only around its outer surface. However, filtration also occurs from the capillary surfaces within the glomerular tuft. Consequently, the total capillary surface area of the glomerulus might be more critical than the 'glomerular surface area'. While there is no absolutely accurate method for determining the capillary surface area (Nash 1931), it is probably proportional to the volume of the glomerular tuft (Marshall 1930). Assuming the glomerulus to be approximately spherical, glomerular volume $GV=\frac{4}{3}(\pi r^3)$, where r is the mean glomerular radius for the entire kidney. Total glomerular volume (*TGV*) for the whole fish is the product of *GV* and *GF*.

Body surface area

The surface area of an isometric body increases in proportion to its volume to the power of $\frac{2}{3}$ (Schmidt-Nielsen 1984). Hence, some authors (Morgulis 1915; Gray 1953; Mehrotra & Jha 1962) who investigated the relationship between body surface area (*BSA*) and body mass (*M*) in teleost fish, assumed a value of $\frac{2}{3}$ for the exponent, b, in the allometric equation $BSA=aM^b$.

It is unlikely that this isometric relationship can be extended to most organisms because their retention of a constant shape or density is improbable (Gould 1971; Sweet 1980; Roth *et al.* 1990). Body surface area of the small gobiid species examined does not scale isometrically with body mass; the regression exponents are significantly smaller than 0.667 (Table 1 and Fig. 1). These exponents are lower than corresponding exponents for juvenile and adult stages of many other, larger species (Table 1). The gobiids have relatively large mass-specific body surface areas, which is indicated when these values are calculated for fish of mid-range body mass (using ranges in Table 1; Fig. 2).

Low exponents and high mass-specific body surface areas are also reported for the pre- and post-larval stages of *Clupea harengus, Pleuronectes platessa* and *Cyprinus carpio* (De Silva 1974; Oikawa & Itazawa 1985) (Table 1 and Fig. 2). Hence, similar allometric trends in juvenile or adult specimens of the gobiids *Aphia minuta*, *Gobiopterus chuno*, and *Pomatoschistus microps* might be a paedomorphic character. Although it was not possible to make comparisons with adult *Sicydium plumieri*, the low exponent and high mass-specific body surface areas found in post-larval *S. plumieri* may corroborate this hypothesis.

Table 1. Summary of regression analyses for allometric relationships ($Y = aM^b$) of body surface area in mm^2 (Y) to body mass in g (M), for a range of teleosts.

Species	Range in body mass (g)	Regression equation	95% confidence range of *b* (for present study)	Reference
1 *Clupea harengus*	0.01–0.175	$Y = 1106M^{0.608}$		De Silva (1974)
2 *Cyprinus carpio*	0.0018–0.14	$Y = 616M^{0.596}$		Oikawa & Itazawa (1985)
3 *Cyprinus carpio*	0.33–2250	$Y = 719M^{0.664}$		Oikawa & Itazawa (1985)
4 *Clarias batrachus*	15–77	$Y = 565M^{0.743}$		Munshi, Ojha & Sinha (1980)
5 *Saccobranchus fossilis*[b]	10–100	$Y = 851M^{0.684}$		Hughes, Singh, Guha, Dube & Munshi (1974)
6 *Salvelinus namaycush*	Not available	$Y = 617M^{0.731}$		Webb & Skasden (1979)
7 *Oncorhynchus mykiss*	6.8–520	$Y = 744M^{0.633}$		Webb & Skasden (1979)
8 *Oncorhynchus mykiss*	6.3–1072	$Y = 680M^{0.667}$		Webb & Skasden (1979)
9 *Oncorhynchus nerka*	9.9–434	$Y = 782M^{0.65}$		Webb & Skasden (1979)
10 *Amphipnous cuchia*[b]	5–250	$Y = 877M^{0.707}$		Hughes, Singh, Thakur & Munshi (1974)
11 *Ambloplites rupestris*	Not available	$Y = 692M^{0.650}$		Webb & Skasden (1979)
12 *Chaenocephalus aceratus*	342–2240	$Y = 3236M^{0.589}$		Holeton (1976)
13 *Boleophthalmus boddaerti*[b]	1–12	$Y = 601M^{0.687}$		Niva, Ojha & Munshi (1981)
14 *Aphia minuta*	0.032–0.322	$Y = 1054M^{0.601}$	0.561–0.644	Present study
15 *Gobiopterus chuno*	0.010–0.0702	$Y = 727M^{0.617}$	0.587–0.659	Present study
16 *Pomatoschistus microps*	0.1206–1.1485	$Y = 714M^{0.593}$	0.560–0.627	Present study
17 *Sicydium plumieri*	0.0087–0.0514	$Y = 898M^{0.606}$	0.572–0.642	Present study
18 *Pleuronectes platessa*[a]	0.001–0.02	$Y = 258M^{0.402}$		De Silva (1974)

[a] Adapted from the formula given by De Silva (1974: 468) for larval fish.
[b] Air breathing fish. Note that body surface area shows positive allometry with respect to body mass ($b < 0.067$). This is due to the continued importance of the skin for functional respiration throughout ontogeny.

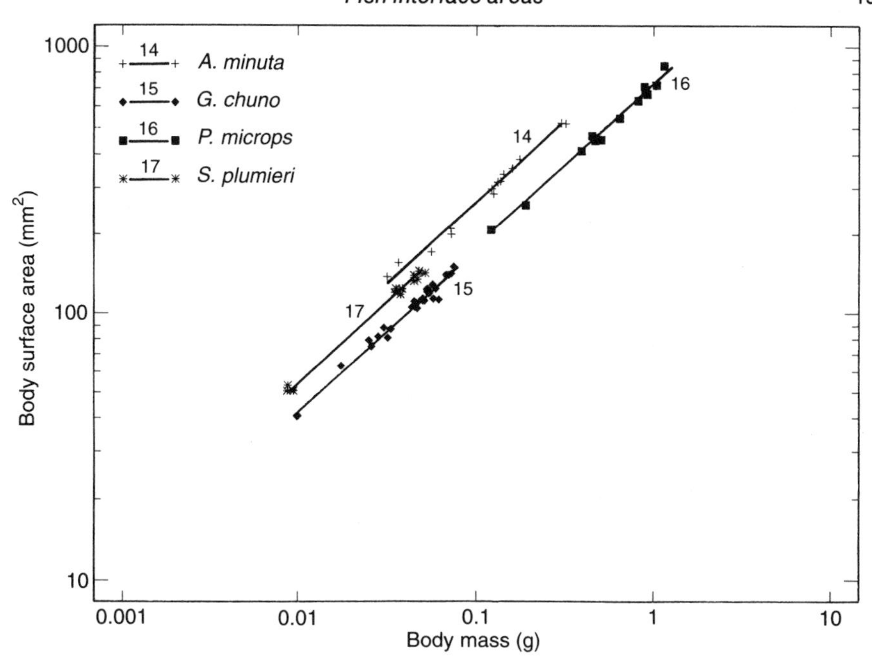

Fig. 1. Allometric relationships of body surface area to body mass for small gobiids. Regression line numbers correspond to numbering of species in Table 1.

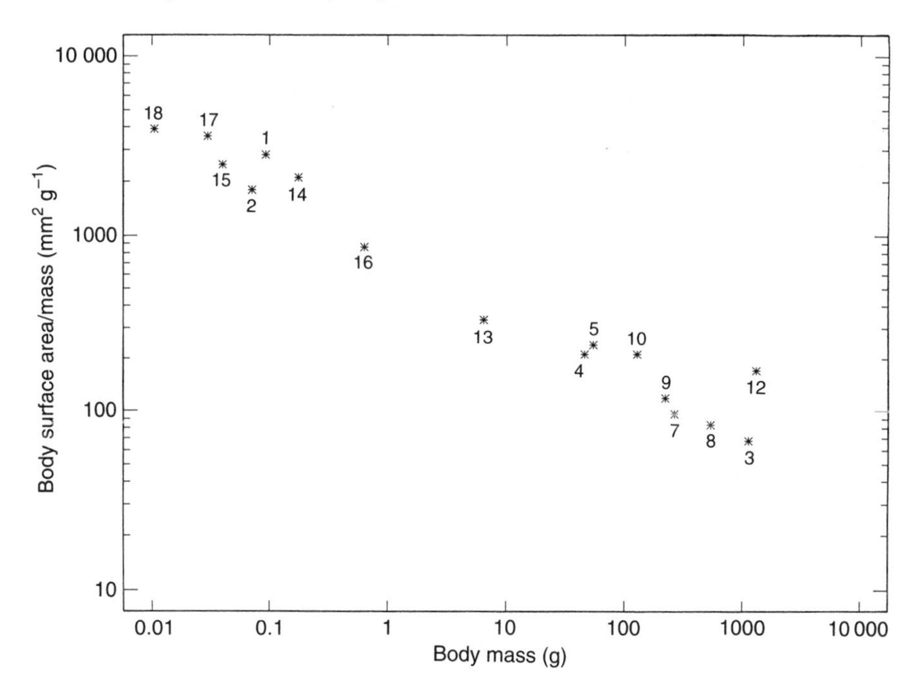

Fig. 2. Allometric relationships of body surface area per unit body mass to body mass for average-sized specimens of various teleost species. Plot numbers correspond to numbering of species in Table 1.

Cutaneous gas exchange has been demonstrated in a variety of fish (Feder & Burggren 1985). The large mass-specific body surface areas of larval fish are probably important sites of gas exchange, complementing the gills while the latter are still developing (De Silva 1974; Oikawa & Itazawa 1985; Al-Kadhomiy & Hughes 1988; see below). Adults of the small gobiid species exhibit skin morphometry comparable to that found in young stages of larger species (see above). Therefore, cutaneous respiration in these paedomorphic adults could be of particular importance.

Cutaneous respiration is thought to be aided by reduced squamation in some small teleosts (Oikawa & Itazawa 1985), such as post-larval *S. plumieri* with relatively large body surface area (Table 1) and limited squamation (Erdman 1961). However, there are numerous small species which are relatively well scaled. The thickness and physiology (e.g. vascularization) of the skin could also control its respiratory function (Feder & Burggen 1985; Al-Kadhomiy & Hughes 1988).

Large fish of some species show a greater degree of euryhalinity than smaller ones; this size-related salinity tolerance and osmoregulatory capability could result from change in surface/volume ratios (Nordlie, Szelistowski, & Nordlie 1982; McCormick & Naiman 1984), perhaps causing greater osmoregulatory demand in small fishes (Miller 1979). Nevertheless, several species of miniature teleosts maintain a high level of osmoregulatory competence over a broad spectrum of conditions, such as the tiny Philippines goby *Pandaka pygmaea* (Norman 1981), and various small cyprinodontids whose euryhalinity is unmatched among other teleost fishes (Nordlie 1985).

The skin's contribution to osmoregulation is probably less than that of the gills because it is less permeable to water and may have a relatively smaller surface area (Parry 1966; W. S. Marshall 1977; Evans 1980; see below). The poor transfer of ions or water across the skin could also be attributed to poorer vascularization and the presence of scales and mucus (Fromm 1968).

Thus, reduced squamation in various small teleosts might be expected to increase permeability and, hence, osmotic stress. However, the skin may show proportionately greater thickening or development of slime glands when squamation is poor (Parry 1966; Mittal & Banerjee 1980). Surface mucus could play an enhanced role in osmoregulation (Mittal & Banerjee 1980), but mucus production is energetically costly (Feder & Burggren 1985), particularly when the mass-specific body surface area is large, as in small teleosts.

Friction drag is the dominant resistance component to locomotion in 'periodic' or 'steady' swimmers (Webb & Blake 1985). Drag force on a moving fish is correlated with its body surface area (Boddeke 1963). Therefore, small fish, with relatively large body surface areas, must experience large surface drag relative to their volume. Consequently, small fish are less efficient steady swimmers than larger ones, requiring greater effort to power themselves through the water (Schmidt-Nielsen 1984; Dabrowski 1986). Miniature South American freshwater fishes are usually found in slow-moving or still waters, perhaps because of cost in maintaining position (Weitzman & Vari 1988).

An alternative to steady swimming is 'transient' or 'burst-and-coast' swimming, with lower frictional drag on a body, briefly accelerating through water, since growth of the water boundary layer does not mature until the fish has travelled a certain distance. External surface mucus further reduces friction drag and can be of particular importance to transient swimmers (Bernadsky, Sar, & Rosenberg 1993). Most energy is expended overcoming the fishes' own inertia (Webb & Blake 1985). Thus, large body surface areas can be advantageous for transient swimming. They promote large initial thrust (Webb 1982a) and, for the short burst-and-coast movement, carry little penalty in terms of drag and, hence, performance (Webb & Blake 1985). Small teleosts would thus seem well suited to transient swimming, which may substantially reduce energy expenditure compared to steady swimming (Weihs 1974; Videler & Weihs 1982; Dabrowski 1986). Transient swimming is also an adaptation for predator avoidance (Webb 1982a, b). However, all but the shortest bursts of swimming could become energetically unfeasible at extreme miniaturization. Moreover, distension of a small abdomen with eggs in female teleosts may further reduce performance (Miller 1979).

The preceding discussion has indicated that increased body surface area relative to mass could facilitate cutaneous respiration but also confers some energetic problems. While the extent of these energetic costs in dwarf teleosts might be limited by compensatory physiology and behaviour, they will probably still be a significant part of overall metabolism. Although metabolism is not directly surface-related, relative metabolic rate is greater in some small fishes (Schmidt-Nielsen 1984; Miller, this volume, pp. 175–199).

The allometric relationship between body surface area and mass shows noticeably less interspecific variation than others, such as gill area–body mass (see below). Hence, despite the obvious variation that exists in body form of teleosts, body surface area is relatively rigidly controlled. The physiological and hydrodynamic factors, which might constrain variation in body form and surface area, are perhaps critical in very small teleosts. Indeed, the greatest variation in body form apparently occurs in medium-sized species rather than minute ones. Dial & Marzluff (1988) presented a similar hypothesis of size-related taxonomic diversity.

Gill morphometry

Fish gills serve for respiration, clearance of nitrogenous waste products, maintenance of acid–base and mineral balances, and passive water diffusion (Maetz 1971; Schmidt-Nielsen 1984; see below). Gill morphometry and functions are related to activity and physicochemical aspects of ecotope (De Jager & Dekkers 1975; Hughes & Al-Kadhomiy 1986).

Previous work (De Silva 1974; Al-Kadhomiy 1985; Oikawa & Itazawa 1985; Hughes & Al-Kadhomiy 1988; Prasad 1988) has shown that gill morphometry changes markedly near metamorphosis. In pre-metamorphic larval fish the allometric exponent of the gill area–body mass relationship is large (1.2–7.1), indicating considerable dimensional changes with increasing body mass (Table 2). A similar condition exists in post-larval *S. plumieri* (Table 2). Post-metamorphic juvenile and adult teleosts typically show much lower exponents (usually approximately 0.8: Muir 1969; Hughes 1970, 1984b; Robotham 1978; Schmidt-Nielsen 1984; see Table 2).

Table 2. Summary of regression analyses for allometric relationships ($Y = aM^b$) of gill parameters (Y) to body mass in g (M) and predicted body surface area in mm^2 (BSA), in selected teleosts.

Species	Range in body mass (g)	Regression equation	95% confidence range of b (for present study)	Reference
1 *Clupea harengus*	0.015–0.175[a]	$L = 11325M^{1.84}$		De Silva (1974)
		$n/2 = 42.7M^{0.201}$		
		$bl = 0.0794M^{1.30}$		
		$GA = 84158M^{3.36}$		
	0.175–3.0[a]	$L = 1071M^{0.57}$		
		$n/2 = 31.7M^{0.035}$		
		$bl = 0.0106M^{0.36}$		
		$GA = 823M^{0.78}$		
2 *Cyprinus carpio*	0.0016–0.33	$L = 835M^{0.740}$		Oikawa & Itazawa (1985)
	0.0016–0.0028	$n/2 = 1.32 \times 10^{11}M^{3.880}$		
		$bl = 1053M^{2.267}$		
		$GA = 6.74 \times 10^{17}M^{7.066}$		
	0.0028–0.33	$n/2 = 40.3M^{0.111}$		
		$GA = 1334M^{1.222}$		
	0.0028–2250	$bl = 0.0185M^{0.336}$		
	0.33–2250	$L = 719M^{0.508}$		
		$n/2 = 32.26M^{-0.0787}$		
		$GA = 846M^{0.794}$		
3 *Oncorhynchus mykiss*	0.068–0.1	$GA = 137187M^{3.443}$		Hughes & Al-Kadhomiy (1988)
	0.1–1274	$GA = 317M^{0.932}$		
	400–2000	$GA = 104M^{1.058}$		Niimi & Morgan (1980)
4 *Aphia minuta*	0.0341–0.5024	$L = 361M^{0.677}$	0.559–0.821	Present study
		$n/2 = 13.1M^{-0.137}$	−0.251–−0.074	
		$bl = 0.036M^{0.512}$	0.362–0.724	
		$GA = 318M^{1.102}$	0.883–1.161	
		$GA = 0.00257BSA^{1.68}$	1.47–1.93	
5 *Crystallogobius linearis*	0.0185–0.4199	$L = 196M^{0.517}$	0.444–0.602	Present study
		$n/2 = 11.8M^{-0.168}$	−0.235–−0.120	
		$bl = 0.072M^{0.697}$	0.560–0.866	
		$GA = 330M^{1.043}$	0.890–1.223	
6 *Gobiopterus chuno*	0.0107–0.106	$L = 298M^{0.499}$	0.399–0.625	Present study
		$n/2 = 17.0M^{-0.127}$	(P = NS)	
		$bl = 0.018M^{0.503}$	0.258–0.980	
		$GA = 186M^{0.876}$	0.728–1.054	
		$GA = 0.0161BSA^{1.42}$	1.18–1.71	
7 *Mistichthys luzonensis*[b]	0.00633–0.01370	$GA = 887M^{0.818}$		Te Winkel (1935)
	0.011–0.03	$GA = 266M^{0.664}$		Pauly (1982)
8 *Pomatoschistus microps*	0.2144–1.4345	$L = 292M^{0.604}$	0.497–0.734	Present study
		$n/2 = 19.2M^{-0.159}$	(P = NS)	

		$bl = 0.028M^{0.751}$	0.437–1.289		
		$GA = 319M^{1.210}$	0.909–1.612		
		$GA = 0.00005BSA^{2.04}$	1.53–2.72		
9 *Pomatoschistus minutus*	0.991–5.425	$L = 299M^{0.406}$	0.324–0.508		Present study
		$n/2 = 22.4M^{-0.152}$	−0.366– −0.063	($P < 0.01$)	
		$bl = 0.031M^{0.530}$	0.384–0.731		
		$GA = 418M^{0.788}$	0.652–0.952		
10 *Sicydium plumieri*	0.0154–0.110	$L = 444M^{0.561}$	0.475–0.661		Present study
		$n/2 = 23.2M^{-0.195}$		($P =$ NS)	
		$bl = 0.025M^{0.625}$	0.502–0.777		
		$GA = 863M^{1.169}$	0.968–1.411		
		$GA = 0.00174BSA^{1.93}$	1.60–2.33		
11 *Colisa fasciatus*[b]	0.010–0.026	$L = 36197M^{1.8271}$			Prasad (1988)
		$n/2 = 28.3M^{-0.1207}$			
		$bl = 0.0468M^{0.727}$			
		$GA = 833551M^{2.4132}$			
	0.030–0.290	$L = 445M^{0.4325}$			
		$n/2 = 34.7M^{-0.0327}$			
		$bl = 0.0163M^{0.4042}$			
		$GA = 499M^{0.8017}$			
12 *Platichthys flesus*	0.001–0.05	$L = 3503M^{1.0985}$			Al-Kadhomiy (1985)
		$n/2 = 19.8M^{-0.0522}$			
		$bl = 0.149M^{1.1832}$			
		$GA = 1.89M^{2.2128}$			
	0.05–200	$L = 410M^{0.502}$			
		$n/2 = 22.9M^{-0.0641}$			
		$bl = 0.034M^{0.386}$			
		$GA = 14.4M^{0.824}$			
13 *Pleuronectes platessa*	0.0015–0.035[a]	$L = 156.5M^{0.88}$			De Silva (1974)
		$n/2 = 28.9M^{0.045}$			
		$bl = 0.0324M^{0.71}$			
		$GA = 1837M^{1.59}$			
	0.035–1.0[a]	$L = 389.2M^{0.59}$			
		$n/2 = 28.3M^{0.040}$			
		$bl = 0.00683M^{0.21}$			
		$GA = 165M^{0.85}$			

[a] Ranges in body mass for *C. harengus* and *P. platessa* are estimates based on length ranges and conversion scales given by De Silva (1974).
[b] Regressions by reduced major axis technique from original data of author.
L, total filament length in mm; *n*/2, average frequency of secondary lamellae per mm on one side of the filament; *bl*, average bilateral surface area of secondary lamellae in mm^2; *GA*, gill area in mm^2. *P* values are given when > 0.005.

However, exponents for juvenile and adult *Aphia minuta*, *Crystallogobius linearis*, and *Pomatoschistus microps* are large (Table 2 and Fig. 3) and those for the relationships of total filament length to body mass and bilateral surface area of the secondary lamellae to body mass are intermediate between the large values obtained for pre-metamorphic fish (Table 2) and the lower values typical of post-metamorphic/adult stages (generally, 0.3–0.5/0.6: Hughes & Iwai 1978; Munshi, Ojha, & Sinha 1980; Hughes & Al-Kadhomiy 1986; Roubal 1987; Kobayashi, Murata, & Harada 1988; Table 2). The aphyines, *A. minuta* and *C. linearis*, are nektonic (Miller 1986) and the activity associated with this mode of life might contribute towards the large exponents, although *P. microps* is epibenthic but has a larger exponent than the aphyines. The similarities between the allometric relationships of *A. minuta*, *C. linearis*, and *P. microps* and the pre-metamorphic stages of other taxa suggest a paedomorphic trait in adults of these gobies. The absence of this in *Pomatoschistus minutus* may be attributed to the larger maximum size in this epibenthic species (Miller 1986).

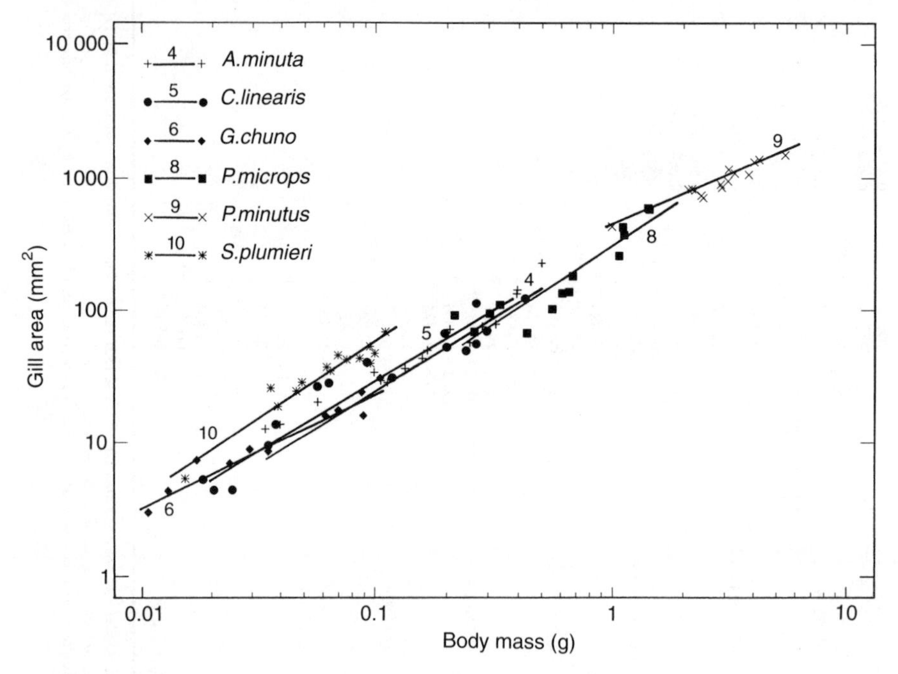

Fig. 3. Allometric relationships of gill area to body mass for small gobiids. Regression line numbers correspond to numbering of species in Table 2.

The smallest (juvenile) *Gobiopterus chuno*, a nektonic species, have gill dimensions comparable to, if not larger than, those of similar-sized specimens of other taxa (Table 3). Hence, mass-specific increases in gill dimensions must have been relatively large in the post-larval and early immature specimens of *G. chuno*. This trait does not continue as a paedomorphic feature of juvenile and adult *G. chuno*. The exponents of the allometric relationships in *G. chuno* are smaller than those found in the other aphyines and *P. microps*, being more or less comparable to those for several larger species (Table 2 and see above). These results might derive from experimental protocol, ecotopic adaptation, or some other morphological constraint in *G. chuno*.

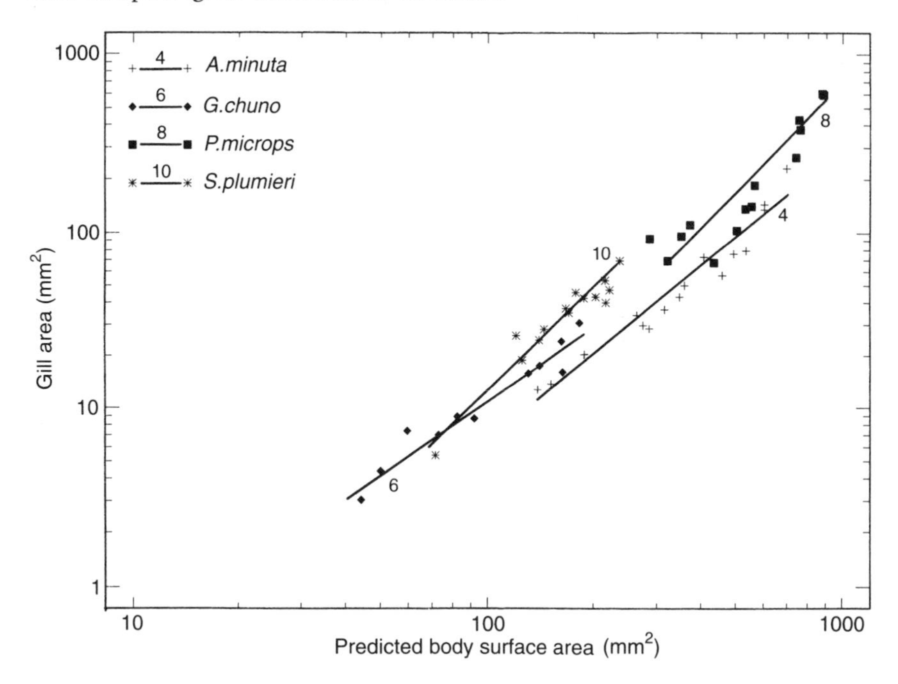

Fig. 4. Allometric relationships of gill area to body surface area for small gobiids, where body surface areas are predicted from the regression equations in Table 1. Regression line numbers correspond to numbering of species in Table 2.

Te Winkel (1935) gave body surface areas and gill areas for *Mistichthys luzonensis* (probably congeneric with *Gobiopterus*), which reaches only 14 mm in adult standard length (Miller 1979). Body mass of her specimens can be estimated by using the body surface area–body mass relationship of *G. chuno* (assuming the species to be morphologically comparable). Accordingly, the two species have similar exponents for gill area–body mass relationships (Table 2). *Mistichthys luzonensis* has relatively well-developed gills, compared to other species (Table 3). (Note, however, that Pauly's (1982) treatment of Te Winkel's data gives slightly different results (Tables 2 and 3), but does not explain how body masses were estimated.)

Available space will obviously determine the size and frequency of individual gill filaments (Galis & Barel 1980). Small teleosts have a relatively large brain and eyes (Wellensiek 1953; Harrison 1987) and the expansion of these organs might be at the cost of a reduced volume of the pharyngobranchial chamber. This 'competition' for head space (Hanken 1983) could account for the shorter total filament lengths found in small gobiids of standardized mass (e.g. 0.03–0.1 g; Table 3); it might similarly account for the lower exponent for total filament length versus body mass in *G. chuno*, compared to the corresponding exponents in other gobiids.

Table 3. Gill dimensions at selected standard body lengths. Abbreviations and regressions as Table 2.

Species	Parameter	Dimensions of parameter at standard body mass 0.011g	0.03 g	0.1 g	Reference
Clupea harengus	*L*	–	17.7	163.7	De Silva (1974)
	n/2	–	21.1	26.9	
	bl	–	0.0008	0.0040	
	GA	–	0.643	36.74	
Cyprinus carpio	*L*	29.7	62.4	151.9	Oikawa & Itazawa (1985)
	n/2	24.4	27.3	31.2	
	bl	0.0041	0.0057	0.0085	
	GA	5.4	18.4	80.0	
Cobitis taenia	*L*	–	–	69.2	Robotham (1978)
	n/2	–	–	22.8	
	bl	–	–	0.0064	
	GA	–	–	55.0	
Oncorhynchus mykiss	*GA*	–	–	49.47	Hughes & Al-Kadhomiy (1988)
	GA	–	–	37.07	
Lipophrys pholis	*L*	–	62.8	116.2	Milton (1971)
	n/2	–	46.4	39.2	
	bl	–	0.0032	0.0057	
	GA	–	38.3	106.9	
Aphia minuta	*L*	–	9.2	75.9	Present study
	n/2	–	21.2	17.9	
	bl	–	0.0060	0.0111	
	GA	–	9.2	30.9	
Crystallogobius linearis	*L*	19.0	32.0	59.6	Present study
	n/2	25.2	21.3	17.4	
	bl	0.0031	0.0062	0.0144	
	GA	3.0	8.51	29.9	
Gobiopterus chuno	*L*	31.4	51.8	94.5	Present study
	n/2	30.1	26.5	22.8	
	bl	0.0019	0.0031	0.0058	
	GA	3.6	8.6	24.7	
Mistichthys luzonensis	*GA*	22.2	–	–	Te Winkel (1935)[a]
	GA	13.3	25.9	–	Pauly (1982)
Sicydium plumieri	*L*	35.4	62.1	122.0	Present study
	n/2	55.9	46.0	36.3	
	bl	0.0015	0.0028	0.0060	
	GA	4.43	14.3	58.5	
Colisa fasciatus	*L*	9.6	97.6	164.3	Prasad (1988)
	n/2	48.7	38.9	37.4	
	bl	0.0018	0.0040	0.0064	
	GA	1.6	30.0	78.8	
Platichthys flesus	*L*	24.7	74.4	129.1	Al-Kadhomiy (1985)
	n/2	25.1	23.8	26.5	
	bl	0.0007	0.0024	0.014	
	GA	0.00009	0.0008	2.2	
Pleuronectes platessa	*L*	3.0	7.2	100.0	De Silva (1974)
	n/2	23.6	24.7	25.8	
	bl	0.0013	0.0027	0.0042	
	GA	1.4	7.0	23.3	

[a] See text for discussion of Te Winkel's data.

A mass-specific increase in bilateral surface area of the secondary lamellae is probably constrained by the size of the filaments bearing the lamellae and the requirement of lamellar rigidity (Galis & Barel 1980). Any considerable mass-specific increase in frequency of secondary lamellae will result in greater resistance to water flow over the gills (Hughes 1966), which increases the energetic cost of respiration. A higher frequency of larger lamellae might also incur more chance of clogging in turbid waters (Niimi & Morgan 1980).

Elaboration of the branchial apparatus (and mass-specific increase in gill area) is only possible if accompanied by a corresponding expansion of the branchial and opercular chambers (Fuiman 1983). The bones of the neurocranium, suspensorium and branchial apparatus are small or poorly ossified in the adults of some of the present gobies, for example *A. minuta*, *C. linearis*, and *P. microps* (Harrison 1987). This might afford a degree of 'plasticity' in development of the head skeleton which would aid the large mass-specific increases in gill dimensions. However, this hypothesis requires further testing; for example, the minute aphyines *G. chuno* and *M. luzonensis* have relatively poorly ossified skeletons (Harrison 1987) but do not exhibit such dramatic mass-specific increases in gill morphometry as some of the other species. It would be interesting to compare gill ontogeny in *Gobiopterus* with that of other miniature gobies which possess more completely ossified skeletons, such as *Eviota*, *Priolepis* and *Pandaka* (Harrison 1987).

Fish may exhibit triphasic relationships between oxygen consumption and body mass; the large mass-specific increases in oxygen consumption in the pre- and post-larval stages of fish are probably correlated with the rapid development of organs with high metabolic activity, for example the brain (Oikawa, Itazawa & Gotoh 1991). These demands will require larger or more efficient surfaces for respiratory exchange. Therefore, it is not surprising to find large mass-specific increases in gill areas of pre- and post-larval fish (see above). A similar explanation might be applied to the large increases in gill areas found in adults of diminutive gobiids. Many small species have relatively large brains (see above) and exhibit reproductive strategies which require considerable energetic investment at early stages of adult development (Miller 1984).

A reduced major axis regression analysis (Harrison 1987) of Rogers's (1986) data for *P. microps* shows that oxygen consumption $= 0.241M^{1.32}$ ($r = 0.725$, $P < 0.001$). The exponent is not significantly different from the exponent for gill area versus body mass (1.21; Table 2). Although mass-specific increases in gill area are inevitably linked with mass-specific increases in oxygen consumption (Muir & Hughes 1969; Holeton 1976; Schmidt-Nielsen 1984; Oikawa & Itazawa 1985), these two parameters are not always proportional to each other (Hughes & Al-Kadhomiy 1988; Prasad 1988). Allometric comparisons of respiratory area and oxygen consumption would be more accurate if they also accounted for skin area (Hughes & Al-Kadhomiy 1988). At small body size the small respiratory exchange area of the gills could be offset by cutaneous respiration across the relatively large body surface area (Oikawa & Itazawa 1985). As the fish grows, the pharyngobranchial chamber and gills rapidly expand relative to body mass and body surface area (Table 2; Figs 3 and 4); this indicates a greater respiratory significance for the gills than for the less efficient skin

(Holeton 1976; Fuiman 1983). These adjustments to the respiratory system of the fish have been noted in several other instances (Iwai & Hughes 1977; McDonald & McMahon 1977; Hughes 1980; Dabrowski 1986), usually in young or metamorphosing stages. However, in the present gobies, changes might not commence until a later stage in ontogeny. A similar situation may exist with gill and body surface areas in the mosquitofish *Gambusia affinis* (Murphy & Murphy 1971).

Fish gills are the principal site for osmotic and ionic exchange, both active and passive, with the surrounding water (Parry 1966; Maetz 1971; Evans 1980). Free exposure of gill surface, high blood flow, and extensive dispersion of blood within the gills means they are a weak link in a fish's defence against osmotic stress (Pauly 1981). Thus, while a large mass-specific increase in gill area might permit greater respiratory and active ionic exchange, it could have the disadvantage of increasing passive osmotic exchange. This might partly explain the smaller mass-specific increases in gill area of the minute gobiids *G. chuno* and *M. luzonensis* (see above).

The osmoregulatory function of the gills could be assisted by other features, such as specialized rodlet cells and microprojections on the 'chloride cells' of *Brachydanio rerio* which could aid osmoregulation and ionic absorption in this small cyprinid with a relatively large gill area (Karlsson 1983). Also, mucus on the secondary lamellae can reduce permeability to water and ions without largely affecting oxygen permeability (Hughes 1970). Ionic permeability of the gills is controlled by prolactin-like hormones in response to changes in salinity or blood concentration (Lahlou 1980). Such endocrine control of gill permeability might support rapid development of the gills in small teleosts.

Alimentary tract morphometry

Gross morphology of the fish gut is determined by phylogenetic constraints, configuration of the body cavity, and adaptation for a particular diet (Suyehiro 1942; Zihler 1982). Zihler's 'spatial demand' seems a particularly important factor regulating gut development in miniature teleosts. With the exception of *P. microps*, all the gobiid species investigated here possess guts which increase in length less than or not significantly different from isometry with body mass (i.e. $b=0.33$, Table 4 and Fig. 5). *Pomatoschistus microps* is more similar to other larger species with positive allometry of gut length relative to body mass (Table 4), possibly helped by larger body-cavity volume. In these species, elongation of the gut usually involves looping and coiling.

In the small aphyine gobies, the gut does not develop much beyond a simple, relatively straight tube, so that a generally isometric relationship exists between gut length and body mass. Convolution of the gut might be functionally unnecessary in these small species. Mass-specific gut lengths (for specimens of mid-range body mass) are certainly greater in diminutive species than in larger ones; this is most evident for the smallest species, *G. chuno* (Fig. 6). However, if phyletic miniaturization occurs to below a certain size, a straight gut might be too short for efficient processing of food. Therefore, an increase in gut length by

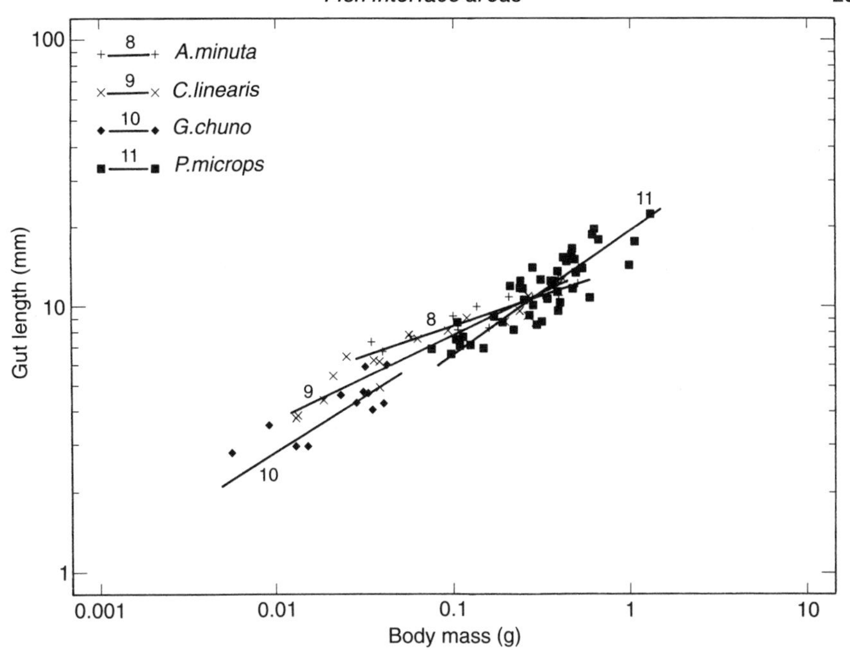

Fig. 5. Allometric relationships of gut length to body mass for small gobiids. Regression line numbers correspond to numbering of species in Table 4.

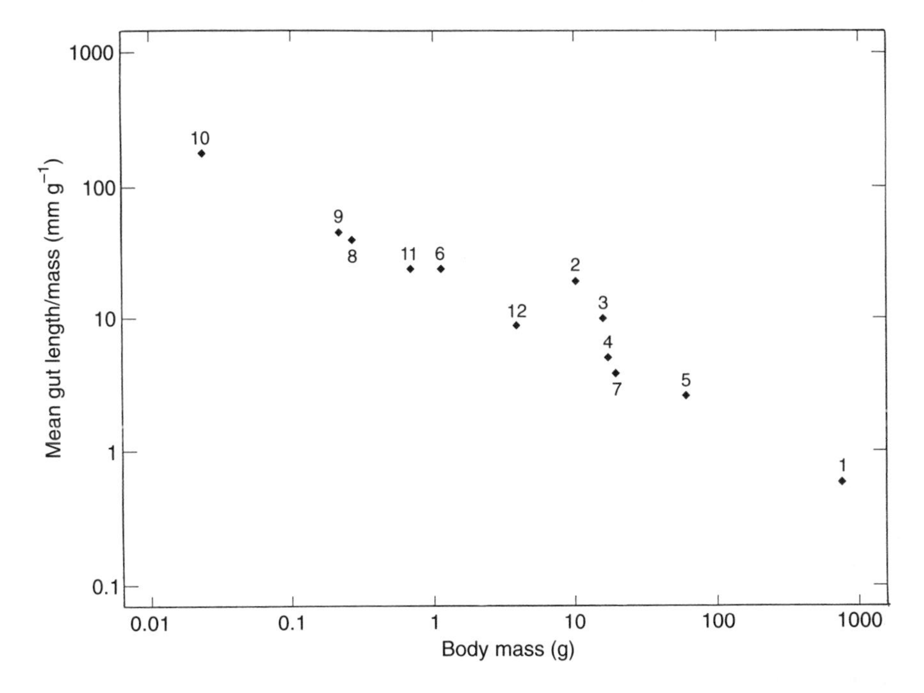

Fig. 6. Allometric relationships of gut length per unit body mass to body mass for mean size specimens of various teleosts. Plot numbers correspond to numbering of species in Table 4.

Table 4. Summary of regression analyses for allometric relationships ($Y = aM^b$) of gut parameters (Y) to body mass in g (M), for various teleosts.

Species	Range in body mass (g)	Regression equation	95% confidence range of b (for present study)	Reference
1 *Abramis brama*	28–1490	$GL = 36.9M^{0.374}$		Siankowa (1966)
		$MA = 481M^{0.713}$		
2 *Carassius auratus*	8.4–12.1	$GL = 81.9M^{0.378}$		Unnithan (1965)
		$MA = 55.9M^{1.38}$		
3 *Cyprinus carpio*	11.5–20.2	$GL = 39.8M^{0.50}$		Al-Hussaini (1949)
		$MA = 290M^{0.948}$		
4 *Gobio gobio*	11–23.2	$GL = 16.9M^{0.568}$		Al-Hussaini (1949)
		$MA = 384M^{0.807}$		
5 *Rutilus rutilus*	14–105.5	$GL = 43.5M^{0.306}$		Al-Hussaini (1949)
		$MA = 373M^{0.799}$		
6 *Glyptosternum reticulatum*	1–1.25	$GL = 24.3M^{0.945}$		Unnithan (1965)
		$MA = 180M^{2.87}$		
7 *Heteropneustes fossilis*	1–37.75	$GL = 21M^{0.418}$		Unnithan (1965)
		$MA = 101M^{0.756}$		
8 *Aphia minuta*	0.0341–0.5024	$GL = 14.5M^{0.228}$	0.179–0.289	Present study
		$GC = 10.7M^{0.445}$	0.374–0.528	
		$MA = 143M^{0.663}$	0.564–0.779	
9 *Crystallogobius linearis*	0.0129–0.4199	$GL = 15.7M^{0.302}$	0.247–0.370	Present study
		$GC = 8.79M^{0.393}$	0.306–0.504	
		$MA = 130M^{0.672}$	0.572–0.788	
10 *Gobiopterus chuno*	0.0056–0.0416	$GL = 18.9M^{0.398}$	0.228–0.695	Present study
	0.0129–0.0416	$GC = 24.9M^{0.654}$	($P < 0.10$)	
		$MA = 568M^{1.11}$	0.553–2.25 ($P < 0.01$)	
11 *Pomatoschistus microps*	0.0754–1.3026	$GL = 19.8M^{0.473}$	0.394–0.568	Present study
	0.0972–1.3026	$GC = 13.4M^{0.538}$	0.349–0.830	
		$MA = 226M^{0.856}$	0.724–1.012	
12 *Trichogaster fasciatus*	3.2–4.5	$GL = 54.9M^{1.35}$		Unnithan (1965)
		$MA = 288M^{1.06}$		

GL, gut length in mm; *GC*, mean gut mucosal circumference in mm; *MA*, gut mucosal surface area in mm^2. P values are given when > 0.005.

convolution would be necessary, but perhaps prevented in adult fish by other vital organs already filling the small body cavity. The limited mass-specific increases in gut lengths of small aphyines might, therefore, be an unavoidable consequence of this physical constraint on gut elongation, rather than the lack of any functional requirement for a longer gut. Gut elongation could be further constrained by development of voluminous gonads. In ripe female *G. chuno*, the ovaries leave little space for the alimentary canal and feeding stops (Pillay & Sarojini 1950; pers. obs.).

When gut length is constrained, the effect may be ameliorated by submucosal collagen fibres enhancing the absorptive capacity (Mishra, Rao, Alim, & Ahmad 1991) and by greater complexity of mucosal folding, for example in some cyprinids (Kapoor, Smit, & Verighina 1975; Moitra & Ray 1977, 1979; Junger, Kotrschal, & Goldschmid 1989). The large mass-specific increase in gut circumference and consequently mucosal surface area of *G. chuno* (Table 4 and Fig. 7) indicates some morphometric compensation for short gut length.

However, phyletic miniaturization below a certain size is also likely to constrain mucosal folding of the gut, which might otherwise obstruct the passage of food (Bari & Nazneen 1984). In turn, food can flatten such folds and reduce their effective surface (Siankowa 1966). One way to increase available space is to reduce the thickness of the intestine wall, as has occurred in some small poecilids (Wellensiek 1953). Nevertheless, the muscle layer of the intestine wall must remain sufficiently thick to permit peristalsis. Consequently, there may be a minimum adult body size below which there is insufficient abdominal space for the development of a functional alimentary tract and other organs.

Increase in mucosal area is isometric with body mass ($b = 0.667$) in the aphyines *A. minuta* and *C. linearis*, whereas it is positively allometric in *P. microps* and larger species (Table 4). The relatively small mass-specific development of the gut surfaces in the two aphyine genera may be due to the constraints on gut length and mucosal circumference discussed above. However, it is surprising that these two species should suffer greater spatial limitation than the smaller *G. chuno*. A more probable explanation is that there is no functional requirement for a pattern of morphological compensation similar to that in *G. chuno*.

In general, carnivorous fishes have short alimentary tracts while herbivorous species have longer guts for digestion of plant material, as exemplified among gobioids (Geevarghese 1983). Al-Hussaini (1947) found that planktivores usually have especially short guts. In keeping with this, the small aphyine gobies *A. minuta*, *C. linearis* and *G. chuno* feed on plankton (Pillay & Sarojini 1950; Wheeler 1969). *Pomatoschistus microps* demonstrates a greater mass-specific development in gut length and, in some cases, mucosal surface area when compared to the aphyines; it has a broader carnivorous diet, feeding on planktonic copepods and benthic invertebrates (Wheeler 1969). In some small fish, limitations of gut length and muscularity probably restrict trophic opportunities to planktivory or carnivory and small herbivores are rare among coral fish (Kotrschal & Thomson 1986). Nevertheless, there are some striking exceptions. Small gobiids such as *Chlamydogobius* and the minute species of *Kellogella* have relatively elongate guts which are probably correlated with herbivory (Hoese 1975; Miller 1987).

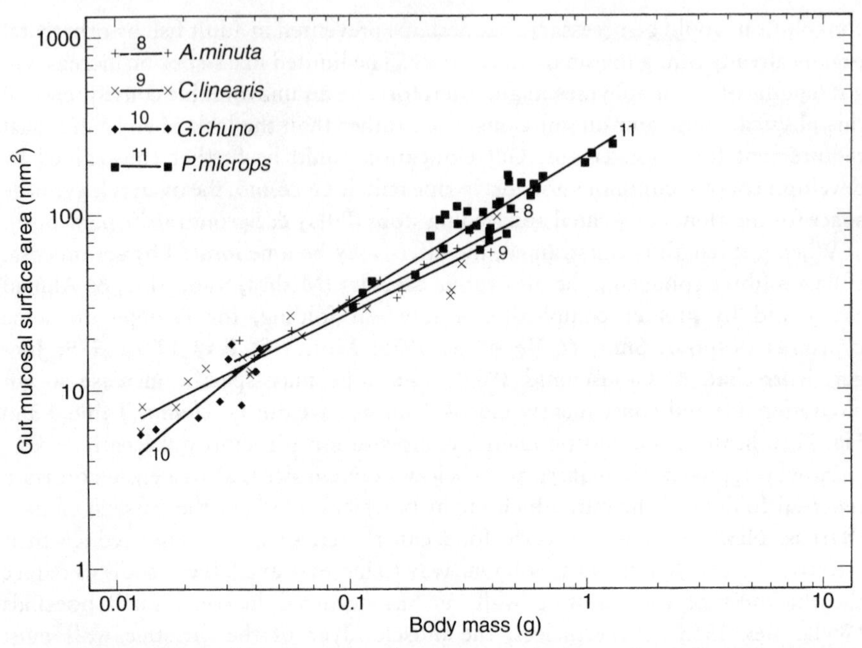

Fig. 7. Allometric relationships of gut mucosal surface area to body mass for small gobiids. Regression line numbers correspond to numbering of species in Table 4.

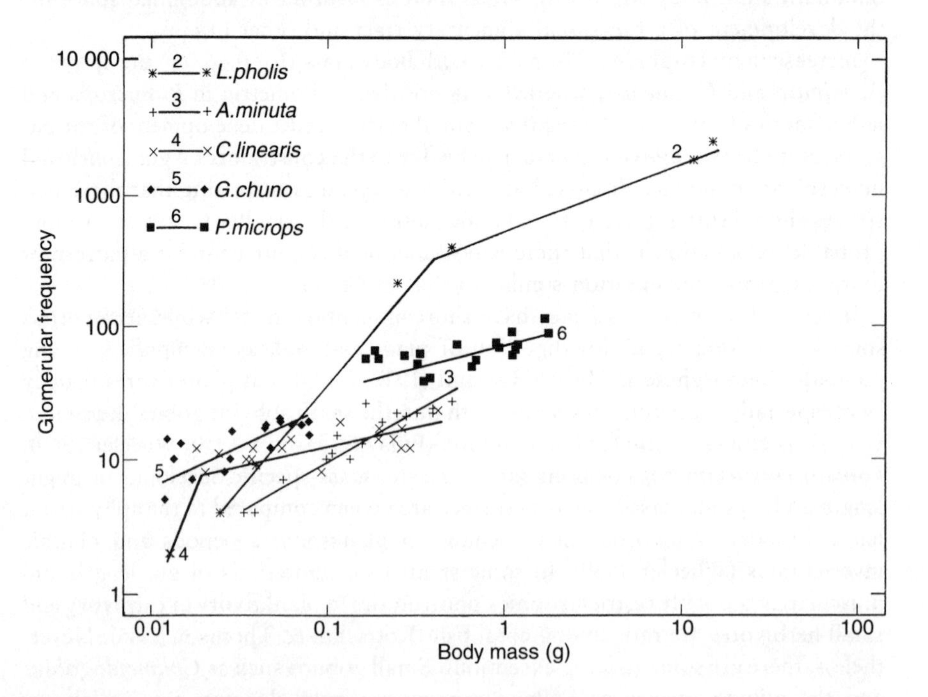

Fig. 8. Allometric relationships of glomerular frequency to body mass for small gobiids and a blenniid. Regression line numbers correspond to numbering of species in Table 5.

Sweet (1980) suggested that the mucosal surface area of the gut is correlated with metabolic demand. Comparison of the gut area–body mass regression exponent for *P. microps* (0.856) with the oxygen consumption–body mass exponent of 1.321 (from Rogers 1986) does not support direct correlation, but further anatomical and physiological studies are required. Similarly, teleostean growth may be correlated with the development of the gut (Klust 1939, 1940; Siankowa 1966) because the gut assimilates the nutrients required for anabolism. Small and larval fish demonstrate ingestion and digestion rates which maximize growth rates (Kapoor *et al.* 1975; Houde & Schekter 1980), with relatively large food intake, rapid passage of food along the gut, and high rates of digestion and assimilation (Govoni, Boehlert, & Watanbe 1986). Norman (1981) suggested that, in the tiny goby *P. pygmaea*, high food requirements but low holding capacity of the gut mean that the fish must expend considerable energy in devouring plankters and processing them in steady succession. Under these conditions, a constrained gut surface area in small adults might be physiologically limiting for anabolism.

Further phyletic reduction in adult body size might not permit development of a gut mucosal surface large enough for adequate assimilation of nutrients. This limitation could be compensated for by an increase in number of goblet cells (Siankowa 1966), as suggested for larval stages of larger teleosts (Albertini-Berhaut 1987). Pauly (1981) stated that anabolism can be independent of the gut surface because energy-rich substances may be stored as fat or liver oil. This allows the fish to maintain anabolic activities long after feeding and nutrient assimilation has ceased. However, in small teleosts, liver size might also be constrained by body cavity space. As a further limitation, Kapoor *et al.* (1975) noted that food intake drops with decreasing temperature as a result of lower metabolic rates and decreased intestinal clearance. The combination of restricted gut mucosal surface and decreased intestinal clearance at lower temperatures might limit the rate of food processing, growth, and gonad maturation in small fish at higher latitudes, where small teleosts are not so frequent.

The gut also functions in osmoregulation in marine teleosts after ingestion of water (Parry 1966; Maetz 1971; Evans 1980). Clearly, an interrelationship between digestive and osmoregulatory functions of the gut is particularly important in small marine teleosts, where the gut mucosal surface area might be morphometrically and, hence, functionally constrained.

Renal glomerular morphometry

The teleostean kidney is almost solely concerned with osmoregulation (Parry 1966; Evans 1980; Nishimura & Imai 1982). While some nitrogen is excreted at the kidneys, the principal excretory compound, ammonia, is easily lost across the gills (Schmidt-Nielsen 1984). Discussion above has indicated that small fish might be subject to considerable osmotic demands, so that renal osmoregulation may be of increased significance.

Glomerular size

Various studies of glomerular morphometry have been made, usually involving analysis of glomerular size (E. K. Marshall & Smith 1930; Nash 1931; Te Winkel 1935; Holliday & Blaxter 1961; Milton 1971; Oikari 1977; Colville *et al.* 1983; Eastman & DeVries 1986). Marine fish, which must conserve water, generally possess smaller glomeruli than freshwater fish, which are faced with an osmotic gain of water from the hypotonic environment (Nash 1931; Gambaryan 1988).

The glomerular diameter increases almost isometrically ($b = 0.33$) relative to body mass in small fish but remains constant in larger fish (Nash 1931; present study, Table 5). Milton's (1971) data for the blenniid *Lipophrys pholis*, while apparently supporting this hypothesis, are too sparse to permit quantitative comparisons of small and large fish.

The isometry in glomerular development remains constant in the small gobiids examined here; the fish do not reach a size where the process stops. This is not a paedomorphic trait because isometric development of glomerular diameter is not restricted only to early ontogeny in larger species (e.g. *Carassius auratus* and *L. pholis*). Furthermore, the small species and particularly the aphyines possess well-developed glomeruli relative to their body size (Table 6; cf. Nash's (1931) values for larger species). The range in mean glomerular diameter obtained for *G. chuno* is virtually identical to that of *P. microps*, despite the former species weighing less than 10% of the latter.

These two species from brackish waters possess large glomeruli probably correlated with the need for high glomerular filtration rates in hypotonic conditions. Adult *A. minuta* and *C. linearis*, which are marine species, also have large glomeruli attaining sizes (70–85 μm; Table 6) greater than Nash (1931) reported for other marine species. Under hypertonic conditions, possible excess water loss in the aphyine gobies could be hormonally regulated (De Ruiter 1981; Nishimura & Imai 1982). The glomerular filtration rate, which is affected by renal blood flow, can also vary according to environmental conditions; hence, the level of 'glomerular recruitment' can be under physiological control (Evans 1980; Colville *et al.* 1983; Gambaryan 1988).

Glomerular frequency

Freshwater teleosts often possess more abundant glomeruli than marine species (E. K. Marshall & Smith 1930; Nash 1931; Daikoku 1978; Colville *et al.* 1983). The glomerular number is reported as varying directly with, but not necessarily in proportion to, body mass (Nash 1931; Holliday & Blaxter 1961). Data presented here (Table 5 and Fig. 8) indicate that glomerular frequency increases uniphasically with body mass in some species and, apparently, diphasically in others, i.e. *C. linearis* and *L. pholis*. However, a small degree of experimental error could easily account for the regression inflexion in *C. linearis*.

The marine species *A. minuta* and *C. linearis* have the lowest glomerular frequencies relative to body mass (Table 6). This reduced glomerular frequency

Table 5. Regression analyses of the allometric relationships ($Y = aM^b$) of glomerular parameters (Y) to body mass in g (M), for a range of teleosts.

Species	Range in body mass (g)	Regression equation	95% confidence range of b (for present study)	Reference
1 *Carassius auratus*[a]	6.2–29.7	$d = 23.9M^{0.280}$		Nash (1931)
	29.7–132.5	$d = 41.5M^{0.009}$		
	18–173	$n = 637M^{0.662}$		
	6.2–132.5	$d = 31M^{0.163}$		
2 *Lipophrys pholis*[a]	0.025–0.5	$n = 1501M^{1.592}$		Milton (1971)
		$TGV = 26.0M^{1.81}$		
	0.5-15	$n = 565M^{0.516}$	($P < 0.10$)	
		$TGV = 12.4M^{1.14}$		
	0.025–15	$d = 39.0M^{0.151}$		
		$n = 291M^{0.985}$		
		$TGV = 8.79M^{1.4}$		
3 *Aphia minuta*	0.0341–0.065	$TGV = 312M^{2.54}$	($P = NS$)	Present study
	0.065–0.5024	$TGV = 9.79M^{1.26}$		
	0.0341–0.5024	$d = 93.3M^{0.352}$	0.170–0.762	
		$n = 57.7M^{0.769}$	0.621–0.95	
		$TGV = 15.9M^{1.58}$	1.15–2.17	
4 *Crystallogobius linearis*	0.0129–0.02	$n = 2.16 \times 10^6 M^{3.19}$	($P < 0.05$)	Present study
		$TGV = 1.39 \times 10^6 M^{3.97}$	($P < 0.10$)	
	0.02–0.4199	$n = 24.6M^{0.272}$	($P < 0.01$)	
		$TGV = 11.4M^{1.01}$		
	0.0129–0.4199	$d = 115M^{0.327}$	0.247–0.405	
		$n = 44.9M^{0.556}$	0.314–0.985	
		$TGV = 17.4M^{1.21}$	0.999–1.47	
5 *Gobiopterus chuno*	0.0123–0.0793	$d = 180M^{0.385}$	0.303–0.489	Present study
		$n = 101M^{0.603}$	0.332–1.09	
		$TGV = 187M^{1.61}$	1.29–2.02	
6 *Pomatoschistus microps*	0.1646–1.7889	$d = 60.5M^{0.357}$	0.292–0.435	Present study
		$n = 76.3M^{0.351}$	0.161–0.763	
		$TGV = 8.47M^{1.35}$	1.05–1.72	

[a] Regressions by reduced major axis technique from original data of author.
d, mean glomerular diameter in μm; n, number of glomeruli in kidney (glomerular frequency); TGV, total glomerular volume in $mm^3 \times 10^{-3}$. P values are given when > 0.005.

Table 6. Glomerular dimensions at selected standard body lengths. Abbreviations and regressions as Table 5.

Species	Parameter	Measured range for parameter	Dimensions of parameter at standard body mass				Reference
			0.03 g	0.07 g	0.4 g	1.0 g	
Carassius auratus	*d*	41.7–68.8	–	–	–	–	Nash (1931)
	n	4316–19307	–	–	–	–	
Lipophrys pholis	*d*	24.9–60	23.0	26.1	34.0	39.0	Milton (1971)
	n	4–2490	5.65	21.8	349.0	565.0	
	TGV	0.0323–274.6	0.046	0.211	4.95	12.4	
Aphia minuta	*d*	21.5–71.3	–	36.6	67.6	–	Present study
	n	4–27	–	10.2	27.0	–	
	TGV	0.0282–5.1243	–	0.343	3.09	–	
Crystallogobius linearis	*d*	21–81.8	36.5	48.2	85.2	–	Present study
	n	2–23	9.48	11.9	19.2	–	
	TGV	0.0232–4.8720	0.330	0.777	4.52	–	
Gobiopterus chuno	*d*	34–76.9	46.7	64.7	–	–	Present study
	n	5–20	12.2	20.3	–	–	
	TGV	0.1529–4.5197	0.661	2.58	–	–	
Mistichthys luzonensis	*d*	40.4–54.4	–	–	–	–	Te Winkel (1935)
Pomatoschistus microps	*d*	33.6–83.9	–	–	43.6	60.5	Present study
	n	38–114	–	–	55.3	76.3	
	TGV	1.152–30.333	–	–	2.46	8.47	

should be advantageous for water conservation. *Pomatoschistus microps* and the blenniid *L. pholis* may be subject to more variable salinities. However, *P. microps* has a smaller mass-specific increase in glomerular frequency than *L. pholis* (Table 5), perhaps resulting from the smaller adult body size attained by *P. microps*. However, confidence ranges for the glomerular frequency–body mass regression in *P. microps* are very broad (Table 5) and so any interspecific comparisons must be treated with caution. *Gobiopterus chuno*, the smallest species examined here, has high relative glomerular frequencies (Table 6). The possession of relatively large and frequent glomeruli might, under the correct physiological control, increase renal versatility (Holliday & Blaxter 1961; see above).

Total glomerular volume

Current results suggest a diphasic relationship between total glomerular volume and body mass, as in *A. minuta*, *C. linearis*, and perhaps *L. pholis* (Table 5 and Fig. 9). With a diphasic relationship, the initial phase of mass-specific development of glomerular volume is greater than the later phase. Some small teleosts, such as *G. chuno*, exhibit large mass-specific increases in total glomerular volume throughout ontogeny. These results suggest the importance of the glomerular filtration surface to small teleosts likely to experience osmoregulatory stress. The relatively large total glomerular volumes of *G. chuno* at standardized body mass (0.03–0.07 g; Table 6), compared to corresponding values for other species, might be a consequence of the relatively large body surface areas over which passive exchange could occur. Moreover, the provision of a relatively large and physiologically adaptable total glomerular volume would be advantageous in this brackish water species. Te Winkel's data (1935: table 2) can be used to calculate a mean total glomerular volume for the freshwater *M. luzonensis* of $0.65 \times 10^{-3} mm^3$, corresponding to a specimen of approximately 0.0106 g (applying body surface area–body mass regression for the related *G. chuno*). With the higher output of urine required in fresh water, nephric reabsorption of ions becomes necessary to prevent loss of electrolytes. While the energetic cost of ionic and osmotic regulation is variable in fishes, it can account for up to 50% of a freshwater fish's total standard metabolism (Febry & Lutz 1987). The cost of ionic reabsorption in small freshwater teleosts with large total glomerular volumes (e.g. *M. luzonensis*) is probably high.

Aphia minuta, *C. linearis*, and *P. microps* might owe their smaller mass-specific total glomerular volumes (Table 6) to selection for life in more saline habitats. The relatively small total glomerular volume of *P. microps* indicates better adaptation to hypertonic conditions. Osmoregulatory problems that this species might experience under hypotonic conditions (e.g. at low tide in an estuary channel) could be overcome by changes in renal physiology and in behaviour, when the species tends to be less active (Rogers 1986). Increased activity would incur a greater water and ion flux across the gills, which would need to be counteracted by increased renal function with higher energetic costs (Wood & Randall 1973; Hofmann & Butler 1979; Febry & Lutz 1987). Similarly, *L. pholis* is least active at low tide (Gibson 1969) and, according to the pooled morphometric data (0.025–15 g) for this species, the allometric relationship between total glomerular volume and body mass is comparable to that of *P. microps* (Table 5).

Some small marine teleosts, such as *A. minuta* or *C. linearis*, might be 'functionally' aglomerular despite being morphologically equipped with glomeruli. This condition has been observed in Antarctic notothenioids (Eastman & DeVries 1986) and will reduce renal water loss and the energetic cost of water reabsorption across the nephric tubules. Nevertheless, the physiological modifications of functional aglomerulism could also be energetically costly. The total glomerular volume–body mass exponent for *P. microps* (1.35) is not significantly different from Rogers's (1986) oxygen consumption–body mass regression exponent (1.32) for that species. One might conclude that rapid development of the glomerular filtration surface is directly correlated with the fish's metabolism. However, a proportional, allometric relationship between glomerular filtration surface and body metabolism requires further testing.

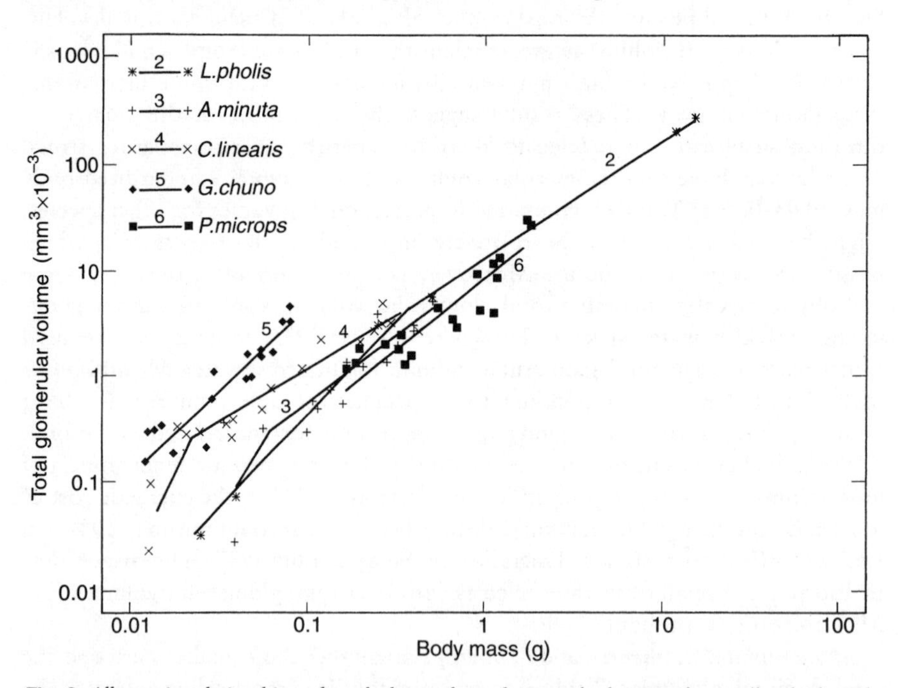

Fig. 9. Allometric relationships of total glomerular volume to body mass for small gobiids and a blenniid. Regression line numbers correspond to numbering of species in Table 5.

The discussion above indicates that phyletic reduction in adult body size might be constrained by prohibitive metabolic demands associated with necessary renal osmoregulation. One could postulate that if juveniles of smaller body size and only partially developed kidneys can survive, then adults existing at similar body size should also be viable. However, juveniles differ in physiology from adults. Ontogenetic changes in the permeability of the skin and gills have been reported for larval fish (Brown & Tytler 1993) and changes also occur in the composition of body fluids as fish mature (Ehrlich 1974; Gill, Weatherley, & Bhesania 1982). These ontogenetic changes in physiology will produce a shift in the osmoregulatory demands to be met by the fish.

Conclusion

Morphological or morphometric specializations of physiologically important interface areas appear to be necessary for functional competence in some extremely small teleosts. These changes might also impose physiological demands which could induce modifications to their ecology and ethology (Harrison 1987). Nevertheless, Hanken & Wake (1993) have noted the dangers of speculating about the functional processes that underpin morphological patterns in miniaturized animals. The present work may be doing exactly that, by using simple allometric data as the basis for discussing extremely complex systems. The possibilities for over-simplification and conjecture are manifold in such a study. Hence, the foregoing information is submitted as a series of 'working hypotheses' for general consideration; the statements are certainly far from dogmatic.

Acknowledgements

I am grateful for the assistance and helpful discussion of various staff and associates at Bristol University (particularly Dr P. J. Miller), Mr G. J. Howes (Natural History Museum), Ms J. K. Skelton, and Dr M. Lugli (Università di Parma, Italy). The research on which this review is based was supported by a SERC studentship.

References

Al-Hussaini, A. H. (1947). The feeding habits and the morphology of the alimentary tract of some teleosts living in the neighbourhood of the marine biological station, Ghardaqa, Red Sea. *Publs mar. biol. Stn Ghardaqa* No.5: 4–61.

Al-Hussaini, A. H. (1949). On the functional morphology of the alimentary tract of some fish in relation to differences in their feeding habits: anatomy and histology. *Q. J. microsc. Sci.* **90**: 109–140.

Al-Kadhomiy, N. K. (1985). *Gill development, growth and respiration of the flounder* (Platichthys flesus). PhD thesis: University of Bristol.

Al-Kadhomiy, N. K. & Hughes, G. M. (1988). Histological study of different regions of the skin and gills in the mudskipper, *Boleophthalmus boddarti* with respect to their respiratory function. *J. mar. biol. Ass. U.K.* **68**: 413–422.

Albertini-Berhaut, J. (1987). L'intestin chez les Mugilidae (Poissons; Téléostéens) à différentes étapes de leur croissance. 1. Aspects morphologiques et histologiques. *Z. angew. Ichthyol.* **3**: 1–12.

Balon, E. K. (1981). Saltatory processes and altricial to precocial forms in the ontogeny of fishes. *Am. Zool.* **21**: 573–596.

Bari, G. A. & Nazneen, S. (1984). Biological studies on the freshwaters of Pakistan. Correlation of the alimentary canal structures of some fishes of Haleji Lake with their herbivorous nature of feeding. *Pakist. J. scient. ind. Res.* **27**: 28–32.

Bernadsky, G., Sar, N., & Rosenberg, E. (1993). Drag reduction of fish skin mucus: relationship to mode of swimming and size. *J. Fish Biol.* **42**: 797–800.

Boddeke, R. (1963). Size and feeding of different types of fishes. *Nature, Lond.* **197**: 714–715.

Brown, J. A. & Tytler, P. (1993). Hypoosmoregulation of larvae of the turbot, *Scophthalamus maximus*: drinking and gut function in relation to environmental salinity. *Fish Physiol. Biochem.* **10**: 475–483.

Colville, T. P., Richards, R. H., & Dobbie, J. W. (1983). Variations in renal corpuscular morphology with adaptation to sea water in the rainbow trout, *Salmo gairdneri* Richardson. *J. Fish Biol.* **23**: 451–456.

Dabrowski, K. R. (1986). Active metabolism in larval and juvenile fish: ontogenetic changes, effect of water temperature and fasting. *Fish Physiol. Biochem.* **1**: 125–144.

Daikoku, T. (1978). Adaptation to seawater environment in guppy *Poecilia reticulata*. *J. Osaka Cy med. Cent.* **27**: 605–636.

De Jager, S. & Dekkers, W. J. (1975). Relations between gill structure and activity in fish. *Neth. J. Zool.* **25**: 276–308.

De Ruiter, A. J. H. (1981). Testosterone-dependent changes *in vivo* and *in vitro* in the structure of the renal glomeruli of the teleost *Gasterosteus aculeatus* L. *Cell Tissue Tes.* **219**: 253–266.

De Silva, C. (1974). Development of the respiratory system in herring and plaice larvae. In *The early life history of fish*: 465–485. (Ed. Blaxter, J. H. S.). Springer-Verlag, Berlin.

Dial, K. P. & Marzluff, J. M. (1988). Are the smallest organisms the most diverse? *Ecology* **69**: 1620–1624.

Eastman, J. T. & DeVries, A. L. (1986). Renal glomerular evolution in Antarctic notothenioid fishes. *J. Fish Biol.* **29**: 649–622.

Ehrlich, K. F. (1974). Chemical changes during growth and starvation of herring larvae. In *The early life history of fish*: 301–324. (Ed. Blaxter, J. H. S.). Springer-Verlag, Berlin.

Erdman, D. S. (1961). Notes on the biology of the gobiid fish *Sicydium plumieri* in Puerto Rico. *Bull. mar. Sci. Gulf Caribb.* **11**: 448–456.

Evans, D. H. (1980). Osmotic and ionic regulation by freshwater and marine fishes: a status report. In *Environmental physiology of fishes:* 93–122. (Ed. Ali, M.A.) Plenum Press, New York. (*NATO adv. Stud. Inst. Ser. A (Life Sci.)* **35**.)

Febry, R. & Lutz, P. (1987). Energy partitioning in fish: the activity-related cost of osmoregulation in a euryhaline cichlid. *J. exp. Biol.* **128**: 63–85.

Feder, M. E. & Burggren, W. M. (1985). Cutaneous gas exchange in vertebrates: design, patterns, control and implications. *Biol. Rev.* **60**: 1–45.

Fromm, P. O. (1968). Some quantitative aspects of ion regulation in teleosts. *Comp. Biochem. Physiol.* **27**: 865–869.

Fuiman, L. A. (1983). Growth gradients in fish larvae. *J. Fish Biol.* **23**: 117–123.

Galis, F. & Barel, C. D. N. (1980). Comparative functional morphology of the gills of African lacustrine Cichlidae (Pisces, Teleostei). *Neth. J. Zool.* **30**: 392–430.

Gambaryan, S. P. (1988). Kidney morphology in sturgeons: a microdissectional and ultrastructural study. *J. Fish Biol.* **33**: 383–398.

Geevarghese, C. (1983). Morphology of the alimentary tract in relation to diet among gobioid fishes. *J. nat. Hist.* **17**: 731–741.

Gibson, R. N. (1969). The biology and behaviour of littoral fish. *Oceanogr. mar. Biol.* 7: 367–410.
Gill, H. S., Weatherley, A. H. & Bhesania, T. (1982). Histochemical characterization of myotomal muscle in the bluntnose minnow, *Pimephales notatus* Rafinesque. *J. Fish Biol.* **21**: 205–214.
Gould, S. J. (1971). Geometric similarity in allometric growth: a contribution to the problem of scaling in size. *Am. Nat.* **105**: 113–136.
Govoni, J. J., Boehlert, G. W., & Watanbe, Y. (1986). The physiology of digestion in fish larvae. *Envir. Biol. Fishes* **16**: 59–77.
Gray, I. E. (1953). The relation of body weight to body surface area in marine fishes. *Biol. Bull. mar. biol. Lab., Woods Hole* **105**: 285–288.
Hanken, J. (1983). Miniaturization and its effects on cranial morphology in plethodontid salamanders, genus *Thorius* (Amphibia, Plethodontidae): II. The fate of the brain and sense organs and their role in skull morphogenesis and evolution. *J. Morph.* **177**: 255–268.
Hanken, J. & Wake, D. B. (1993). Miniaturization of body size: organismal consequences and evolutionary significance. *A. Rev. Ecol. Syst.* **24**: 501–519.
Harrison, I. J. (1987). *The implications of small size in teleost fish, with special reference to the Gobioidei.* PhD thesis: Bristol University.
Hoese, D. F. (1975). A revision of the gobiid fish genus *Kelloggella*. *Rec. Aust. Mus.* **29**: 473–484.
Hofer, R. (1988). Morphological adaptations of the digestive tract of tropical cyprinids and cichlids to diet. *J. Fish Biol.* **33**: 399–408.
Hofmann, E. L. & Butler, D. G. (1979). The effect of increased metabolic rate on renal function in the rainbow trout, *Salmo gairdneri*. *J. exp. Biol.* **82**: 11–23.
Holeton, G. F. (1976). Respiratory morphometrics of white and red blooded Antarctic fish. *Comp. Biochem. Physiol. (A)* **54**: 215–220.
Holliday, F. G. T. & Blaxter, J. H. S. (1961). The effects of salinity on herring after metamorphosis. *J. mar. biol. Ass. U.K.* **41**: 37–48.
Houde, E. D. & Schekter, R. C. (1980). Feeding by marine fish larvae: developmental and functional feeding responses. *Envir. Biol. Fishes* **5**: 315–334.
Hughes, G. M. (1966). The dimensions of fish gills in relation to their function. *J. exp. Biol.* **45**: 177–195.
Hughes, G. M. (1970). A comparative approach to fish respiration. *Experientia* **26**: 113–122.
Hughes, G. M. (1980). Morphometry of fish gas exchange organs in relation to their respiratory function. In *Environmental physiology of fishes*: 33–56. (Ed. Ali, M.A.). Plenum Press, New York. (*NATO adv. Stud. Inst. Ser. A (Life Sci.)* **35**.)
Hughes, G. M. (1984a). Measurement of gill area in fishes: practices and problems. *J. mar. biol. Ass. U.K.* **64**: 637–655.
Hughes, G. M. (1984b). General anatomy of the gills. In *Fish physiology* **10.** *Gills, Part A. Anatomy, gas transfer, and acid–base regulation*: 1–72. (Eds Hoar, W.S. & Randall, D.J.). Academic Press, New York.
Hughes, G. M. & Al-Kadhomiy, N.K. (1986). Gill morphometry of the mudskipper, *Boleophthalmus boddarti*. *J. mar. biol. Ass. U.K.* **66**: 671–682.
Hughes, G. M. & Al-Kadhomiy, N. K. (1988). Changes in scaling of respiratory systems during the development of fishes. *J. mar. biol. Ass. U.K.* **68**: 489–498.
Hughes, G. M. & Iwai, T. (1978). A morphometric study of the gills in some Pacific deep-sea fishes. *J. Zool., Lond.* **184**: 155–170.
Hughes, G. M., Singh, B. R., Guha, G., Dube, S. C., & Munshi, J. S. D. (1974). Respiratory surfaces of an air-breathing siluroid fish, *Saccobranchus (=Heteropneustes) fossilis* in relation to body size. *J. Zool., Lond.* **172**: 215–232.

Hughes, G. M., Singh, B. R., Thakur, R. N., & Munshi, J. S. D. (1974). Areas of the air breathing surfaces of *Amphipnous cuchia* (Ham.). *Proc. Indian natn. Sci. Acad. (B)* **40**: 379–392.

Hutchinson, G. E. (1959). Homage to Santa Rosalia or why are there so many kinds of animals? *Am. Nat.* **93**: 145–159.

Huxley, J. (1943). *The uniqueness of man*. Chatto and Windus, London.

Iwai, T. & Hughes, G. M. (1977). Preliminary morphometric study on gill development in black sea bream (*Acanthopagrus schlegeli*). *Bull. Jap. Soc. scient. Fish.* **43**: 929–934.

Junger, H., Kotrschal, K., & Goldschmid, A. (1989). Comparative morphology and ecomorphology of the gut in European cyprinids (Teleostei). *J. Fish Biol.* **34**: 315–326.

Kapoor, B. G., Smit, H., & Verighina, I. A. (1975). The alimentary canal and digestion in teleosts. *Adv. mar. Biol.* **13**: 109–239.

Karlsson, L. (1983). Gill morphometry in the zebrafish, *Brachydanio rerio* (Hamilton-Buchanan). *J. Fish Biol.* **23**: 511–524.

Klust, G. (1939). Über Entwicklung, Bau und Funktion des Darmes beim Karpfen (*Cyprinus carpio* L.). *Int. Rev. Hydrobiol.* **39**: 498–536.

Klust, G. (1940). Über Entwicklung, Bau und Funktion des Darmes beim Karpfen (*Cyprinus carpio* L.). *Int. Rev. Hydrobiol.* **40**: 89–173.

Kobayashi, H., Murata, O., & Harada, T. (1988). Some aspects of gill measurement in relation to the growth of the yellowtail *Seriola quinqueradiata*. *Bull. Jap. Soc. scient. Fish.* **54**: 49–54.

Kotrschal, K. & Thomson, D. A. (1986). Feeding patterns in eastern tropical Pacific blennioid fishes (Teleostei: Trypterygiidae, Labrisomidae, Chaenopsidae, Blenniidae). *Oecologia* **70**: 367–378.

Lahlou, B. (1980). Les hormones dans l'osmorégulation des poissons. In *Environmental physiology of fishes*: 201–240. (Ed. Ali, M.A.). Plenum Press, New York. (*NATO adv. Stud. Inst. Ser. A (Life Sci.)* **35**.)

McCormick, S. D. & Naiman, R. J. (1984). Osmoregulation in the brook trout, *Salvelinus fontinalis*. II. Effects of size, age and photoperiod on seawater survival and ionic regulation. *Comp. Biochem. Physiol. (A)* **79**: 17–28.

McDonald, D. G. & McMahon, B. R. (1977). Respiratory development in Arctic char *Salvelinus alpinus* under conditions of normoxia and chronic hypoxia. *Can. J. Zool.* **55**: 1461–1467.

Maetz, J. (1971). Fish gills: mechanisms of salt transfer in fresh water and sea water. *Phil. Trans. R. Soc. (B)* **262**: 209–249.

Marshall, E. K. (1930). Function of the glomerular and aglomerular kidneys. *Am. J. Physiol.* **94**: 1–10.

Marshall, E. K. & Smith, H. W. (1930). The glomerular development of the vertebrate kidney in relation to habitat. *Biol. Bull. mar. biol. Lab., Woods Hole* **59**: 135–153.

Marshall, L. G. & Corruccini, R. S. (1978). Variability, evolutionary rates, and allometry in dwarfing lineages. *Paleobiology* **4**: 101–119.

Marshall, W. S. (1977). Transepithelial potential and short-circuit current across the isolated skin of *Gillichthys mirabilis* (Teleostei: Gobiidae), acclimated to 5% and 100% seawater. *J. comp. Physiol.* **144**: 157–165.

Mehrotra, P. N. & Jha, V. S. (1962). On the relation between body weight and surface area in certain fresh-water fishes. *Ichthyologica, Kanpur* **1**: 71–72.

Miller, P. J. (1979). Adaptiveness and implications of small size in teleosts. *Symp. zool. Soc. Lond.* No. 44: 263–306.

Miller, P. J. (1984). The tokology of gobioid fishes. In *Fish reproduction: strategies and tactis*: 119–153. (Eds Potts, G. W. & Wootton, R. J.). Academic Press, London.

Miller, P. J. (1986). Gobiidae. In *Fishes of the north-eastern Atlantic and the Mediterranean* 3: 1019–1085. (Eds Whitehead, P. J. P., Bauchot, M.-L., Hureau, J.-C., Nielsen, J. & Tortonese, E.). UNESCO, Paris.

Miller, P. J. (1987). Affinities, origin and adaptive features of the Australian desert goby *Chlamydogobius eremius* (Zietz, 1896) (Teleostei: Gobiidae). *J. nat. Hist.* **21**: 687–705.

Milton, P. (1971). Oxygen consumption and osmoregulation in the shanny, *Blennius pholis*. *J. mar. biol. Ass. U.K.* **51**: 247–265.

Mishra, K. P., Rao, N. V. A., Alim, A., & Ahmad, M. F. (1991). Functional peculiarities of the alimentary canal of *Lepidocephalichthys guntheа* (Hamilton). *J. inld Fish. Soc. India* **23**: 59–63.

Mittal, A. K. & Banerjee, T. K. (1980). Keratinization versus mucus secretion in fish epidermis. *Linn. Soc. Symp. Ser.* No. 9: 1–12.

Moitra, S. K. & Ray, A. K. (1977). Morpho-histology of the alimentary canal of an Indian fresh-water perch, *Colisa fasciata* (Bloch) in relation to food and feeding habits. *Anat. Anz.* **141**: 37–58.

Moitra, S. K. & Ray, A. K. (1979). The comparative morpho-histology and anatomy of the digestive system of two Indian freshwater perches, *Ambassis nama* (Ham.) and *Ambassis ranga* (Ham.) in relation to their food and feeding habits. *Zool. Jb. (Anat.)* **102**: 142–169.

Morgulis, S. (1915). The body surface of flounders and its relation to the gaseous metabolism. *Am. J. Physiol.* **36**: 207–216.

Muir, B. S. (1969). Gill dimensions as a function of fish size. *J. Fish. Res. Bd Can.* **26**: 165–170.

Muir, B. S. & Hughes, G. M. (1969). Gill dimensions for three species of tunny. *J. exp. Biol.* **51**: 271–285.

Munshi, J. S. D., Ojha, J., & Sinha, A. L. (1980). Morphometrics of the respiratory organs of an air-breathing catfish, *Clarias batrachus* (Linn.) in relation to body weight. *Proc. Indian natn. Sci. Acad. (B)* **46**: 621–635.

Murphy, P. G. & Murphy, J. V. (1971). Correlations between respiration and direct uptake of DDT in the mosquito fish *Gambusia affinis*. *Bull. envir. Contam. Toxicol.* **6**: 581–588.

Nash, J. (1931). The number and size of glomeruli in the kidneys of fish with observations on the morphology of the renal tubules of fishes. *Am. J. Anat.* **47**: 425–445.

Nelson, J. S. (1994). *Fishes of the world.* (3rd edn). John Wiley & Sons, New York.

Niimi, A. J. & Morgan, S. L. (1980). Morphometric examination of the gills of walleye, *Stizostedion vitreum vitreum* (Mitchill) and rainbow trout, *Salmo gairdneri* Richardson. *J. Fish Biol.* **16**: 685–692.

Nishimura, H. & Imai, M. (1982). Control of renal function in freshwater and marine teleosts. *Fed. Proc.* **41**: 2355–2360.

Niva, B., Ojha, J., & Munshi, J. S. D. (1981). Morphometrics of the respiratory organs of an estuarine goby, *Boleophthalmus boddaerti*. *Jap. J. Ichthyol.* **27**: 316–326.

Nordlie, F. G. (1985). Osmotic regulation in the sheepshead minnow *Cyprinodon variegatus* Lacépède. *J. Fish Biol.* **26**: 161–170.

Nordlie, F. G., Szelistowski, W. A., & Nordlie, W. C. (1982). Ontogenesis of osmotic regulation in the striped mullet, *Mugil cephalus* L. *J. Fish Biol.* **20**: 79–86.

Norman, A. (1981). The smallest fish there is. *Freshwat. mar. Aquarium* **4**: 26–27.

Oikari, A. (1977). Renal adaptations of Baltic *Myoxocephalus scorpius* (L.) and *M. quadricornis* (L.) to hypotonic brackish water. *Annls zool. fenn.* **14**: 162–172.

Oikawa, S. & Itazawa, Y. (1985). Gill and body surface areas of the carp in relation to body mass, with special reference to the metabolism–size relationship. *J. exp. Biol.* **117**: 1–14.

Oikawa, S., Itazawa, Y., & Gotoh, M. (1991). Ontogenetic change in the relationship between metabolic rate and body mass in a sea bream *Pagrus major* (Temminck & Schlegel). *J. Fish Biol.* **38**: 483–496.

Parry, G. (1966). Osmotic adaptation in fishes. *Biol. Rev.* **41**: 392–444.

Pauly, D. (1981). The relationships between gill surface area and growth performance in fish: a generalization of von Bertalanffy's theory of growth. *Meeresforschung Rep. mar. Res.* **28**: 251–282.

Pauly, D. (1982). Further evidence of a limiting effect of gill size on the growth of fish: the case of the Philippine goby, *Mistichthys luzonensis*. *Kalikasan* **11**: 379–383.

Pillay, T. V. R. & Sarojini, K. K. (1950). On the larval development of the Indian transparent goby, *Gobiopterus chuno* (Hamilton) with observations on its bionomics. *Proc. natn. Inst. Sci. India* **16**: 181–187.

Prasad, M. S. (1988). Morphometrics of gills during growth and development of the air-breathing habit in *Colisa fasciatus* (Bloch and Schneider). *J. Fish Biol.* **32**: 367–381.

Robotham, P. W. J. (1978). The dimensions of the gills of two species of loach, *Noemacheilus barbatulus* and *Cobitis taenia*. *J. exp. Biol.* **76**: 181–184.

Rogers, S. I. (1986). *Reproductive energetics of the common goby,* Pomatoschistus microps *(Teleostei: Gobiidae)*. PhD thesis: University of Bristol.

Roth, G., Rottluff, B., Grunwald, W., Hanken, J. & Linke, R. (1990). Miniaturization in plethodontid salamanders (Caudata: Plethodontidae) and its consequences for the brain and visual system. *Biol. J. Linn. Soc.* **40**: 165–190.

Roubal, F. R. (1987). Gill surface area and its components in the yellowfin bream, *Acanthopagrus australis* (Günther) (Pisces: Sparidae). *Aust. J. Zool.* **35**: 25–34.

Schmidt-Nielsen, K. (1984). *Scaling: why is animal size so important?* Cambridge University Press, Cambridge.

Siankowa, L. (1966). The surface area of the intestinal mucosa in bream—*Abramis brama* (L.). *Studia Soc. Sci. torun. (Zool.)* **8**: 20–74.

Suyehiro, Y. (1942). A study on the digestive system and feeding habits of fish. *Jap. J. Zool.* **10**: 1–303.

Sweet, S. S. (1980). Allometric inference in morphology. *Am. Zool.* **20**: 643–652.

Te Winkel, L. E. (1935). A study of *Mistichthys luzonensis* with special reference to conditions correlated with reduced size. *J. Morph.* **58**: 463–535.

Unnithan, R. R. (1965). On the mucosal coefficient of a few Indian fresh water fishes. *Sci. Cult.* **31**: 387–388.

Videler, J. J. & Weihs, D. (1982). Energetic advantages of burst-and-coast swimming of fish at high speeds. *J. exp. Biol.* **97**: 169–178.

Webb, P. W. (1982a). Locomotor patterns in the evolution of actinopterygian fishes. *Am. Zool.* **22**: 329–342.

Webb, P. W. (1982b). Fast-start resistance of trout. *J. exp. Biol.* **96**: 93–106.

Webb, P. W. & Blake, R. W. (1985). Swimming. In *Functional vertebrate morphology*: 110–128. (Eds Hildebrand, M., Bramble, D. M., Liem, K., & Wake, D. B.). Harvard University Press, Massachusetts.

Webb, P. W. & Skadsen, J. M. (1979). Reduced skin mass: an adaptation for acceleration in some teleost fishes. *Can. J. Zool.* **57**: 1570–1575.

Weihs, D. (1974). Energetic advantages of burst swimming of fish. *J. theor. Biol.* **48**: 215–229.

Weitzman, S. H. & Vari, R. P. (1988). Miniaturization in South American freshwater fishes; an overview and discussion. *Proc. biol. Soc. Wash.* **101**: 444–465.

Wellensiek, U. (1953). Die Allometrieverhältnisse und Konstructionsänderungen bei dem kleinsten Fisch im Vergleich mit etwas grösseren verwandten Formen. *Zool. Jb. (Anat.)* **73**: 187–228.

Wheeler, A. (1969). *The fishes of the British Isles and north-west Europe.* Macmillan, London.

Wood, C. M. & Randall, D. J. (1973). The influence of swimming activity on water balance in the rainbow trout *Salmo gairdneri. J. comp. Physiol.* **82**: 257–276.

Zihler, F. (1982). Gross morphology and configuration of digestive tracts of Cichlidae (Teleostei, Perciformes): phylogenetic and functional significance. *Neth. J. Zool.* **32**: 544–571.

Symp. zool. Soc. Lond. (1996) No. 69: 47–61

Miniaturization in tetrapods: consequences for skull morphology

OLIVIER RIEPPEL

Department of Geology
The Field Museum
Roosevelt Road at Lake Shore Drive
Chicago, IL 60605–2496, USA

Synopsis

Miniaturization and its consequences for skull structure in amphibians, reptiles, and mammals are reviewed. The smallest tailed tetrapod is a salamander; physiological constraints related to endothermy prevent size in mammals decreasing to the degree observed in amphibians and reptiles. With decreasing head size, the brain and sense organs increase in relative size and become dominant factors in the shaping of the skull of miniaturized tetrapods. Skull morphology may be further modified by the mode of life of the organisms, as it is in miniaturized and fossorial / burrowing tetrapods (caecilians, squamates), where physiological constraints of miniaturization (minimal size of brain and labyrinth organs) combine with adaptational needs in a radical change of skull structure. Paedomorphosis is always involved in changes of skull morphology in miniaturized lizards, but in no case is it also a sufficient explanation for all the observed modifications.

Introduction

Miniaturization has been defined as '. . . the evolution of extremely small body size within a lineage' (Hanken & Wake 1993: 502). By that definition, miniaturization is understood as a process of developmental heterochrony which can be discovered only with reference to a well-corroborated hypothesis of phylogenetic relationships indicating relatively large size as the ancestral character. 'Extremely' small size is identified as the size class at which '. . . important physiological or ecological functions . . . are affected' (Hanken & Wake 1993: 502), sometimes resulting in major morphological changes. Because functional and/or physiological constraints may trigger fundamental morphological changes in miniaturized clades, miniaturization has also been viewed as a key to the understanding of the origin of higher taxa. In that sense, miniaturization reflects the inversion of Cope's rule (Gould 1977: 285) which states that overall body size generally increases

ZOOLOGICAL SYMPOSIUM No. 69
ISBN 0–19–857787–7

through time within a clade, its earliest representatives typically being smaller than later forms.

The widespread occurrence of miniaturization throughout the animal kingdom has recently been reviewed by Hanken & Wake (1993). The present paper will focus on morphological consequences in the skull of selected miniaturized tetrapods. Skull morphology is particularly susceptible to morphological changes as a consequence of miniaturization because it relates to such important organs and functions as the brain, the sense organs and the feeding mechanism. The brain generally shows a relative size increase in smaller animals, as do sense organs unless they are reduced (as are the eyes in burrowing tetrapods). Miniaturization may require fundamental changes in jaw adductor musculature in order to preserve an effective muscle fibre length, since the range of action of a muscle fibre within a relatively optimal range of the length–tension curve depends on the absolute length of the muscle fibre. Selection pressure for small size may therefore result in physiological and/or functional constraints necessitating morphological changes in the skull which are not part of the primary adaptive strategy of the particular organism.

Amphibia

The plethodontid salamander genus *Thorius* has been identified as the smallest extant tailed tetrapod (Hanken 1983, 1984). This genus illustrates the consequences of relative size increase of the sense organs and brain in the head of miniaturized tetrapods. The skull loses several bones present in related and larger genera, while other elements remain poorly developed and may fail to articulate with neighbouring bones. The skull has 'in effect shrunken around the now predominant brain and sense organs' (Hanken 1984: 67). The jaws remain poorly developed and many species are virtually toothless. Prey capture is effected by an 'elaborate tongue projecting mechanism' (Hanken 1984: 68) which involves complex changes in the hyobranchial skeleton and the associated musculature. Several aspects of miniaturization in *Thorius* can be explained as a consequence of paedomorphosis, and may indeed be related to progenesis (Roth, Rottluff, & Linke 1988; Roth, Rottluff, Grunwald *et al.* 1990), such as the loss of cranial elements or their reduced ossification. However, other characters of miniaturized morphology in *Thorius* result from accelerated development, such as ossification of mesopodial elements and degree of ossification of the long bones in the limbs (Hanken 1982) or from processes compensating for the relatively increased cell size, particularly in the eye and the optic central nervous system (Roth, Rottluff, & Linke 1988; Roth, Rottluff, Grunwald *et al.* 1990). Paedomorphosis likewise does not account for the emergence of evolutionary novelties in the limb skeleton of *Thorius* (Hanken 1985). Miniaturization is thus shown to be a complex phenomenon for which paedomorphosis may be a necessary, but certainly not a sufficient explanation.

In frogs, again, there is no consistent correlation between relative degree of paedomorphosis and adult body size (Davies 1989). Loss of cranial bones is also common in miniaturized frogs, the lost elements generally corresponding to those appearing late in ontogeny in larger and related forms (Trueb & Alberch 1985), but the relative degree of ossification of dermal skull elements shows no direct correlation with miniaturization (Trueb & Alberch 1985).

The skull of a miniaturized microsaur from the Lower Permian of Texas is well ossified, but again shows profound changes in the ethmoidal, orbital, and otico-occipital regions due to the relative size increase of the nasal and otic capsules and of the eyes (Carroll 1990). Modification of skull structure in this miniaturized microsaur results in convergent similarities of *Quasicaecilia* with extant caecilian amphibians, a group of limbless aquatic and/or burrowing Lissamphibia. In particular the burrowing forms will be under a selection pressure favouring reduction of the skull diameter. Owing to the relative increase of the eye and of the otic capsules in small skulls, the size of the adductor chamber housing the jaw adductor musculature is severely reduced in caecilians. This is compensated for by a particular reorientation of the interhyoideus muscle which now inserts into the retroarticular process of the lower jaw, functioning as a jaw adductor (Nussbaum 1977; Bemis, Schwenk, & Wake 1983). One of the smallest species of caecilians is *Idiocranium russeli* and it shows a number of paedomorphic features in skull structure due to the early cessation of development (Wake 1986). Truncation of development is invoked in the explanation of the reduction of the frontals, the persistence of a large dorsal fontanelle, the small vomers, and the absence of sutures or overlap among many dermal elements (Wake 1986). However, the extensive mineralization of the sphenethmoid and mesethmoid components of the neurocranium is unusual, and is viewed as a compensation for the reduction of the dermatocranium in correlation with the burrowing mode of life of this species.

Reptilia

The stem reptiles (captorhinomorphs) are characterized by an anapsid skull, and the neurocranium located within the dermatocranium. The temporal area of the dermatocranium is closed and the jaw adductor musculature is packed into the space between neurocranium and dermatocranium. To maximize the physiological cross-section of the jaw adductor musculature within the limited space defined by skull architecture, the jaw adductors show a complex pinnate structure in relation to a multipartite internal tendinous skeleton. The generalized lizard skull is of a modified diapsid type (Rieppel 1993). Of the neurocranium, only the otico-occipital segment ossifies; the ethmoidal and orbitotemporal components generally remain cartilaginous. The ossified neurocranium is movably suspended within the dermatocranium: points of articulation are between the basipterygoid processes and the pterygoid bones (palatobasal articulation), between the distal tips of the paroccipital processes and the squamosal bones, and between the ascending process of the supraoccipital and the parietal bone (metakinetic joint). The dermatocranium has lost the lower temporal arch and comprises a mesokinetic line of flexion (at the fronto-parietal suture) correlated with a hypokinetic line of flexion in the dermal palate. The jaw adductor musculature originates from the upper temporal arch, from the lateral margin of the parietal, from the margins of the post-temporal fossa (parietal and paroccipital process), and from the lateral wall of the ossified braincase.

Modifications of skull morphology in miniaturized lizards have been reviewed by Rieppel (1984a, 1988, and references therein). In lizards, miniaturization is generally correlated with body elongation (Griffith 1990) and limb reduction. Profound changes in cranial structure are not a by-product of miniaturization only, but are correlated with fossorial and/or burrowing habits. Miniaturized yet non-fossorial lizards may retain a skull structure closely similar to the generalized condition described above, as seen in sphaerodactylid geckos and small Lacertidae such as *Algyroides fitzingeri* (Rieppel 1984a). Again, fossorial and/or burrowing habits will favour reduction of skull diameter, since this will decrease the energy required during the burrowing action of elongated limbless lizards, but burrowing may also require compensations for paedomorphic changes in order to strengthen the skull. Modifications of skull structure in miniaturized and fossorial and/or burrowing lizards are similar in unrelated clades (e.g. acontine skinks within the Scincomorpha, *Anniella* within the Anguimorpha, and *Dibamus* and Amphisbaenia within the Autarchoglossa *incertae sedis*) and, hence, excellent examples of convergent evolution due to physical and physiological constraints (Fig. 1).

The skull of miniaturized fossorial and/or burrowing lizards is generally characterized by the loss of the upper temporal arch, a marked increase of the size of the otic capsules (related to physiological constraints set by the function of the semicircular canals: Jones & Spells 1963), reduction of the eyes and telescoping of the snout. The relative size of the braincase increases with increasing relative size of the brain and the braincase combines with the dermatocranium in the formation of a closed cranial box, correlated with the loss of meso- and metakinesis, and closure of the post-temporal fenestrae. The lateral wall of the braincase is closed by ventral extension of the parietal, and an anterior extension (alary process) of the prootic.

The jaw adductor musculature expands in a posterodorsal direction across the parietal and prootic to a position dorsal to the otic capsules (the 'temporalis' position of Säve-Söderbergh 1945). Miniaturization of the skull reduces the distance between the top of the skull roof and the dorsal tip of the coronoid process, and hence the maximum absolute length of anterior, vertically orientated muscle fibres. To reduce passive stretching of muscle fibres and to increase their individual length, the musculature becomes obliquely orientated, trending in a posterodorsal direction (for further discussion of these parameters, see below). The loss of mechanical advantage incurred by such rearrangement is at least partially compensated for by an increased height of the coronoid process, which is particularly obvious in *Dibamus* and Amphisbaenia (Fig. 1b, c).

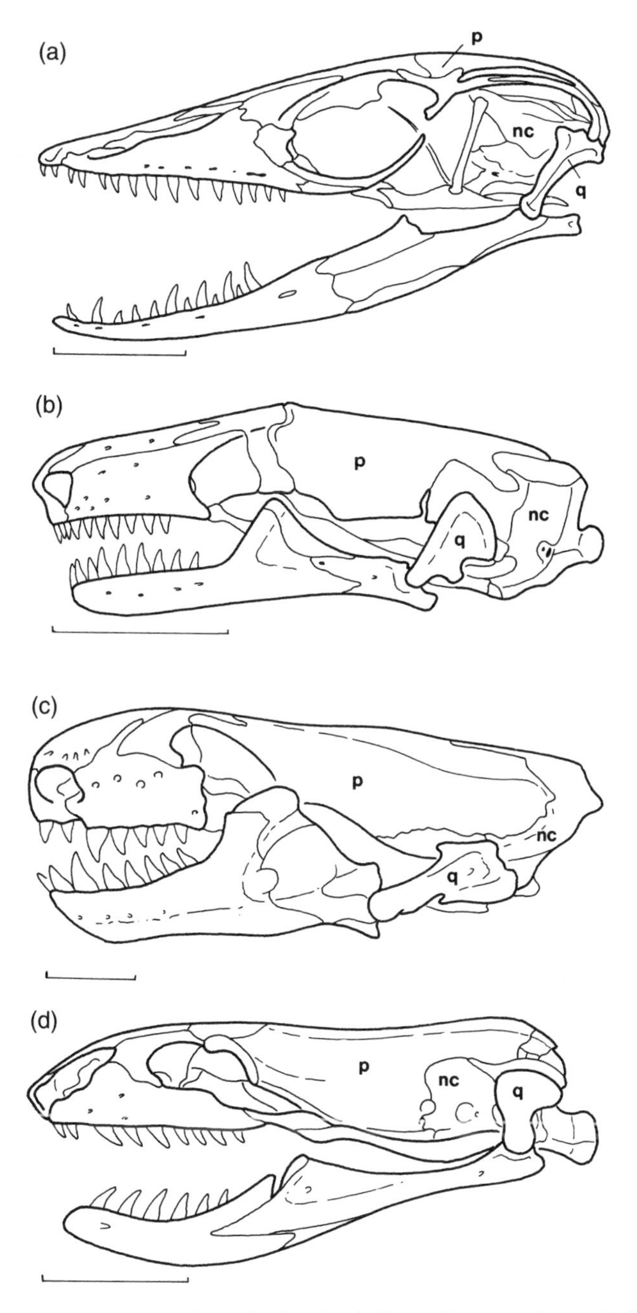

Fig. 1. Convergence in miniaturized lizard and snake skulls. (a) *Varanus salvator*, skull in left lateral view (scale bar equals 20 mm); (b) *Dibamus novaeguineae*, skull in left lateral view (scale bar equals 2 mm); (c) *Trogonophis wiegmanni*, skull in left lateral view (scale bar equals 5 mm); (d) *Cylindrophis rufus*, skull in left lateral view (scale bar equals 5 mm). Abbreviations: nc, neurocranium; p, parietal; q, quadrate. (Redrawn after various sources.)

Some of the changes observed, such as the loss of the upper temporal arch (Fig. 2; Rieppel 1982), are related to paedomorphosis. Other modifications result from peramorphosis, such as the closure of the lateral wall of the braincase by parietal downgrowths (Rieppel 1984b). This complex pattern of heterochronic changes is reflected in the ontogeny of the dermatocranium. The general pattern of ossification of the dermatocranium is fairly constant for generalized (i.e. non-miniaturized) lizards and may reflect the plesiomorphic pattern for the group (Rieppel 1992, in press, and references therein). The first element to ossify is the pterygoid, and there is a general predominance of the ossification of the bones associated with the dermal palate and jaws. The parietal invariably has a paired origin in squamates. In generalized lizards, its ossification starts in the marginal zones along the lateral and, later, posterior edges in the embryo, whereas the skull table generally ossifies during early post-embryonic growth only. The fronto-parietal suture therefore becomes defined during post-embryonic growth only. This delayed ossification of the dermal skull table is a marked contrast to turtles and crocodiles, and may be related to the evolution of mesokinesis in lizards (Rieppel in press). In fact, the skull of a hatchling lizard is a highly kinetic framework composed of all elements necessary for prey capture and ingestion (Fig. 3). The jaws and dermal palate required to seize and ingest prey are fully

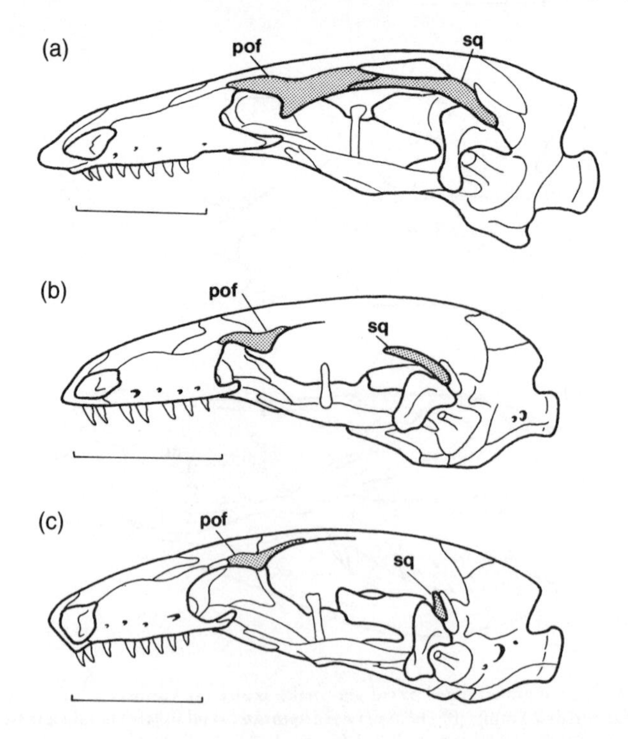

Fig. 2. The loss of the upper temporal arch in the genus *Typhlosaurus* (Scincidae, Acontinae). (a) *Typhlosaurus lineatus*; (b) *Typhlosaurus vermis*; (c) *Typhlosaurus cregoi*. Scale bar equals 2 mm. Abbreviations: pof, postorbitofrontal; sq, squamosal. (Redrawn after Rieppel 1982.)

ossified, as are the lateral wall of the braincase, the marginal areas of the parietal and the upper temporal arch, required as sites for muscle attachment (Rieppel 1984b). However, lack of ossification in the skull leaves large areas of the brain uncovered by bone. Whether or not this relates to particular patterns of brain growth remains unknown.

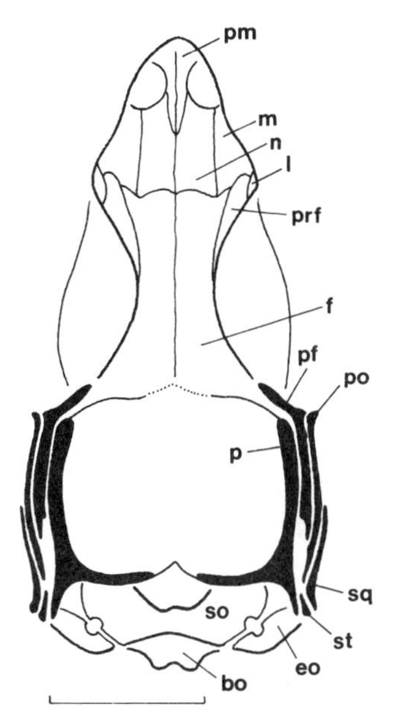

Fig. 3. Ossification of the parietal skull table in a hatchling *Podarcis muralis* (Lacertidae). Scale bar equals 2 mm. Abbreviations: bo, basioccipital; eo, exoccipital; f, frontal; l, lacrimal; m, maxilla; n, nasal; p, parietal; pf, post-frontal; pm, pre-maxilla; po, post-orbital; prf, pre-frontal; so, supraoccipital; sq, squamosal; st, supratemporal. (Redrawn after Rieppel 1984b.)

In fossorial/and or burrowing miniaturized lizards, the timing of ossification in the dermatocranium is changed. The upper temporal arch and some circumorbital elements are reduced or lost through paedomorphosis (Rieppel 1984b), but the parietal shows precocious ossification relative to the generalized pattern. Again ossifying in the lateral areas first, i.e. in the lateral parietal flanges providing the secondary lateral braincase wall, the two parietals meet along the dorsal midline and may have fused in the hatchling (*Acontias*: Brock 1941). Peramorphosis of the parietal correlates with a loss of cranial kinesis as is required in a skull used for burrowing. The frontoparietal suture may still be braced by a rudimentary postorbitofrontal bone in some groups, but the element is lost in others such as *Dibamus* and Amphisbaenia (Figs 1, 2). The latter two taxa also develop different interlocking mechanisms at the frontoparietal suture (Rieppel 1984c). The palatobasal articulation generally persists, although the joint may become synarthrotic, and the basipterygoid processes reorientated so as to brace the dermal palate against stress resulting from burrowing actions.

An interesting aspect of the restructuring of the skull in fossorial and/or burrowing lizards is that it results in a morphology closely comparable to the ophidian skull (Fig. 4). Snakes share a number of characters with miniaturized lizards, such as the elongated body, loss of limbs, loss of upper temporal arch, reduction or loss of some circumorbital elements, relative size increase of the braincase, and combination of the braincase with the peramorphotic parietal to form a closed cranial box, in fact a central strut from which the movable jaws are suspended. These convergences tie in with earlier theories postulating an origin of snakes from some fossorial and/or burrowing squamate clade (Walls 1940; Bellairs & Underwood 1951; Rieppel 1988). Whereas the sister group of snakes has not yet been firmly established (see Cundall, Wallach & Rossmann 1993, for a discussion), miniaturization of a fossorial and/or burrowing squamate may provide an explanatory model for origin of the ophidian bauplan (Rieppel 1988). It is interesting to note that heterochrony is involved in the miniaturization of the lizard skull, and the same has been postulated for the origin and further evolution of the ophidian skull (Irish 1989). Indeed, the transition from a lizard-like to an ophidian squamate may be the only example among extant tetrapods illustrating the role of miniaturization in the origin of a higher taxon.

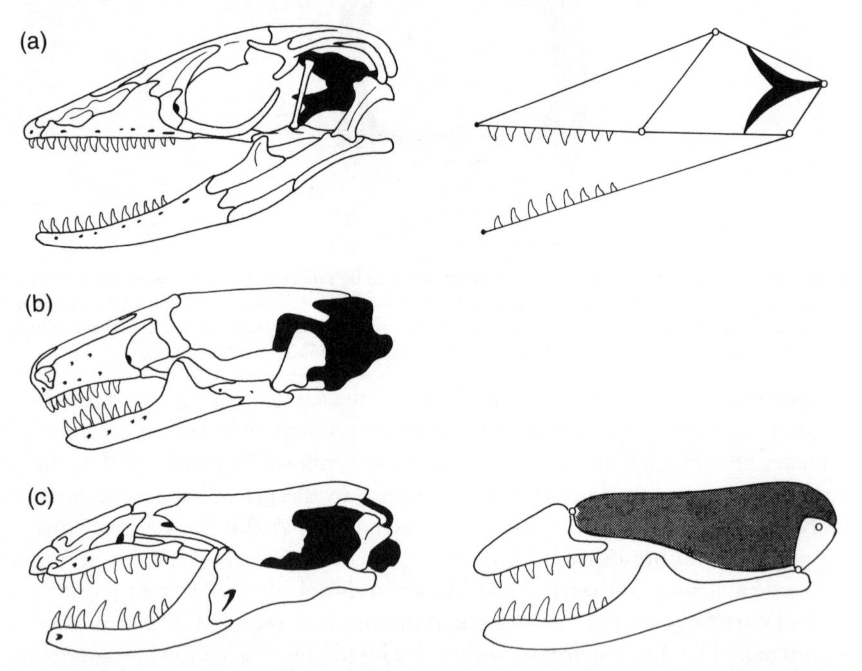

Fig. 4. The transition from the lacertilian to the ophidian skull. (a) Left, *V. salvator*; right, schematic representation of cranial kinesis in lizards. (b) *Dibamus novaeguineae*. (c) Left, *C. rufus*: right, schematic representation of cranial kinesis in snakes. Not to scale. (Redrawn after various sources.)

Mammalia

The Soricidae (shrews) offer an excellent example of the complex integration of size-constraining factors (Rieppel 1985) in endotherms. The smallest shrew, *Suncus etruscus*, is among the smallest living mammals, yet still distinctly larger than the smallest amphibians or reptiles. The relatively high metabolic rate constrains minimal size for endotherm organisms, since their surface increases with decreasing body mass (Alexander this volume, pp. 3–14). To maintain a relatively high body temperature at small size, *S. etruscus* has exploited maximal physiological capacities of heart musculature (Hoppeler *et al.* 1984).

Shrews are unique among mammals in a number of characters relating to skull structure (Fig. 5). The shrew skull is low and elongate and has no zygomatic arch, the lower jaws have a double mandibular joint, the coronoid process of the lower jaw is very high relative to the skull and bears a 'fossa temporalis interna' on its medial side, and the jaw adductor musculature is essentially horizontally orientated (Dötsch & Dantuma 1989). These parameters are the same in relatively large (*Suncus murinus*) as well as in small species (*S. etruscus*) and may therefore have made miniaturization possible rather than being a consequence thereof. Subtle differences can be noted, however, in the comparison of the skulls of these two species.

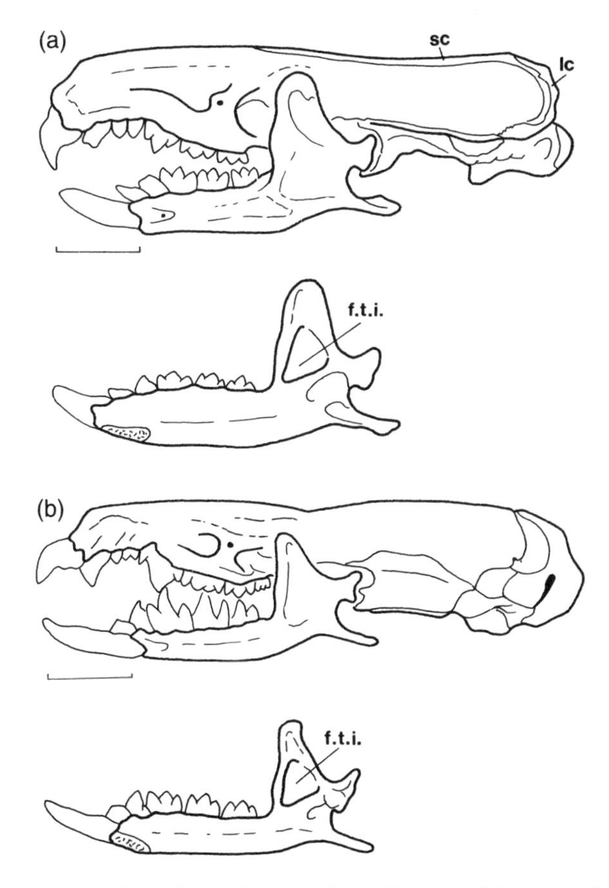

Fig. 5. (a) *Suncus murinus*, skull in left lateral view and mandible in medial view. Scale bar equals 5 mm. (b) *Suncus etruscus*; skull in left lateral view and mandible in medial view. Scale bar equals 2 mm. Abbreviations: f.t.i., fossa temporalis interna; lc, lambdoidal crest; sc, sagittal crest. (Redrawn after Rieppel 1985.)

The skull of *S. murinus* (condylobasal length, 32 mm) shows the characteristic low but elongated shape of the skull. The jaw adductor musculature spreads across the temporal region of the skull, reaching a low sagittal crest along the dorsal midline, and a distinct lambdoidal crest which separates the temporal region from the occiput posteriorly and posterolaterally. Basic skull structure is the same in *S. etruscus* (condylobasal length: 12.5 mm), although the relative size of the braincase has increased owing to a relative size increase of the brain and of the otic capsules. The temporal region protrudes from behind the parietal, and the lambdoidal crest is lost. Rearrangement of the neurocranium relative to the dermatocranium results in a slight expansion of the area of origin for the temporalis musculature.

The temporalis, masseter, and pterygoideus musculature of shrews show a complex pinnate architecture described by Gasc (1963), Dötsch (1982, 1983), and Rieppel (1985). The general orientation of the muscle fibres is horizontal. Passive stretching of muscle fibres during jaw opening depends on the angle of insertion of the fibre relative to the long axis of the lower jaw (decreasing with decreasing angle), and on the distance of the insertion from the fulcrum (increasing with increasing distance) (Rieppel & Gronowski 1981). The horizontal fibre orientation in the jaw adductor musculature of *Suncus* reflects the low and elongated skull shape and reduces the passive stretching of fibres during jaw opening, but at the same time results in poor mechanical advantage. This can only be compensated for by an increased height of the coronoid process (DeMar & Barghusen 1972).

Increasing the height of the coronoid process causes problems for the anterior, more vertically positioned portion of the temporalis muscle, which also comprises the fibres with greatest mechanical advantage. The high coronoid process severely limits the distance between its tip and the top of the skull, yet the fibres spanning that distance are also those which undergo the greatest degree of stretching during jaw opening. The solution to that mechanical problem is the differentiation of a 'fossa temporalis interna' on the medial side of the coronoid process, allowing the anterior temporalis fibres to insert into the base of the coronoid process rather than into its tip.

Discussion and conclusions

Miniaturization in tetrapods involves complex heterochronic changes in skull development. Paedomorphosis is a likely pattern to be expected, since miniaturization may result from truncation of development (progenesis: Gould 1977). Paedomorphosis results in the loss or reduction of elements such as bones in the skull of amphibians and reptiles, or cranial crests in the skull of mammals. It may also result in a reduction of the general degree of ossification in the skull, as in salamanders (Hanken 1983, 1984), although this is not a general pattern and is not evident in frogs (Trueb & Alberch 1985; Davies 1989) or fossorial and/or burrowing squamates (Rieppel 1984a). Indeed, in all groups of miniaturized tetrapods studied so far heterochronic changes did not involve paedomorphosis only, but also peramorphosis of certain skeletal components as a compensatory mechanism for reduction, as well as the emergence of evolutionary novelty.

One of the most obvious consequences of miniaturization, resulting in convergent changes of skull morphology throughout tetrapods, relates to the relative size increase of the brain and sense organs (particularly the otic capsules) with decreasing overall size. The basic structure of the gnathostome skull corresponds to two cylinders, one positioned inside the other. The inner cylinder corresponds to the neurocranium, the outer to the dermatocranium, and the jaw adductor musculature relating to the splanchnocranium is accommodated between the two. The splanchnocranium is highly modified in tetrapods (palatoquadrate, stapes, Meckel's cartilage, hyobranchial skeleton), but the neurocranium remains movably suspended within the dermatocranium in lower tetrapods (Rieppel 1978) and the jaw adductor muscles remain situated between the two components. Fenestration or emargination of the temporal region of the dermatocranium in tetrapods may expose jaw musculature in superficial view, but does not change the basic topological relations of the skull components.

In most extant lower tetrapods (but see the discussion of caecilians above), only the posterior (otico-occipital) parts of the neurocranium ossify (if at all). With a relative size increase of the neurocranium in miniaturized forms (due to the relative size increase of the brain and semicircular canals), the neurocranium 'bulges' out from below and behind the dermatocranium and combines with the dermatocranium to form a closed cranial box from which the jaws are suspended. This constitutes a fundamental change in the topological relations of neurocranium and dermatocranium, which can be observed in lissamphibians (caecilians) as well as in reptiles (miniaturized fossorial and/or burrowing lizards, amphisbaenians, and snakes). One consequence is the loss of kinesis between the two components. The other consequence is that the jaw adductor musculature now lies superficial to the skull and is free to expand across the surface of the cranium to what Säve-Söderbergh (1945) called the temporalis position.

Such expansion of the area of origin of the jaw adductor musculature may be necessary in miniaturized forms to preserve a functional muscle fibre length. If miniaturization is, indeed, a valid model for the evolution of the ophidian bauplan, the expansion of the jaw adductor musculature to the temporalis position had important consequences for the late evolutionary success of the group. Combining neurocranium and dermatocranium in a solid and akinetic cranial box provided a central strut from which the highly mobile jaws characteristic of macrostomate snakes could be suspended (Rieppel 1988). Reduction of the internal tendinous skeleton in the jaw adductor musculature maximized individual fibre length, thus providing the fibres with the range of action required for the highly kinetic jaws.

The most fundamental change in the skull of miniaturized tetrapods relates to the relative size increase of the neurocranium, which combines with the dermatocranium to form a closed cranial box across the surface of which the jaw adductor musculature expands to the 'temporalis position' above the otic capsule.

With respect to these features, miniaturization induces changes in the lower tetrapod skull which mimic modifications and characterize the evolution of the avian and mammalian skull. In the Synapsida and the birds, the relative increase in size of the neurocranium is not correlated with miniaturization, but with a phylogenetic size increase of the brain. The evolution of the mammalian skull in particular involves changes superficially similar to those seen in some miniaturized squamates: closure of the cranium, expansion of the musculature to the temporalis position, and increasing height of the coronoid process (Barghusen 1968, 1972; DeMar & Barghusen 1972). It is perhaps because of this historical background that miniaturization in mammals (as seen in *S. etruscus*) does not affect skull structure to the same degree as it does in non-mammalian tetrapods. The skull of mammals is already built upon a scheme which lends itself to significant size reduction without any fundamental change, whereas the degree of size reduction is constrained by physiological constraints as they result from endothermy.

Summary

All miniaturized tetrapods investigated show a certain degree of paedomorphosis in skull morphology, but never is paedomorphosis the only factor influencing restructuring of the skull. Reduction or loss of cranial elements is often compensated for by peramorphosis of other parts of the skull as required by the mode of life of the animals in question. In the miniaturized salamander genus *Thorius*, reduction of the dermatocranium and dentition is correlated with a complex tongue-projecting mechanism for food capture (Hanken 1983, 1984). This results in the dominance of the eyes and optic centres in the brain in the shaping of the miniaturized head (Roth, Rottluff, & Linke 1988; Roth, Rottluff, Grunwald *et al.* 1990). Reduction of the dermatocranium in a miniaturized caecilian is compensated for by an increased ossification of the neurocranium, rendering the skull sturdy and rigid enough for burrowing (Wake 1986). Fossorial and/or burrowing squamates again show the paedomorphic reduction of the dermatocranium to be compensated for by peramorphosis of other elements resulting in an akinetic cranium suitable for subterranean locomotion without limbs (Rieppel 1984b).

Heterochronic changes in the development of miniaturized tetrapods are also frequently correlated with the emergence of evolutionary novelties (Hanken 1985). However, although miniaturization has been linked to the origin of a new bauplan, the only actual example for tetrapods which documents the possible involvement of miniaturization in the origin of a higher taxon is the transition from a miniaturized fossorial and/or burrowing lizard-like squamate to snakes (Rieppel 1988). Although miniaturization seems to offer an explanatory model for the origin of the ophidian bauplan, this example continues to be problematical because sister-group relations of snakes among other squamates have not yet been firmly established (Cundall *et al.* 1993).

Acknowledgements

I thank Dr Peter Miller of the University of Bristol for his invitation to participate in this symposium. The shrew material discussed in this study was kindly made available by Dr Peter Vogel, University of Lausanne.

References

Barghusen, H. R. (1968). The lower jaw of cynodonts (Reptilia, Therapsida) and the evolutionary origin of mammal-like adductor jaw musculature. *Postilla* No. 116: 1–49.

Barghusen, H. R. (1972). The origin of the mammalian jaw apparatus. In *Morphology of the maxillo-mandibular apparatus*: 26–32. (Ed. Schumacher, G.-H.). VEB Thieme, Leipzig.

Bellairs, A. d'A. & Underwood, G. (1951). The origin of snakes. *Biol. Rev.* **26**: 193–237.

Bemis, W. E., Schwenk, K. & Wake, H. M. (1983). Morphology and function of the feeding apparatus in *Dermophis mexicanus* (Amphibia: Gymnophiona). *Zool. J. Linn. Soc.* 77: 75–96.

Brock, G. T. (1941). The skull of *Acontias meleagris*, with a study of the affinities between lizards and snakes. *J. Linn. Soc. (Zool.)* **41**: 71–88.

Carroll, R. L. (1990). A tiny microsaur from the Lower Permian of Texas: size constraints in Palaeozoic tetrapods. *Palaeontology* **33**: 893–909.

Cundall, D., Wallach, V., & Rossman, D. A. (1993). The systematic relationships of the snake genus *Anomochilus*. *Zool. J. Linn. Soc.* **109**: 275–299.

Davies, M. (1989). Ontogeny of bone and the role of heterochrony in the myobatrachine genera *Uperoleia, Crinia* and *Pseudophryne* (Anura: Leptodactylidae: Myobatrachinae). *J. Morph.* **200**: 269–300.

DeMar, R. & Barghusen, H. R. (1972). Mechanics and the evolution of the synapsid jaw. *Evolution, Lawrence, Kans.* **26**: 622–637.

Dötsch, C. (1982). Der Kauapparat der Soricidae (Mammalia, Insectivora). Funktionsmorphologische Untersuchungen zur Kaufunktion bei Spitzmäusen der Gattungen *Sorex* Linnaeus, *Neomys* Kaup und *Crocidura* Wagler. *Zool. Jb. (Anat.)* **108**: 421–484.

Dötsch, C. (1983). Morphologische Untersuchungen am Kauapparat der Spitzmäuse *Suncus murinus* (L.), *Soriculus nigrescens* (Gray) und *Soriculus caudatus* (Horsfield) (Soricidae). *Säugetierk. Mitt.* **31**: 27–46.

Dötsch, C. & Dantuma, R. (1989). Electromyography and masticatory behavior in shrews (Insectivora). *Fortschr. Zool.* **35**: 146–147.

Gasc, J.-P. (1963). La musculature céphalique chez *Suncus* Ehr., *Crocidura* Wag., *Sylvisorex* Thom., *Myosorex* Gr. (Insectivores). *Mammalia* **27**: 582–601.

Gould, S. J. (1977). *Ontogeny and phylogeny*. Harvard University Press, Cambridge, Ma.

Griffith, H. (1990). Miniaturization and elongation in *Eumeces* (Sauria: Scincidae). *Copeia* **1990**: 751–758.

Hanken, J. (1982). Appendicular skeletal morphology in minute salamanders, genus *Thorius* (Amphibia, Plethodontidae): growth regulation, adult size determination, and natural variation. *J. Morph.* **174**: 57–77.

Hanken, J. (1983). Miniaturization and its effects on cranial morphology in plethodontid salamanders, genus *Thorius* (Amphibia: Plethodontidae): II. The fate of the brain and sense organs and their role in skull morphogenesis and evolution. *J. Morph.* **177**: 255–268.

Hanken, J. (1984). Miniaturization and its effects on cranial morphology in plethodontid salamanders, genus *Thorius* (Amphibia: Plethodontidae): I. Osteological variation. *Biol. J. Linn. Soc.* **23**: 55–75.

Hanken, J. (1985). Morphological novelty in the limb skeleton accompanies miniaturization in salamanders. *Science* **229**: 871–874.

Hanken, J. & Wake, D. B. (1993). Miniaturization of body size: organismal consequences and evolutionary significance. *A. Rev. Ecol. Syst.* **24**: 501–519.

Hoppeler, H., Lindstedt, S. L., Claassen, H., Taylor, C. R., Mathieu, O., & Weibel, E. R. (1984). Scaling mitochondrial volume in heart to body mass. *Respirat. Physiol.* **55**: 131–137.

Irish, F. (1989). The role of heterochrony in the origin of a novel bauplan: evolution of the ophidian skull. *Géobios Mém. spéc.* No. 12: 227–333.

Jones, G. M. & Spells, K. E. (1963). A theoretical and comparative study of the functional dependence of the semicircular canal upon its physical dimensions. *Proc. R. Soc. (B)* **157**: 403–419.

Nussbaum, R. A. (1977). Rhinatrematidae: a new family of caecilians (Amphibia: Gymnophiona). *Occ. Pap. Mus. Zool. Univ. Mich.* **682**: 1–30.

Rieppel, O. (1978). The phylogeny of cranial kinesis in lower vertebrates, with special reference to the Lacertilia. *Neues Jb. Geol. Paläont. Abh.* **156**: 353–370.

Rieppel, O. (1982). The phylogenetic relationships of the genus *Acontophiops* Sternfeld (Sauria: Scincidae), with a note on mosaic evolution. *Ann. Transv. Mus.* **33**: 241–257.

Rieppel, O. (1984a). Miniaturization of the lizard skull: its functional and evolutionary implications. *Symp. zool. Soc. Lond.* No. **52**: 503–520.

Rieppel, O. (1984b). The upper temporal arcade of lizards: an ontogenetic problem. *Rev. suisse Zool.* **91**: 475–482.

Rieppel, O. (1984c). The cranial morphology of the fossorial lizard genus *Dibamus* with a consideration of its phylogenetic relationships. *J. Zool., Lond.* **204**: 289–327.

Rieppel, O. (1985). Miniaturization of the tetrapod head: muscle fibre length as a limiting factor. In *Konstruktionsprinzipien lebender und ausgestorbener Reptilien*. Konzepte SFB 230 (4): 121–138. (Eds Riess, J. & Frey, E.). University of Tübingen, Tübingen.

Rieppel, O. (1988). A review of the origin of snakes. *Evol. Biol.* **22**: 37–130.

Rieppel, O. (1992). Studies on skeleton formation in reptiles. III. Patterns of ossification in the skeleton of *Lacerta vivipara* Jacquin (Reptilia, Squamata). *Fieldiana (Zool.)* (N.S.) No. **68**: 1–25.

Rieppel, O. (1993). Patterns of diversity in the reptilian skull. *The skull* 2: 344–390. (Eds Hanken, J. & Hall, B. K.). The University of Chicago Press, Chicago.

Rieppel, O. (In press). Studies on skeleton formation in reptiles. Patterns of ossification in the skeleton of *Lacerta agilis exigua* Eichwald (Reptilia, Squamata). *J. Herpet.*

Rieppel, O. & Gronowski, R. W. (1981). The loss of the lower temporal arcade in diapsid reptiles. *Zool. J. Linn. Soc.* **72**: 203–217.

Roth, G., Rottluff, W., Grunwald, W., Hanken, J., & Linke, R. (1990). Miniaturization in plethodontid salamanders (Caudata: Plethodontidae) and its consequences for the brain and visual system. *Biol. J. Linn. Soc.* **40**: 165–190.

Roth, G., Rottluff, B. & Linke, R. (1988). Miniaturization, genome size and the origin of functional constraints in the visual system of salamanders. *Naturwissenschaften* **75**: 297–304.

Säve-Söderbergh, G. (1945). Notes on the trigeminal musculature in non-mammalian tetrapods. *Nova Acta R. Soc. Scient. upsal.* (4) **13** (7): 1–59.

Trueb, L. & Alberch, P. (1985). Miniaturization and the anuran skull: a case study of heterochrony. *Fortsch. Zool.* **30**: 113–121.

Wake, M. H. (1986). The morphology of *Idiocranium russeli* (Amphibia: Gymnophiona), with comments on miniaturization though heterochrony. *J. Morph.* **189**: 1–16.

Walls, G. L. (1940. Ophthalmological implications for the early history of the snakes. *Copeia* **1940**: 1–8.

Symp. zool. Soc. Lond. (1996) No. 69: 63–81

Energetics and the evolution of body size in small terrestrial mammals

J. R. SPEAKMAN

Department of Zoology
University of Aberdeen
Aberdeen AB9 2TN, UK

Synopsis

Many species of very small terrestrial mammals exhibit clinal patterns in their body size, often with larger individuals predominating in populations from higher latitudes (Bergmann's law). Classically, this pattern has been explained as reflecting the advantageous surface-to-volume ratio of larger animals, which provides them with a thermoregulatory advantage in colder regions, because it reduces their energy demands. Attempts to expand on this energetics explanation are, however, fraught with difficulty. A major problem relates to the fact that it is not possible to speculate on the direction of selection on body size by considering costs alone. However, when both costs and gains are considered there is a range of potential effects, which depend critically on the interspecific scaling relationships of energy gains and energy costs. Since these traits may vary in time, in space, and across species, general predictions from the energetics hypothesis of the direction of selection on body size are therefore not possible, and a wide range of directional effects of latitude on size is therefore predicted (and observed). It is generally the case that, within a species, mass explains less than 50% of the individual variation in metabolic energy expenditure. This low explained variation raises the question of why some individuals sustain high rates of metabolism, when this would be likely to make them more susceptible to failure in achieving an energy balance. In wood mice, *Apodemus sylvaticus*, we have found a correlation between basal metabolic rate (BMR) and the thermogenic capacity (noradrenaline-induced metabolic rate). These latter data suggest that animals which might be at a disadvantage in terms of achieving an energy balance might be at an advantage in terms of achieving a heat balance under severe temperature conditions. This indicates a more complex involvement of energetics in the phenomenon of overwinter survivorship than is addressed by models of energy balance alone.

Introduction

Miniature terrestrial mammals exhibit a high degree of individual variability in their body masses, with coefficients of variation, across populations of a given species, often exceeding 30% of the mean mass. Such high coefficients of variation

ZOOLOGICAL SYMPOSIUM No. 69
ISBN 0–19–857787–7

mean that the smallest mature individuals of a given species are generally less than one-third the size of the largest. This variability in body size, however, does not occur at random. For example, discrete populations, with extreme sizes, both large and small, are frequently found on islands. The mature adult body mass of the mainland European population of the common vole (*Microtus arvalis*), for instance, varies between 15 and 30 g. However, a population on the Orkney islands, off the northern coast of Britain (60°N), and probably isolated from the remainder of the European population for approximately 4000–6000 years (Gorman 1991), has adult masses in the range 35–70 g. Within small island populations of miniature rodents, large deviations in body mass from the species average may, in part, be due to founder effects and drift, since inter-island transfer of small rodents may be rare (Redfield 1976; but see Ebenhard 1990).

Continental populations of small mammals, however, also deviate systematically from the mean mass of the species. A frequently observed trend in body sizes, in these groups of animals, involves size clines, with body mass increasing with increase in latitude. This trend is familiar as Bergmann's law, which was first formulated in 1847 (Blaxter 1989). Rensch (1936) suggested that 81% of North American mammals, including many of the smallest terrestrial mammals, conform to Bergmann's law. Since founder effects and drift are unlikely to be important factors influencing the systematic patterns of variation in body mass across large continental areas, it is appropriate to consider how these patterns might come about as a consequence of natural selection.

A major, and obvious, difference between environments, as one moves from the tropical regions to the poles, is the development of temperature-based seasonality, the year becoming separated into a warm summer and a cold winter. The severity and duration of the cold winter period become progressively greater as latitude increases. Many small mammals, in temperate and Arctic regions, experience high mortality during this cold period. It is not exceptional for more than 50% of individuals in small terrestrial mammal populations to perish during winter. For example, the population of house mice (*Mus musculus domesticus*) on the small Welsh island of Skokholm (52°N), where mortality can be separated from emigration, routinely falls from a zenith of approximately 2000 individuals at the end of each summer to a nadir of 100–200 individuals at the start of each spring (Berry 1968).

Given the profound impact of winter temperature conditions on mortality of small terrestrial animals, it is not unreasonable to infer that the directional variability in body masses of these animals is somehow linked to the abilities of the animals to survive periods spent in the cold. Supporting this view, several studies of small mammals have demonstrated that, during periods of cold, larger individuals have a greater probability of surviving than smaller individuals (e.g. Fleming & Rauscher 1978; Boonstra & Krebs 1979; Murie & Boag 1984; Sauer & Slade 1985; Campbell & Slade 1993). Murie & Boag (1984), for example, found that among juvenile ground squirrels (*Spermophilis columbianus*) heavy individuals were almost twice as likely to survive the winter as light individuals.

Given this background, it is widely believed that Bergmann's law describes an adaptive response, which has an explanation rooted in energetics. In this paper I will explore several theoretical attempts to elaborate on the link between energetics and the evolution of clines in body size, in these small mammals. This analysis will illustrate the complexity of deriving predictions from the energetics hypothesis, and the key roles played by several factors, for which, in most circumstances, we have no information. In the second section of the paper I shall present some empirical data which illustrate the difficulties of generating realistic theoretical models, and suggest that the role of energetics during overwinter survival may be considerably more complex than these models suggest.

Theoretical models

Models based on surface-to-volume effects on energy demands

If I take a kettle of boiling water and pour it into two cups, one of which holds 10 times more than the other, and then allow them to cool down, the temperatures in both cups will decline exponentially towards the ambient temperature. However, the water in the larger cup will cool more slowly than the water in the small cup. The reason for this difference is that heat is exchanged over the surface of the vessels and the larger cup has a lower surface-to-volume ratio. This means that the heat loss, relative to the volume of water, is lower in the large cup and, thus, it cools down more slowly. If we reverse this argument, and consider instead how much heat we would need to put into the two different-sized vessels in order to maintain them at a constant temperature—like an animal attempting to maintain a constant body temperature—the greater surface-to-volume ratio of the small vessel means that for each gram of water in that cup, we need to put in more heat than for the larger cup. This difference gets greater as ambient temperature declines (Fig. 1a).

Measurements made on animals of their thermoregulatory responses, in relation to declining temperature, support this simple Newtonian cooling model. For example, in Fig. 1(b), I present data on resting energy demands as a function of ambient temperature for two black mice (strain: C57/BL10), one of which was 47% heavier than the other. The pattern of energy demands and the differences between the smaller and larger individual match exactly the expectation from the Newtonian cooling model (Fig. 1a). The gradient of the line relating metabolic energy demands (W/g) to temperature, is called the whole-body thermal conductance. Several studies have demonstrated that the mean whole-body thermal conductance of populations of animals increases as they get smaller. Brown & Lee (1969), for example, examined 10 populations of wood rats (genus *Neotema*) and found a large decrease in whole-body thermal conductance with increased size of the subjects (see Fig. 1c).

The physical heat balance model presented above is beguilingly simple to accept. For a long period it was considered that the energetics explanation of

Bergmann's law was that larger animals had more favourable surface-to-volume ratios, which gave them a selective advantage over smaller individuals, in terms of heat balance, as it became colder. This effect was therefore believed to result in the observed size clines. For example, Brown & Lee (1969: 337) interpreted their observed trend in thermal conductance of wood rats (Fig. 1c) in this way, when they stated 'Large wood rats have a selective advantage in cold climates . . . because their smaller surface-to-mass ratio . . . permit[s] them to conserve metabolic heat.'

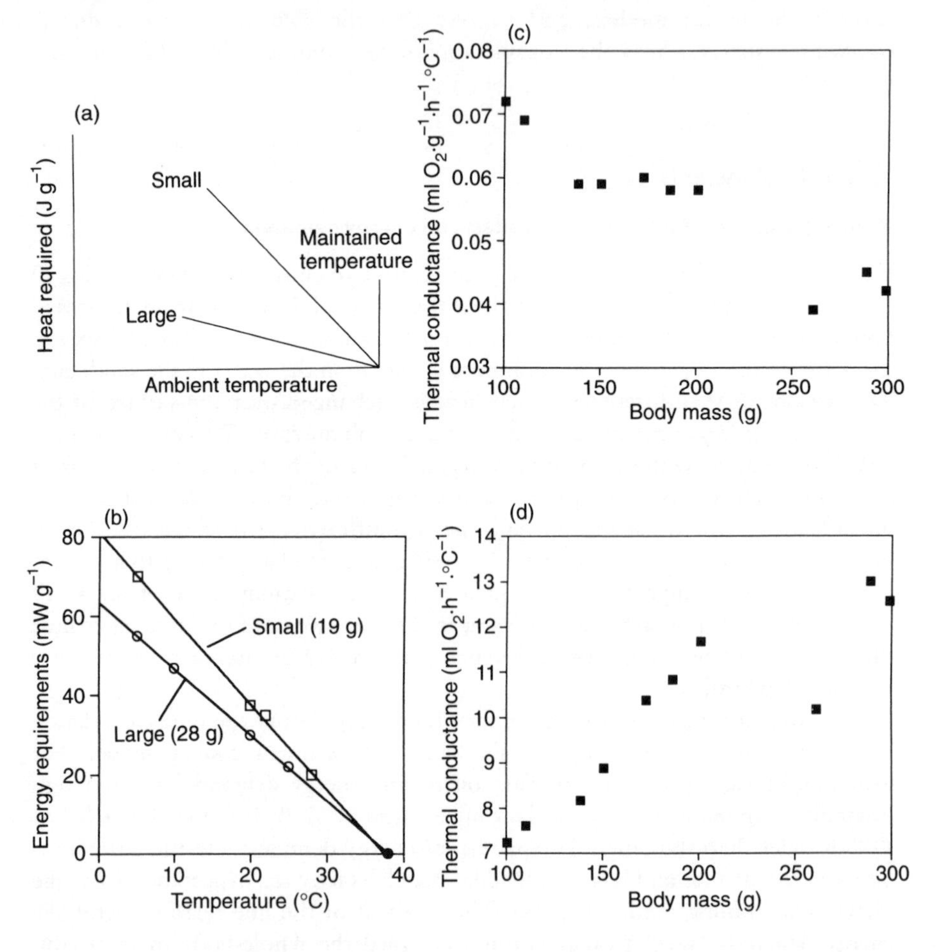

Fig.1. (a) Theoretical energy requirements to sustain the temperature of two vessels of water which have different volumes, and hence surface-to-volume ratios, as a function of ambient temperature. The larger vessel, with smaller surface-to-volume ratio, requires less energy per gram to maintain its temperature. (b) Energy requirements of two mice of the same strain which differ in their body mass, as a function of temperature. Observed demands match closely the theoretical prediction from the heat balance model (in 1a). (c) Thermal conductance measured for wood rats (*Neotema*) in relation to body mass (from Brown & Lee 1969). (d) Thermal conductances of wood rats recalculated on a whole-animal rather than on a per-gram basis.

Despite the fact that there are at least three problems with this explanation, elaborated below, which have been known since at least 1950 (Scholander, Walters, Hock, & Irving 1950; McNab 1971) it is still possible to find references in the literature to size clines and to Bergmann's law being the consequence of favourable surface-to-volume ratios of larger individuals (e.g. Davenport 1992).

The three problems with this explanation are as follows. Firstly, it is unclear from the model why there is ever a selective advantage to being small. If we examine the metabolic energy demands illustrated in Fig. 1b, the line for the small animal never falls below the line for the larger animal. Although the selection favouring large size will be less intense at warmer temperatures, it is clear from these curves that selection will never favour being small. Energetically this model predicts that it is always better to be bigger, whatever the temperature. The second problem is that energy demands are expressed in the model on a per-gram basis (cf. Fig. 1a–c). The rationale for expressing the energy expenditure per gram is that this supposedly normalizes the data for the effects of body mass. Even ignoring the fact that simply dividing a dependent variable by body mass will only normalize the mass effect if the gradient of the scaling relationship is 1.0 (Packard & Boardman 1987), the key question to be considered in this context is why this normalization is being performed at all. The question one is attempting to address, in the context of Bergmann's law, is the effect of body size on energy costs. It makes no sense, therefore, to perform before analysis a procedure which aims to remove the effect one is attempting to reveal! Moreover, the energy budget of an animal must be balanced on the basis of an entire animal, not per gram of animal. As McNab (1971: 846) pointed out 'Although weight-specific expressions are often convenient, it is important to realise that an animal does not live on a per-gram basis, but lives rather as an intact individual.' To illustrate the importance of this effect for the interpretation of thermoregulation data, consider the data for wood rats in Fig. 1(c) again. If these data are recalculated on a per-animal rather than on a per-gram basis, the direction of the relationship is completely reversed (Fig. 1d). Larger animals have greater whole-body thermal conductances. On a whole-animal basis, these data suggest that it would be energetically *disadvantageous* to be larger, which is the complete opposite of the interpretation by Brown & Lee (1969: see quote above) of the same data. The positive relationship between total energy costs and body size, despite larger animals having lower surface-to-volume ratios and generally better surface insulation, and the consequent inapplicability to surface-to-volume arguments in the explanation of Bergmann's law, were pointed out in the early 1950s (Scholander *et al.* 1950).

The third problem with this model is that it considers only energy demands. If energy balance is important to the selection of body size, it is also important to recognize that energy balance is the result not only of how much energy the animal is expending, but also how much energy it can acquire to meet those demands. Models, such as the surface-to-volume arguments presented above, which rely only on examining the effects of mass differences on energy costs, effectively assume that body mass has no effect on the energy gains that the animals can achieve. This is likely to be a naive assumption.

Models based on energy costs and gains

Models which address both the energy demands and the energy gains of animals can be divided into two groups: those which consider the situation when food is present, which, to my knowledge, have not previously been explicitly modelled, and those which consider the situation when food is absent, which have been extensively considered. I will treat each of these situations separately.

When food is present

Imagine a small animal species, individuals of which vary in their body mass. Relatively few studies have examined the effects of intraspecific variation in body mass on energy expenditure. The few studies which have been performed suggest that the effect of intraspecific differences in mass on basal or resting energy expenditure (per whole animal) is positive (Daan, Masman, Strijkstra, & Verhulst 1989; Earle & Lavigne 1990; Millar & Hickling 1990; Hayes, Garland, & Dohm 1992). Interspecifically there is some evidence that in small mammals and birds the total daily energy requirements follow a trend similar to that of basal metabolic rate, as a function of mass, but elevated approximately three to seven times above it (Drent & Daan 1980; Peterson, Nagy, & Diamond 1990; Bryant & Tatner 1991; Weiner 1992). We will assume that this linkage also pertains intraspecifically, although as yet no studies I am aware of have empirically verified this possibility. The basal and total daily energy costs as a function of body mass therefore follow two lines with a fixed ratio (Fig. 2). The exact position of the total cost line might be anticipated to vary with environmental conditions, so that when it is colder, for example, thermoregulatory costs would increase, and the line would be elevated. Conversely, when it is warmer, the total cost line would be lower. This is a simplistic view of the role of environmental variability on energetics. Intuitively it is attractive, because it accords with laboratory studies of the impact of temperature on resting energy demands: lower temperatures lead to greater resting energy demands (cf. Fig. 1b). However, studies in small birds suggest the link between daily energy demands and ambient temperature in the field may be considerably more complex. In the absence of similar information for small mammals, however, I will assume that decreases in temperature will elevate thermoregulatory demands and, thus, elevate total daily energy expenditure.

Consider now some different scenarios for the effects of intraspecific variation in body mass on energy intake. It is possible to imagine many different relationships between energy intake and body size. For example, increases in mass may have a very steep positive effect on energy gains, because, in competition over food resources, larger individuals may have a distinct advantage over smaller individuals. However, where food is abundant, and competition less likely, gains may be independent of mass. Finally, there may be situations where smaller individuals have greater gains than larger individuals. For example, small mustelids may be

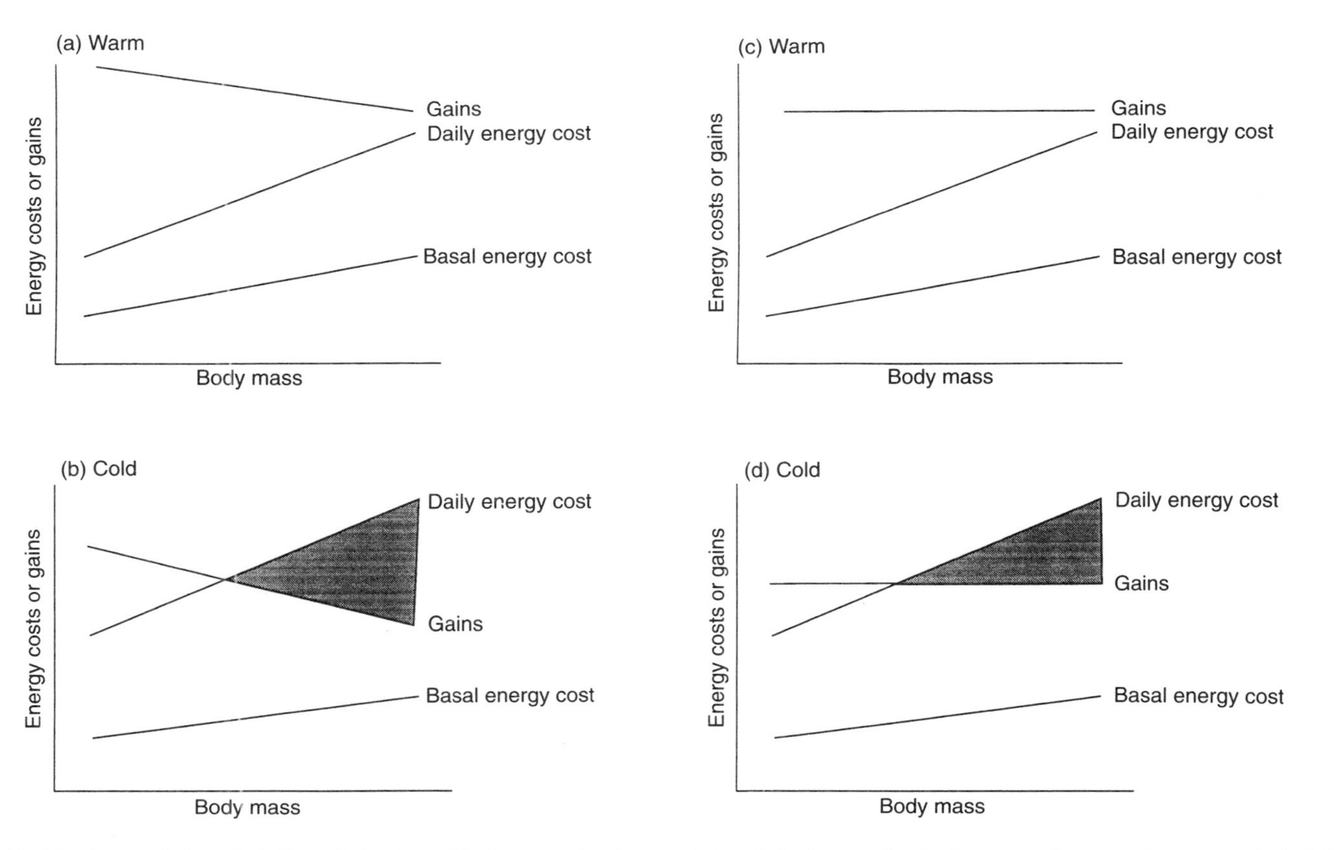

Fig. 2. Models of energy balance including relationships of both costs and gains to variations in body mass. In (a) it is warm and energy gains are negatively linked to body mass. In (b) it is cold and the gain line falls below the cost line for the larger individuals. This would lead to selection for small size. In (c) and (d) the gains are independent of body size. The same effects occur as when the relationship between size and gains is negative.

able to get down burrows with narrower entrances, and gain access to a prey resource which is not accessible to larger individuals (King 1991). In addition to these diverse effects of body mass, we might also imagine that environmental variation would play an important role in influencing the energy gains that animals might achieve. These effects might occur directly or they might occur indirectly because of effects of the changing environment on food availability.

For each different scenario, for the direction of the effect of body mass on energy gains, there would be a series of cost and gain curves the precise location of which would depend on the environmental conditions. The juxtaposition of these cost and gain functions would define whether an individual, of a given body mass, would make an energy balance or not, in that environment, in those particular conditions. For simplicity let us imagine two contrasting environmental conditions: when it is warm and when it is cold. We will assume that these two conditions have opposite effects on energy costs and gains—warmth decreasing costs and increasing potential gains and cold increasing costs but decreasing gains. If we overlay the energy gain patterns on the variation in energy costs we can reveal the situations in which animals are likely to achieve energy balance, fail to achieve energy balance or make an energy surplus. From these patterns of net gain, and net loss, we could thus infer the likely direction of selection on body mass, as a result of energetics (Fig. 2a–d).

In Fig. 2(a) and (b), energy gains are negatively related to body mass. In warm conditions (Fig. 2a) the gain curve always lies above the cost curve. Consequently, in this situation, both large and small animals could make an energy surplus, which would be somewhat larger for the smaller animal. However, in cold conditions (Fig. 2b) the small animals continue to make an energy balance (gains above costs) but large animals fail to do so (gain line falls below the cost line: shaded area). If these cold conditions persisted for any period of time the larger animals in this population would perish. Selection would favour the smaller individuals. In Fig. 2(c) and (d), the energy gains are independent of body size. The pattern of change in costs and gains in this situation is similar to that when gains are negatively related to body size and the consequences are the same: selection will favour smaller individuals. Several previous studies have also suggested that the lower absolute energy costs of small individuals (implicitly combined with an inferred independence of energy gains from body size, as in Fig. 2c) would favour small size in winter (Merritt & Merritt 1978; Ure 1984; Millar & Hickling 1990; Merritt & Zegers 1991). This effect has been suggested to explain the observation that many small mammals reduce their body size during winter (Dehel's phenomenon).

In Fig. 3(a) and (b), the relationship between energy gains and body mass is very steeply positive. This situation might pertain, for example, if there was strong competition for food resources and larger individuals competed favourably. In this situation, it is the smaller individuals which make a loss when it is cold (Fig. 3b: gain line below cost line) and the larger individuals which continue to achieve a positive energy balance. Selection in this situation would be likely to favour larger individuals. Finally, in Fig. 3(c) and (d), the gain line is also positive, but this

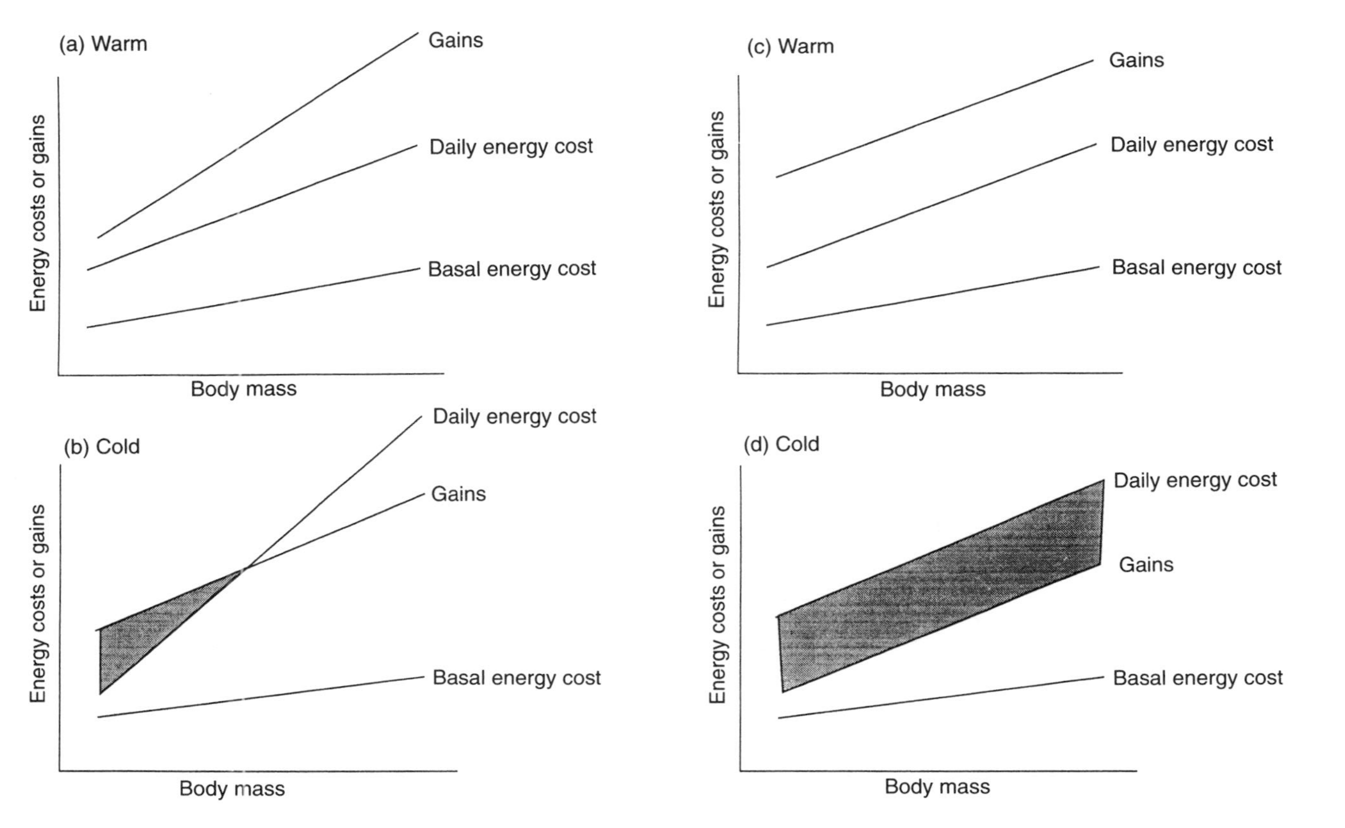

Fig. 3. Models of energy balance including relationships of both costs and gains to variations in body mass. In these examples the gains are positively related to body mass. In (a) it is warm and energy gains are strongly positively linked to body mass. In (b) it is cold and the gain line falls below the cost line for the smaller individuals. This would lead to selection for large size. In (c) and (d) the gains are positively related to body size and increase at the same rate as costs. Consequently net gains are independent of body size.

time has the same gradient as the cost line. In this circumstance, in warm conditions (Fig. 3c) all the animals would make an energy gain, but in cold conditions (Fig. 3d) all the animals would make an energy loss. Consequently, there would be no clear selection on body size as a result of the energetics.

Given this framework, where the effects of changes in environmental conditions affect the elevation of the relationships between energy (costs and gains) and body mass, it is only a small step to imagine more complex scenarios where the gradients of the relationships are also affected by environmental conditions. For example, when it is warm, gains might be negatively related to mass and selection might favour smaller individuals. However, when it is cold, intense competition might result, favouring the larger individuals. This model could lead to size clines as a function of ambient temperature and, hence, latitude, with larger individuals dominating populations where it is colder. It is important to point out that this is one of many potential theoretical effects. There is no reason to favour this latter model over other theoretical treatments where exactly the opposite trends are predicted: with larger animals selected in the warm and smaller individuals favoured in the cold.

These theoretical models suggest that selection on body size will depend critically on the relationships of energy costs *and* energy gains to body mass and how these relationships change with changing environmental conditions. As far as I am aware, no study has yet been performed on a mammal which has elucidated these relationships. The complexity of these models suggests that there is no *a priori* reason to expect from energetics that body mass would increase with increases in latitude. A myriad of different patterns of size-selection as a function of latitude are possible, depending on the cost and gain functions, and how these alter with latitudinal changes in the environment.

When food is absent

When animals have no food available to them they must rely on energy-conservation mechanisms to reduce their energy costs, and their stored fat reserves to supply this energy. As animals will put energy-conservation mechanisms into effect, energy costs will be generally lower than when food is present and they are actively feeding. The amount of time that animals could remain alive while functioning at this low level (called their fasting endurance) would depend on how large their fat reserves were, in relation to how rapidly they were burning them up. The critical factor influencing fasting endurance is therefore the balance of the relationship between energy costs and body size, and the relationship between fat storage and body size. Many studies have observed that the amount of fat that animals can store in their bodies increases disproportionately with body mass (Calder 1984). It has been generally suggested that this increase with size in potential fat storage exceeds the increased energy costs of maintaining a larger body. Consequently, there is a positive relationship between the potential fasting endurance and body size (Rosenzweig 1968; Searcy 1980; Calder 1984; Lindstedt & Boyce 1985; Millar & Hickling 1990). This positive relationship suggests that, in the absence of food, selection will favour larger individuals. Periodic absence of food has been suggested to be a phenomenon linked to the severity of winter. Therefore the fasting endurance model appears to provide a potential mechanism explaining the size clines of increasing body mass with latitude (reference cited above).

Complete absence of food, however, is unlikely to be a prolonged occurrence. Even for animals which must endure frequent fasts, there must also be times when food is available, between the fasting periods. The actual direction of selection on body size, for animals which must cope with periodic complete fasts, will therefore be a trade-off between the selection pressures during the times when food is available and those when it is not available (Speakman 1992, 1993). As I have indicated above, we can expect virtually any pattern of selection when food *is* available, depending on the exact dynamics of the cost and gain curves. Consequently, even when food is periodically absent and the fasting endurance hypothesis suggests that selection would favour larger individuals, the total direction of selection, including periods when food is and is not available, could be positive or negative or could favour some intermediate size where the benefits in the fasting phase are not offset by the disadvantages in the feeding phase (e.g. as suggested by Campbell & Slade 1993 for cotton rats, *Sigmodon hispidus*, in Kansas). There is no unambiguous pattern in the direction of selection which can be anticipated, without specific information on the energy costs and gain relationships, as a function of body mass, for a particular situation.

Millar & Hickling (1990) used the fasting endurance model to generate predictions of the likely patterns of body size selection, as a function of periodic food shortage. I have suggested, however, that these predictions are flawed, because they take into account only the direction of selection during the periods of time when food is not available (Speakman 1992, 1993). The intervening periods between fasts are considered only to the extent that there must be sufficient time for all animals to replenish their fat reserves to the maximum capacity. This scenario, however, is only likely if the effect of mass on energy balance during interfasting periods is neutral (Fig. 3b). If other patterns of costs and gains occur, it is possible to envisage selection during the fasting period being strongly reinforced (e.g. if the pattern was like that in Fig. 3a) or, conversely, completely reversed (e.g. if the pattern was like that in Fig. 2a or c).

Hickling & Millar (1993) have suggested that my critiques (Speakman 1992, 1993) of their predictions from the fasting endurance model (Millar & Hickling 1990) are unscientific. They argue that they have developed a hypothesis and derived predictions from that hypothesis, which are open to testing and falsification, as required by the hypothetico-deductive scientific method (Popper 1962). In contrast, they suggest that my argument is that, because the predictions may be falsified, they are untenable. However, this is a misinterpretation of my criticisms. My critique of their predictions is not that they might be falsified, but that they do not derive from the hypothesis. The hypothesis which overlies the fasting endurance model is that energetics has played a role in the evolution of body size. From this hypothesis, Millar & Hickling (1990) derived the fasting endurance model and generated a series of predictions. However, because they assumed a particular model for what pertains in the periods when food is available, the predictions they generated do not necessarily follow from the original hypothesis concerning the role of energetics in body-size evolution. My argument is that because these predictions do not reflect the original hypothesis, any attempt to falsify them will also not test the original hypothesis.

An analogy may clarify what I mean. Imagine a hypothesis that the earth is not a flat plate, but is actually spherical, like a ball. To test this hypothesis, using the hypothetico-deductive method, the falsifiable prediction might be made that, if the Earth is spherical, all the people living in Australia will have to walk around on their heads. A visit to Australia, to test this prediction, reveals that they do not. What does this mean? I have developed a hypothesis, generated a falsifiable prediction from it, tested the prediction against data, and falsified it. Therefore the Earth must be flat. In this analogy it is very clear what is wrong. The prediction does not stem from the hypothesis. Although it is less obvious, this is the same problem with the Millar & Hickling (1990) predictions. The predictions they generate do not test the hypothesis that energetics has played a role in the evolution of body size, or even the hypothesis that fasting endurance leads to selection in favour of larger body size. It is possible to collect data which completely falsify the predictions made by Millar & Hickling (1990) and yet both these hypotheses may still be correct, because selection on size may be dominated by what is happening when food *is* available, rather than what happens when it is not.

In summary, even when food is periodically absent, the expected direction of selection on body mass as a result of energetics is not immediately obvious. Various patterns of body mass, as a function of latitude, could be expected to occur and trends of increasing size with latitude are possible, but not particularly anticipated, from the energetics models. Since Rensch (1936) suggested that 81% of North American mammals conform to Bergmann's law, this might indicate that energetics has, in fact, not played a dominant role in the evolution of body size—despite the intuitive appeal of such interpretations. This was also suggested by Scholander (1955), who suggested that the observed variation in body size in most clines was physiologically unimportant for heat balance.

More recent reviews, however, have cast doubt on the figure of 81% derived by Rensch (1936). McNab (1971), for example, found that only 32% of species conformed to Bergmann's law and that the opposite trend, of animals becoming smaller at greater latitudes, was observed as frequently in the data he reviewed. Close examination of the data compiled by McNab (1971) reveals many complex trends. To illustrate these, consider the shrew *Blarina brevicauda*. At latitudes less than 30°N this species has a negative trend of body mass with latitude (opposite to Bergmann's law). Between 30 and 45°N the trend is positive, strongly supporting Bergmann's law. However, at latitudes above 45°N, there is no significant relationship. Geist (1987) also re-examined data on body size and latitude in mammals and came to the conclusion that Bergmann's law was generally inapplicable to the trends observed, which were more often than not complex functions of latitude showing both positive and negative relationships over different latitudinal ranges. Although the existence of these complex patterns could not be used as evidence to support the energetics models, they are consistent with the expectation from the energetics models, whereas rigid conformity with Bergmann's law would not be.

Empirical data

The models derived above, relating energy costs and gains to body mass, all implicitly assume that body mass has a significant positive effect on basal and daily energy expenditure, explaining the majority of the variation in energy demands under standard conditions. This implicit assumption derives from studies which describe the effect of body mass on energy demands across species (e.g. Kleiber 1961). These interspecific relationships are generally very strong, with variations in mass typically explaining more than 90% of the variation in energy expenditure. Although there have been many hundreds of studies in which metabolic rates of several individuals of a given species have been measured and in which the individuals must have varied in their body masses, these studies seldom present the relationship between the measured metabolic rate and body mass, preferring to eliminate the mass effect by using mass-specific values or a derived scaling exponent, before seeking some other effect. Where relationships are presented, however, they reveal that individual variation in body mass, within a species, explains far less of the variation in energy expenditure than is routinely found in interspecific studies. This is partly because of the range of masses found within a species, compared with that across species. Nevertheless, when plots of energy expenditure against body mass are examined, it is the residual variation in energy demands across animals of varying body mass that is far more striking than the effect of mass itself. Four previously unpublished plots of basal energy expenditure as functions of individual variations in body mass across three different species illustrate this point (Fig. 4). In these plots the variations in body mass explain less than 50% of the variation in energy demands. This is not an effect unique to my laboratory. Reviewing data from several other studies of small mammals reveals similar low coefficients of determination (e.g. Glazier 1985; Earle & Lavigne 1990; Hayes *et al.* 1992; Konarzewski & Diamond in press).

This high degree of variation between individuals, which is not related to body mass, raises some interesting questions in the context of the models attempting to relate selection on body size to energetic costs and benefits. Consider the data for the Orkney vole illustrated in Fig. 4(c). For Orkney voles of intermediate mass (35–40 g) some individuals have basal metabolic rates (0.7 W) which are over twice as great as the basal metabolic rates of other individuals of the same body mass (0.3 W). Variations in body mass, at this level, appear to have much less effect on the basal metabolic rate than other intrinsic factors. Focusing attention

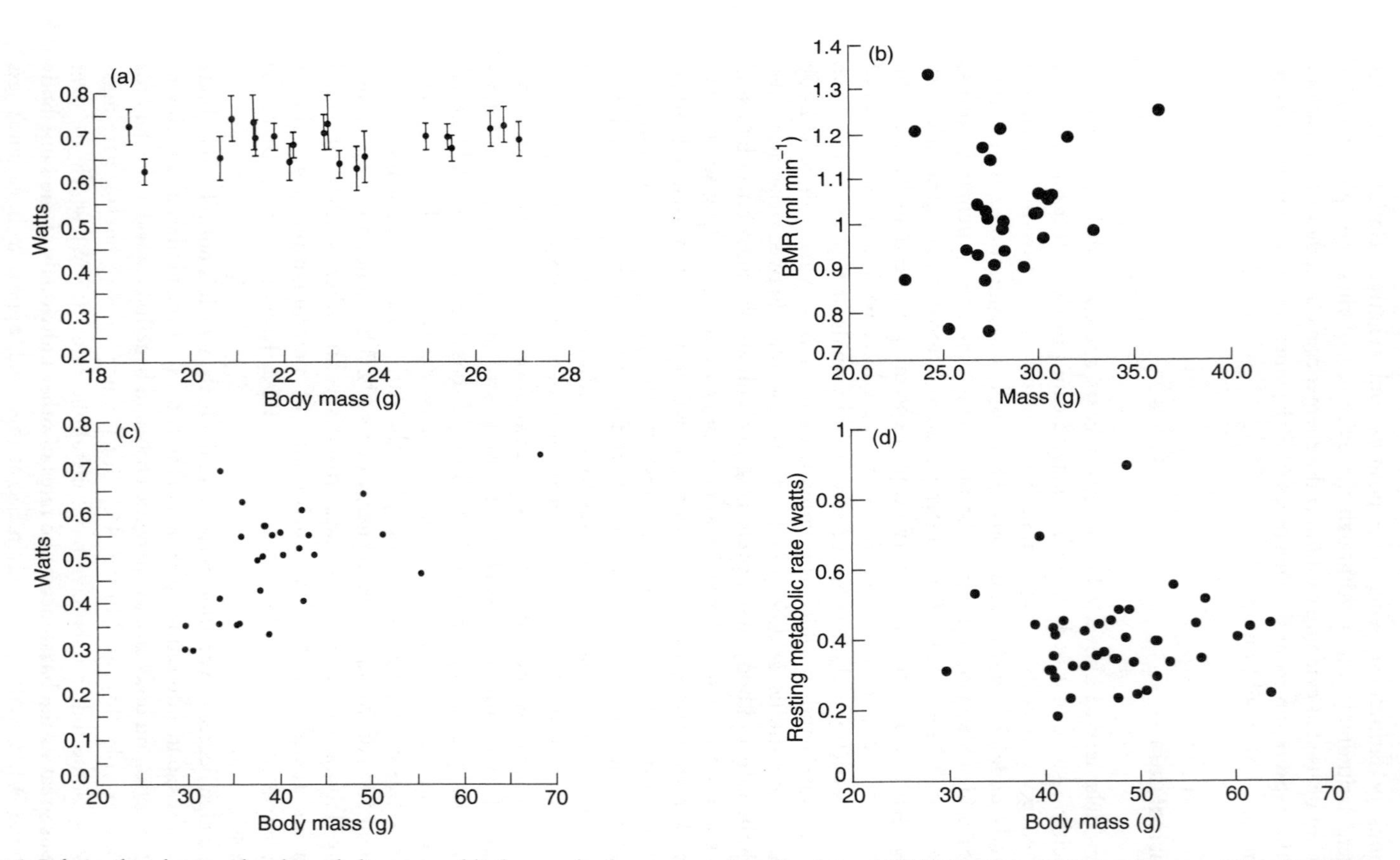

Fig. 4. Relationships between basal metabolic rates and body mass for four groups of small rodents. (a) C57/BL10 mice (J. R. Speakman unpubl.), (b) MF1 mice (S. C. Thomson, M. Johnson, & J. R. Speakman unpubl.), (c) Orkney voles (*M. arvalis*) (J. R. Speakman unpubl.) and (d) Djungarian hamsters (*Phodopus sungorus*) (J. Garvin & J. R. Speakman unpubl.). In all cases body mass does not explain a large amount of the variation in the basal metabolic rate.

on the effects of body mass (as the only intrinsic variable incorporated into the above models) may therefore obscure more interesting questions. It is pertinent to ask, for example, what benefit is derived by the individuals of a given mass which maintain very high basal energy demands? Intuitively, one might anticipate that high levels of BMR would be disadvantageous because they would commit the animal in question to a high rate of food intake. If there was a food shortage, therefore, the animals with lower metabolic rates would appear more likely to achieve an energy balance.

There are, however, a whole series of potential benefits which might be associated with a high basal metabolic rate and which might offset this hypothetical disadvantage. Animals with greater basal metabolism at a given mass might be more able to take in and digest food and, thus, have elevated gains as well as elevated costs. They may be more aggressive and, thus, more likely to compete favourably for food as it becomes scarce. In some small birds, for example, it has been suggested that there is a link between metabolic rate and dominance, independent of the effects of mass, but studies of the link between basal metabolism and dominance in small mammals are currently lacking.

In the last part of this paper I want to consider one further potential energetic benefit which may be linked to individual variation in basal metabolic rate. That is the link of basal metabolism to thermogenic capacity. When animals are placed in the cold, in the laboratory, they increase their thermogenic capacity. A direct consequence of this enhanced thermogenic capacity is that if the animals are subsequently exposed to extremely cold temperatures they survive longer than animals which have had no pre-exposure. Another effect which occurs in synchrony with the enhanced thermogenic capacity, when animals are exposed to the cold, is that their BMR increases. In short-tailed field voles (*Microtus agrestis*) we have shown recently that the increase in BMR is not linked with hypertrophy of the alimentary tract, but rather appears to be more closely linked to changes in the concentration of the uncoupling protein in the brown adipose tissue (McDevitt & Speakman 1994), which is the principal locus of the enhanced thermogenesis (Foster & Frydman 1979).

If enhanced thermogenesis during cold acclimation is linked to increased BMR, is it possible that natural variation in BMR is linked with variation in thermogenic capacity? One measure of thermogenic capacity is the increase in the metabolic rate following injection of noradrenaline (NA) which activates brown adipose tissue. We have recently collected some data in wood mice (*Apodemus sylvaticus*) which indicates that this is indeed the case. Individual variation in the basal metabolism of wood mice is possibly linked to the NA-induced metabolic rate of the same individuals (Fig. 5a: McDevitt, Haim, Thomson, & Speakman in prep.). This effect is found, not only in the raw relationship, which could reflect the covariation of both factors with body mass, but also in the residuals of both factors to body mass (Fig. 5b). Therefore animals of a given mass, which have a higher than expected basal metabolic rate for their mass, appear also to have a greater than expected thermogenic capacity for their mass.

These observations suggest that individual animals may pursue different thermoregulatory strategies when entering the winter. At one extreme animals may opt for a low BMR/low thermogenic capacity option. This might be predicted to be a favourable strategy when conditions are relatively mild, food relatively abundant, and populations low so that competition is reduced. These animals might be most likely to make an energy balance because of their low energy demands. However, if there was a very cold period these animals would die, not from an inability to meet an energy balance, but rather from a lack of metabolic power. In contrast, at the other extreme, animals might opt for the high BMR but high thermogenic capacity strategy. This might make it hard for them to achieve an energy balance, particularly under conditions of food shortage, but they would be more likely to have sufficient metabolic power to survive periods of intense cold. Since cold acclimation takes approximately 10–20 days to develop in the laboratory, switching between strategies would not be a viable option for an animal suddenly confronted with a prolonged period of intense cold.

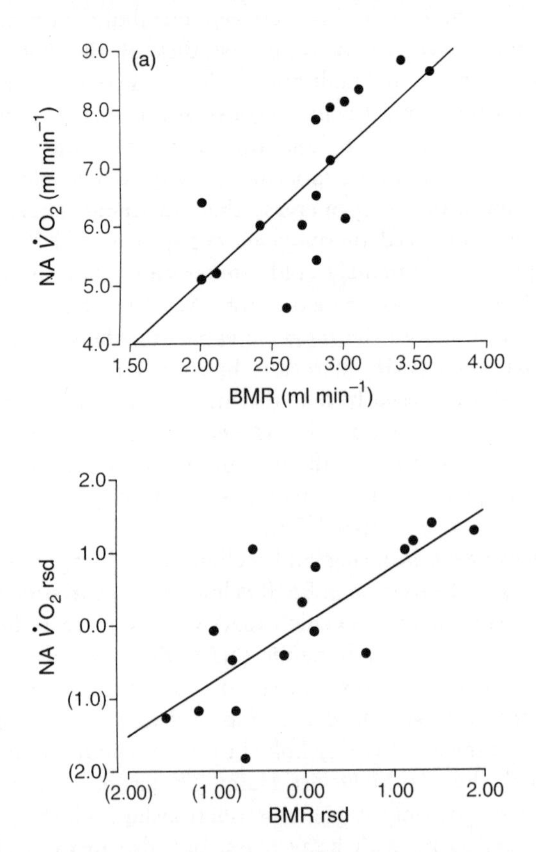

Fig. 5. The relationship between noradrenaline induced metabolic rate (NA $\dot{V}O_2$) and basal metabolic rate (BMR) in wood mice (*A. sylvaticus*) (from R. M. McDevitt, A. Haim, S. C. Thomson, & J. R. Speakman unpubl.). In (a) raw values for both parameters are used and in (b) the residuals of the parameters to the relations with body mass are plotted. In both cases there was a significant positive relationship.

These findings add a novel dimension to the role of energetics as a factor influencing winter survival of small rodents. Studies which have attempted to elucidate how energetics might impact on winter survival have, to date, concentrated entirely on the problem of achieving energy balance. A second problem, however, may be that of achieving heat balance and, thus, of having sufficient metabolic power to generate adequate internal heat to achieve heat balance under extreme conditions. Our data for the wood mouse (Fig. 5 a and b) indicate that the solutions to these two problems may interact negatively. This negative interaction leads to some extremely interesting possibilities for trade-offs in energetic strategies adopted by animals as they enter the winter. It also indicates that the role of energetics as a factor influencing overwinter survival may be considerably more complex than models based on energy balance alone would predict.

Conclusions

Models of energy balance based on costs alone cannot predict the effects of energetics on the selection of body size. Using both costs and gains it is possible to envisage a variety of effects of energetics on body mass which might result in selection for larger or smaller individuals depending on the exact relationships at any particular site. At present the crucial parameters necessary to construct a prediction of the likely effects of energetics on size selection are not available for any small terrestrial mammal. Size clines as a function of latitude can be predicted by using energetics models, but the direction of the cline is not always expected to be positive. A diversity of patterns can be predicted, and this corresponds with the trends observed in nature. Observations of the relationship between metabolic rate and size within species reveals that mass variations generally explain less than 50% of the variation in metabolism. Having a high metabolic rate for one's mass would superficially appear disadvantageous in terms of meeting an energy balance. Animals with higher than anticipated metabolic rates may, however, derive other advantages, one of which could be enhanced thermogenic capacity. There may therefore be a trade-off between capacity to achieve an energy balance and capacity to achieve a heat balance. This suggests that the relationship between energetics and overwinter survival may be considerably more complex than models based on energy balance alone predict.

Acknowledgement

I am grateful to Peter Miller for the opportunity to speak on this topic at the symposium.

References

Berry, R. J. (1968). The ecology of an island population of the house mouse. *J. Anim. Ecol.* 37: 445–470.

Blaxter, K. (1989). *Energy metabolism in animals and man.* Cambridge University Press, Cambridge, UK.

Boonstra, R. & Krebs, C. J. (1979). Viability of large- and small-sized adults in fluctuating vole populations. *Ecology* **60**: 567–573.

Brown, J. H. & Lee, A. K. (1969). Bergmann's rule and climatic adaptation in wood rats (*Neotema*). *Evolution* **23**: 329–338.

Bryant, D. M. & Tatner, P. (1991). Intraspecies variation in avian energy expenditure: correlates and constraints. *Ibis* **133**: 236–245.

Calder, W. A. (1984). *Size, function, and life history*. Harvard University Press, Cambridge, Ma. & London.

Campbell, M. T. & Slade, N. A. (1993). Effect of mass on seasonal survivorship of northern cotton rats. *J. Mammal.* **74**: 971–981.

Daan, S., Masman, D., Strijkstra, A., & Verhulst, S. (1989). Intraspecific allometry of basal metabolic rate: relations with body size, temperature, composition and circadian phase in the kestrel, *Falco tinnunculus*. *J. biol. Rhythms* **4**: 267–283.

Davenport, J. (1992). *Animal life at low temperature*. Chapman & Hall, London.

Drent, R. H. & Daan, S. (1980). The prudent parent: energetic adjustments in avian breeding. *Ardea* **68**: 225–252.

Earle, M. & Lavigne, D. M. (1990). Intraspecific variation in body size, metabolic rate and reproduction of deer mice (*Peromyscus maniculatus*). *Can. J. Zool.* **68**: 381–388.

Ebenhard, T. (1990). A colonization strategy in field voles (*Microtus agrestis*): reproductive traits and body size. *Ecology* **71**: 1833–1848.

Fleming, T. H. & Rauscher, R. J. (1978). On the evolution of litter size in *Peromyscus leucopus*. *Evolution* **32**: 45–55.

Foster, D. O. & Frydman, M. L. (1979). Non-shivering thermogenesis in the rat: II. Measurement of blood flow with microspheres points to brown adipose tissue as the dominant site of the calorigenesis induced by noradrenaline. *Can J. Physiol. Pharmacol.* **56**: 110–122.

Geist, V. (1987). Bergmann's rule is invalid. *Can. J. Zool.* **65**: 1035–1038.

Glazier, D. S. (1985). Energetics of litter size in five species of *Peromyscus* with generalization for other mammals. *J. Mammal.* **66**: 629–642.

Gorman, M. L. (1991). Orkney and Guernsey voles *Microtus arvalis*. In *The handbook of British mammals* (3rd edn): 208–211. (Eds Corbet, G. B. & Harris, S.). Blackwell Scientific Publications, Oxford.

Hayes, J. P., Garland, T. & Dohm, M. R. (1992). Individual variation in metabolism and reproduction of *Mus*: are energetics and life history linked? *Funct. Ecol.* **6**: 5–14.

Hickling, G. J. & Millar, J. S. (1993). The fasting endurance model: a testable hypothesis. *Funct. Ecol.* **7**: 135.

King, C. M. (1991). Body size–prey size relationships in European stoats *Mustela erminea*: a test case. *Holarct. Ecol.* **14**: 173–185.

Kleiber, M. (1961). *The fire of life: an introduction to animal energetics*. John Wiley and Sons, New York & London.

Konarzewski, M. & Diamond, J. (In press). Evolution of metabolic rate and organ masses in laboratory mice. *Evolution*.

Lindstedt, S. L. & Boyce, M. S. (1985). Seasonality, fasting endurance, and body size in mammals. *Am. Nat.* **125**: 873–878.

McDevitt, R. M. & Speakman, J. R. (1994). Central limits to sustained metabolic rate have no role in cold acclimation of the short-tailed field vole (*Microtus agrestis*). *Physiol. Zool.* **67**: 1117–1139.

McNab, B. K. (1971). On the ecological significance of Bergmann's rule. *Ecology* **52**: 845–854.

Merritt, J. F. & Merritt, J. M. (1978). Population ecology and energy relationships of *Clethrionomys gapperi* in a Colorado subalpine forest. *J. Mammal.* **59**: 576–598.

Merritt, J. F. & Zegers, D. A. (1991). Seasonal thermogenesis and body mass dynamics of *Clethrionomys gapperi*. *Can. J. Zool.* **69**: 2771–2777.

Millar, J. S. & Hickling, G. J. (1990). Fasting endurance and the evolution of mammalian body size. *Funct. Ecol.* **4**: 5–12.

Murie, J. O. & Boag, D. A. (1984). The relationship of body weight to overwinter survival in Columbian ground squirrels. *J. Mammal.* **65**: 688–690.

Packard, G. C. & Boardman, T. J. (1987). The misuse of ratios to scale physiological data that vary allometrically with body size. In *New directions in ecological physiology*: 216–239. (Eds Feder, M. E., Bennett, A. F., Burggren, W. W. & Huey, R. B.). Cambridge University Press, Cambridge.

Peterson, C. C., Nagy, K.A., & Diamond, J. (1990). Sustained metabolic scope. *Proc. natn. Acad. Sci. USA* **87**: 2324–2328.

Popper, K. (1962). *Conjectures and refutations*. Basic Books, New York.

Redfield, J. A. (1976). Distribution, abundance, size, and genetic variation of *Peromyscus maniculatus* on the gulf island of British Columbia. *Can. J. Zool.* **54**: 463–474.

Rensch, B. (1936). Studien über klimatische Parallelität der Merkmalsausprägung bei Vögeln und Säugern. *Arch. Naturgesch.* (N.F.) **5**: 317–363. [Cited in McNab (1971).]

Rosenzweig, M. L. (1968). The strategy of body size in mammalian carnivores. *Am. Midl. Nat.* **80**: 299–315.

Sauer, J. R. & Slade, N. A. (1985). Mass-based demography of a hispid cotton rat (*Sigmodon hispidus*) population. *J. Mammal.* **66**: 316–328.

Scholander, P. F. (1955). Evolution of climatic adaptation in homeotherms. *Evolution* **9**: 15–26.

Scholander, P. F., Walters, V., Hock, R., & Irving, L. (1950). Body insulation of some Arctic and tropical mammals and birds. *Biol. Bull. mar. biol. Lab., Woods Hole* **99**: 225–236.

Searcy, W. A. (1980). Optimum body sizes at different ambient temperatures: an energetics explanation of Bergmann's rule. *J. theor. Biol.* **83**: 579–593.

Speakman, J. R. (1992). Evolution of animal body size: a cautionary note on assessments of the role of energetics. *Funct. Ecol.* **6**: 495–496.

Speakman, J. R. (1993). Evolution of body size: predictions from energetics. *Funct. Ecol.* **7**: 134.

Ure, D. C. (1984). Autumn mass dynamics of red-backed voles (*Clethrionomys gapperi*) in Colorado in relation to photoperiod cues and temperature. *Spec. Publs Carnegie Mus. nat. Hist.* **10**: 193–199.

Weiner, J. (1992). Metabolic constraints to mammalian energy budgets. *Acta theriol.* **34**: 3–35.

Symp. zool. Soc. Lond. (1996) No. 69: 83–109

Biomechanical constraints on size in flying vertebrates

JEREMY M. V. RAYNER

School of Biological Sciences
University of Bristol
Woodland Road
Bristol BS8 1UG, UK

Synopsis

As a consequence of the aerodynamic demands of flight, birds occupy a limited range of body mass and individual species experience tight constraints on size. In birds as a whole variation in wing shape or wing size independent of body size is relatively small. Few flying birds are lighter than 0.010 kg: the smallest bird species are all hummingbirds (Trochilidae), with mass down to 0.002 kg. The smallest passeriform bird is appreciably larger, with a mass of 0.004 kg. The largest contemporary flying birds in a number of orders have mass between 10 and 15 kg, although extinct flying birds may have reached masses up to 100 kg.

Allometric models used to describe size-variation of wing design and flight behaviour (including wingbeat kinematics) are discussed and limitations of the allometric scaling approach are considered. Allometry of mechanical (design) adaptations is the result of adaptation within size-dependent constraints. The size-scaling of wingbeat frequency is discussed as an example. As the result of a trade-off between aerodynamic and energetic factors, bird wing size is positively allometric.

Two mechanical constraints, on wingbeat frequency, to provide sufficient lift and thrust, and on power output from the flight muscles, have been proposed as responsible for the upper limit to size in birds, but the largest birds in the modern fauna are much smaller than some extinct giant fliers, and the dominant constraint may be imposed by ecological and environmental factors. Aerodynamic and energetic constraints do not limit the smallest size of birds (compare the wing morphology of larger flying insects), but muscle performance may limit size by setting a maximum wingbeat frequency.

In hummingbirds, wing area and wingspan do not scale geometrically, but vary so that wing loading and aspect ratio are independent of body mass. As a result flight speeds in the smallest individuals are not excessively low, but size is constrained by wingbeat frequency and lift generation.

Introduction

Many aspects of adaptation vary with body size, and rarely do the various parameters determining physiological, ecological, behavioural, or biomechanical

ZOOLOGICAL SYMPOSIUM No. 69
ISBN 0–19–857787–7

adaptive domains vary in the same way with size. The constraints which are critical in determining the viable range of adaptations in small animals may be very different from those which are dominant on larger representatives of the same clade. Any discussion of the limits to design should take account of a broad spectrum of interacting constraints with differing functional relationships, before concluding that any one factor is dominant. However, certain modes of locomotion are especially demanding, because the animal must be able to find sufficient energy to move, and must be sufficiently strongly built to tolerate substantial locomotory forces. Such constraints have had far-reaching effects on adaptation and radiation, and intuitively it is reasonable to expect that in these modes mechanical and physiological factors associated with locomotion have had a major influence on design.

Calder (1984) and Alexander (this volume, pp. 3–14) have elucidated a range of domains in which mechanical factors may impose constraints on small size in vertebrates. Within a typical vertebrate clade of intermediate size, locomotion may play only a weak role in determining the viable range of size or body mass which is suitable for that design, but in such clades locomotor-related constraints may have a significant effect on shape, that is in proportions independent of size. This may be illustrated with an example relevant to the ideas developed in this paper: bird and bat species of average size typically have a tightly bounded range of body masses. In any population, individuals with mass far from the population mean are rare. The range of masses is probably determined in most cases by trophic and/or physiological factors: how much food can be obtained, and how much energy is required for survival and reproduction, for territorial defence, and so on. At any body size, parameters which have the greatest influence on flight performance, such as wing size and shape, are determined by a mosaic of constraints, among which aerodynamic and biomechanical factors predominate (see p. 91–98 below; Norberg & Rayner 1987; Rayner 1988; Norberg 1994). Similarly, limb proportions in terrestrial mammals reflect adaptation to environment (Howell 1944), usually for efficient and/or economical locomotion, but apart from the largest animals locomotion tends to have only indirect influence on body mass in any population.

By quantifying biomechanical (and other) constraints on adaptation, and by determining the likely safety factors (*sensu* Alexander 1981, 1992) or tolerances to the deleterious effects of proximity to these constraints, it should be possible to identify the factor or factors which play the most significant role in the evolution and radiation of any clade. Constraints reflecting locomotion will probably be most significant in those clades in which locomotion is intensive, is highly specialized, or represents a significant component of energy consumption, and may be recognized when locomotion has had a major effect on constructional *Bauplan* and on a broad spectrum of aspects of adaptation. In this respect birds, which are defined primarily by their powers of flight, are an obvious example of a clade in which biomechanical constraints on design and behaviour may be expected to be significant. In this paper I define conceptually some of the aerodynamic and mechanical factors which might act to limit size range of flying birds, and consider some of the adaptations adopted by small birds in response to constraints delimiting the lower extreme of their size range.

The size range of flying vertebrates

Animals are not uniformly distributed across their size range. Each species evolves within a combination of functional (e.g. mechanical and physiological) and ecological (e.g. trophic and environmental) constraints. These constraints act differently on organisms of different sizes, and some constraints—or combinations of constraints—act to delineate the absolute extreme sizes in any clade.

Figure 1 shows the distribution of body masses for a representative sample of living birds and mammals. The distribution for birds (apart from hummingbirds) has an abrupt cut-off at 0.0005–0.01 kg: only a few species (mainly insectivores) are smaller than this. There is a long tail at the upper end of the range: few flying species are bigger than 1 kg, the largest flying birds fall in the range 10–15 kg, and a small number of heavier flightless species belonging to the Spheniscidae and Palaeognathae extend to a maximum at around 100 kg in the ostrich *Struthio*. (For further discussion of the size distribution of birds, see Blackburn & Gaston (1994).)

I have treated hummingbirds (Apodiformes, Trochilidae) separately, because their specialized flight morphology and behaviour and their small size set them apart (Cotton, this volume, p. 239–258). The smallest members of this family are considerably smaller than all other birds, with a lower limit under 0.002 kg. The hummingbirds' size range is narrow, with an abrupt cut-off at 0.01 kg, and only one larger species, *Patagona gigas* (0.02 kg). The reason for this gap is unclear: *Patagona* behaves very much like its smaller relatives.

Although bats are often regarded as comparable with birds, their size distribution is somewhat different. It is highly skewed to the smaller sizes occupied by insectivorous species, with a modal mass of 0.01 kg and a smallest size around 0.002 kg in the bumble-bee bat *Craseonycteris thonglongyai*, comparable with the smallest hummingbirds. Few bat species are heavier than 0.1 kg; the majority of larger species are frugivorous pteropodids, with an upper limit at around 1.5 kg. Rayner (1981) and Norberg & Rayner (1987) argued that this limit is probably determined by behavioural or trophic factors: bats cannot colonize the niches available to larger flying birds because their nocturnality largely rules out soaring flight, because scavenging is exceptionally rare in bats, and no bats are granivorous or grazing herbivores. Bats may also be constrained by their hanging roosting posture, and time-constraints imposed by echolocation may limit size in insectivorous microchiropterans (Jones, this volume, pp. 111–128). Since the flight systems of birds and bats are aerodynamically (Rayner 1987) and physiologically (Thomas 1987) similar, there appears not to be an aerodynamic limit setting upper size in bats, although the possibility cannot be ruled out that the structure of the flight complex, involving a membraneous wing supported by both fore and hind limbs, imposes a constraint on the maximum lift force which the wings can generate.

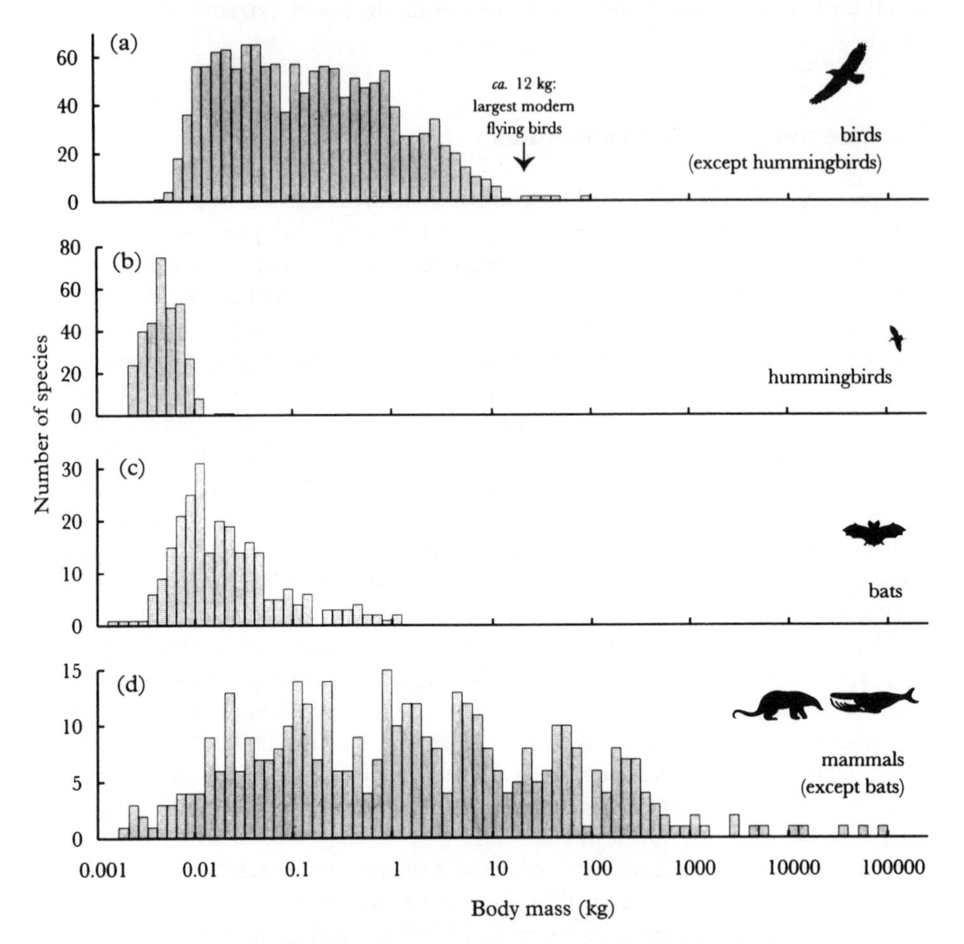

Fig. 1. Distribution of body masses of birds and bats, showing non-flying mammals for comparison, with body mass (kg) shown on logarithmic scales. (a) Birds, $n = 1390$, from flight morphology dataset of Rayner (1988), with addition of flightless species from Dunning (1993). (b) Hummingbirds, $n = 324$, from Dunning (1993) and Greenewalt (1962, 1975). (c) Bats, $n = 255$, from Norberg & Rayner (1987). (d) Mammals other than bats, $n = 411$, from Eisenberg (1981).

A sample of non-flying mammals is shown for comparison (Fig. 1). Mammals occupy a much broader range of trophic niches and, because they are predominantly terrestrial quadrupeds, encompass a wider range of body forms. The data set is limited, but there is a relatively uniform distribution between masses of 0.01 and 100 kg and an appreciable number of larger species which reach up to 1000 times the mass of the largest living flying birds and some 10 000 times that of the largest bats. These largest mammals are of course aquatic.

From this comparison it is evident that flight facilitates small size in homeothermic vertebrates, especially in insectivores and nectarivores. At the same time, factors related directly or indirectly to flight impose a strict upper limit on size within organisms which remain relatively small: the largest modern flying bird, the wandering albatross *Diomedea exulans*, has a wingspan of some 3.5 m, but—with a mass of 10–12 kg—weighs approximately the same as an average domestic cat.

Some caution must be exercised in deducing the presence of absolute limits to size from the distribution of modern fliers. It is well known that the other group of flying vertebrates, the extinct pterosaurs, reached very large sizes towards the end of their radiation in the mid to late Cretaceous (Wellnhofer 1991). The body mass of *Quetzalcoatlus northropi*, the largest known species, has been estimated in the range 80–90 kg, with a wingspan of 11–12 m (Lawson 1975; Langston 1981). The majority of pterosaurs were comparable in size to birds, and some flying juveniles were very small indeed, with masses perhaps as low as 0.01 kg. The Jurassic pterosaur communities of what is now southern Germany seem comparable to those of modern Charadriiformes and other marine birds (Hazelhurst & Rayner 1992). Unfortunately the paucity of reliable mass estimates for pterosaurs and their supposed indeterminate growth rule out a direct comparison with birds and bats. But given the morphological and aerodynamic similarity of pterosaurs to birds (Padian & Rayner 1993) it is likely that their size ranges were broadly comparable. Although we have no record of tiny pterosaurs comparable to the smallest insectivorous birds or to hummingbirds, there is good evidence that in the past the size range of birds was much greater than that of modern species. A number of lineages included giant flying representatives. The best known is the teratorn *Argentavis magnificens* (Ciconiiformes, Cathartidae, a relative of modern New World vultures) from the Miocene of Argentina, which is estimated to have had a body mass of 70–80 kg and a wingspan of 6–8 m (Campbell & Tonni 1983; Campbell & Marcus 1992). The giant albatross *Gigantornis eaglesonei* from the Eocene of Nigeria attained a wingspan of some 6.5 m, twice that of the largest modern species *D. exulans* (Halstead & Middleton 1976).

Wing design in flying animals

The usual approach to the problem of co-variation of morphological or other parameters with size is the method of allometry, which computes the index of the scaling relation as the gradient of a line of best fit through a logarithmically transformed scatter plot of measurements (see, for example, Pedley 1977; McMahon & Bonner 1983; Calder 1984; Schmidt-Nielsen 1984); Rayner (1985a) and Sokal & Rohlf (1994: 541) describe methods. Deductions about the factors controlling adaptation may then be drawn from the values of the scaling indices.

Scatter diagrams of wing dimensions against body mass on logarithmic axes for a comprehensive sample of flying animals were first published by Greenewalt (1962, also 1975) (Fig. 2). Greenewalt (1962) did not derive scaling relations

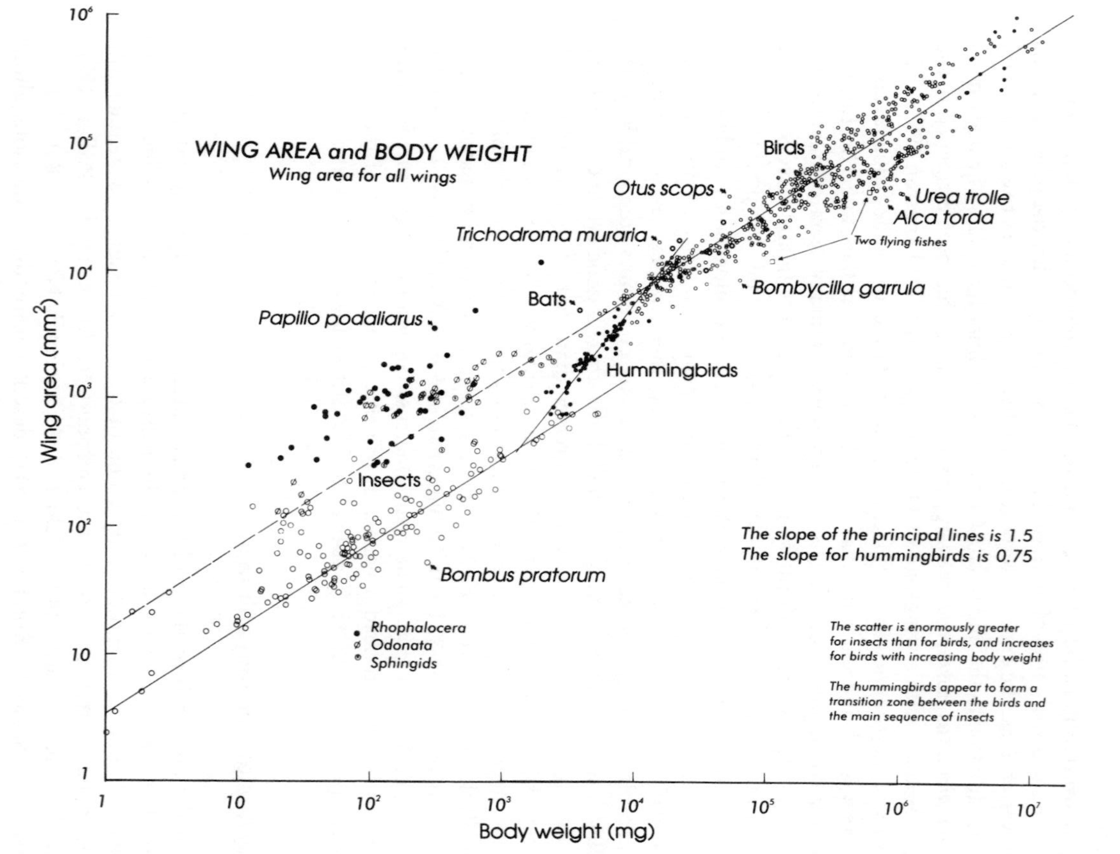

Fig. 2. Scaling of wing area with body weight (*sic*) in flying animals, from Greenewalt (1962). Two lines are shown for insects, divided somewhat arbitrarily. The line drawn through larger winged insects and birds has a slope of 2/3, representing geometric similarity; if this line is extended to much higher masses it passes through the region occupied by modern transport aircraft (Greenewalt 1975 etc.). A steeper scaling applies to hummingbirds; Greenewalt claimed this has a slope of 4/3, but in fact the slope does not differ significantly from 1. The Figure has been relettered for legibility, preserving the original wording.

statistically, but obtained a fit by eye to lines equivalent to geometric similarity, which he treated as a predetermined 'law'. His demonstration that most insects, birds, and some aircraft fitted a single line was persuasive evidence that common aerodynamic factors apply to all flying organisms. This, however, should not be surprising: early aircraft designers employed scaling formulae derived from birds to determine the size of their proposed heavier-than-air flying machines (Winter 1895; Fullerton 1911, etc.). Greenewalt (1975) used regression methods to compute exact slopes for various groups of birds, but, since these groups had neither phylogenetic nor behavioural basis, his gradients have little biological significance. Subsequently (Rayner 1988), I recalculated these relations for a large sample of flying birds (Fig. 3), using where appropriate the reduced major axis (rma, or structural relation) model (Rayner 1985a), to explore how wing dimensions co-vary with body mass and with flight behaviour and ecology.

In his 1962 analysis, Greenewalt implicitly adopted the hypotheses that all linear dimensions increase in proportion with one another as size increases and that mean tissue density is constant, a situation normally referred to as isometric scaling or geometric similarity. Thus, wingspan B should be proportional to body mass $M^{1/3}$ and wing area S should vary as $M^{2/3}$. He found the visual fit acceptable, but had he tested the isometric hypothesis formally he would have found that it failed. In the pooled sample of all flying birds, wingspan and wing area show significant positive allometry, increasing as $M^{0.394}$ and $M^{0.722}$, respectively (Rayner 1988); scaling is also positively allometric within most (but not all) bird and bat families. This illustrates an important pitfall of the method of allometry: when a broad size range is present, visual inspection of the scatter diagram can be misleading, and lines which are poor statistical fits may give good visual agreement with the data cluster. Moreover, the high statistical significance of the relations (Pearson bivariate correlation coefficient $\rho_{1100}=0.962$ in Fig. 3b) can hide considerable variation: at a body mass of 1 kg, wing area in birds varies by a factor of eight, from 0.035 m^2 in Alcidae (auks and puffins) to 0.28 m^2 in Laridae (gulls and terns). It is interesting to note that alcids and larids are widely disparate in wing morphology, although they are closely related, probably as sister groups (Sibley & Ahlquist 1990); this arises largely because of the small wing area associated with underwater swimming in auks.

Greenewalt's hypothetical scaling 'law' exemplifies one of the most important and simplest of similarity rules (e.g. Gould 1971; Weis-Fogh 1977; Economos 1982; Günther & Morgado 1982). These are simple models which summarize how size-dependent parameters vary on the assumption that a certain critical factor remains constant. Geometric similarity or isometry, as applied by Greenewalt (1962), is so-called because in this case it is shape (and, if relevant, density) which remains constant. The similarity methodology does not imply the existence of some underlying developmental constraint which forces the particular design model; we must always expect that under some selective pressures the result of adaptation would be to evade the constraint if it were convenient and feasible to do so and if a viable design solution could be evolved. Moreover, it is rarely possible to describe all features of an organism or adaptive domain by one single rule. None the less, it is convenient to test data sets against an appropriate similarity model and to interpret statistical agreement as confirmation that in this group adaptation is constrained by that similarity hypothesis. Other common similarity rules in studies of biomechanics include dynamic similarity, under which limb elements move at the same rate relative to speed of locomotion (Alexander 1982a), stress similarity, under which muscles produce, and bones and tendons experience, equal stresses (Hill 1950), and elastic similarity, under which compressive buckling stresses in long bones are size independent (McMahon 1973; Alexander 1982a); LaBarbera (1986) lists some examples of other mechanical similarity rules. Life-history invariants (Charnov 1993; Miller, this volume, pp. 175–199) represent the application of similar concepts to aspects of life history and behaviour.

Although helpful as a means of identifying mechanical factors which have a significant influence on design and a convenient way of estimating parameters for an organism of any given size or design, I argue that the approach of deriving a single bivariate allometric relation is over-simplistic, especially when applied to comparative or interspecies samples. It implies that design is determined solely by the factor modelled by the hypothesis under consideration, and it assigns excessive importance to the allometric line, although this may have little more than statistical meaning. Last, and perhaps most seriously, it dilutes the importance of adaptation independent of size in individual clades, which may be the biological phenomena we aim to study, because bivariate allometry implies that variance away from the allometric relation is little more than noise. (A more extensive critique of other aspects of allometry is given by Harvey & Pagel (1991).)

To a considerable extent these problems may be resolved by careful interpretation of the data set and the conclusions drawn from it, and by increasing the number of variates under consideration by one of a number of multivariate generalizations of allometry (Rayner 1985a; see Norberg & Rayner (1987) and Rayner (1988) for examples applied to wing design in bats and birds).

An alternative, and potentially very powerful, approach is to view the individual records in a data set as the result of adaptation within a mosaic of size-dependent constraints, which combine to impose the apparent allometry. This concept is the subject of the following section of this paper.

In birds as a whole, wing design diverges only slightly—but significantly—from geometric similarity. Large birds have relatively longer and larger wings than if all bird wings were the same shape and size relative to the body; equally, smaller birds have relatively shorter and smaller wings. However, in some avian clades wing area does scale geometrically with body mass (e.g. Rayner 1985a). If there is some—unhypothesized—reason for geometrical similarity, then why is it possible for there to be such a broad spread of wingspan at the same mass? Applied strictly, the geometric model would imply that all variation away from the line of best fit was simply statistical noise or was anadaptive (and the commonly high correlation coefficients which result from the broad size range would support this interpretation). However, no biologist could accept this hypothesis, since the variation 'away from the line' is clearly explicable as adaptive: auks have short wings because they 'fly' underwater, while terns have long wings to favour both their extensive migration flights and their use of wind hovering in hunting. Clearly, different factors constrain wingspan in terns and in auks and these factors reflect the different behavioural and ecological adaptations of the two groups. In the same way, it is hard to see how factors which might determine wing shape in a small insectivorous woodland passerine could also apply to wing shape in the largest albatrosses or vultures.

Allometry and constraints

Each species evolves within constraints which depend on the ecology, behaviour, and design of that species. Sampling a population is equivalent to sampling the regions of an adaptive landscape for an individual species where that population is fittest (Rayner 1985a). Unfit organisms are unlikely to survive and therefore are unlikely to be sampled: the geometry of the adaptive landscape is determined by constraints which limit the region in which an organism is fit, and this should be reflected in the distribution of sampled data. Thus, the scatter plot for a clade represents the pooled set of adaptive landscapes for the members of that clade, and the shape of the scatter plot reflects similarities in constraints between the members of the clade.

The scatter plot of wingspan against body mass (Fig. 3a) provides a useful example. The region occupied by flying birds is elongated and the variation in wingspan at any range of body mass is much less than the overall variation in wingspan. This must be the result of strict constraints on wingspan. Rayner (1988) summarized some of the most obvious of these constraints in birds (Fig. 4). Mechanical constraints which affect all birds must be size dependent, corresponding to the boundary of the scatter plot. If the wings are too short they will not generate sufficient lift, while if they are too long they will experience substantial spanwise bending loads. Both of these constraints place limits on wingspan which vary isometrically, as $M^{1/3}$, and therefore it is not surprising that the scaling of wingspan is close to isometry. However, no factor determines this scaling alone and many species will be constrained by factors which are specific to that species. In addition, because species may adapt in different ways in response to selection, there is no reason to think that geometric similarity *must* apply to all flying birds. Hummingbirds deviate from this rule (Fig. 3), and this can be interpreted in terms of their small size and their use of hovering flight. Further, there is no clear mechanism whereby geometric similarity might operate in a similar fashion on organisms which have a range of different habitats, flight styles, etc.

It is, however, more than coincidence that bird wing scalings show only weak positive allometry. The constraints on wingspan (and other design parameters) vary with size for clear mechanical reasons. These confine the variable adaptations of flying birds within a narrow band; for individual species the viable band is generally much narrower still.

This concept that allometry is the result of radiation and adaptation within size-dependent constraints was used to explain limits on wingbeat frequency in hovering flight by Lighthill (1977); see also Weis-Fogh (1977) and Alexander

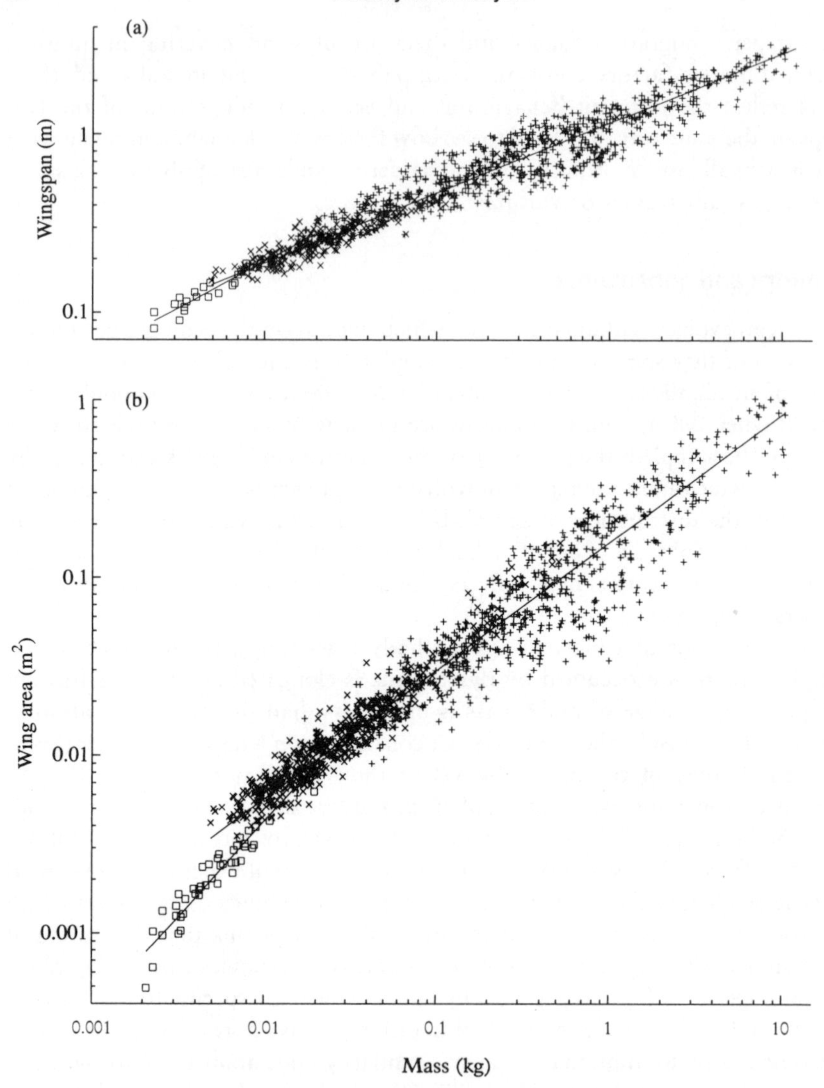

Fig. 3. Scaling of (a) wingspan B (m) and (b) wing area S (m^2) with body mass, after Rayner (1988). +, non-passeriform birds (except hummingbirds); ×, passeriform birds; □, hummingbirds. Reduced major axis (rma) relations as follows: birds (except hummingbirds) wingspan $B = 1.166M^{0.394}$, $\rho_{922} = 0.954$; wing area $S = 0.1576M^{0.722}$, $\rho_{1100} = 0.962$; hummingbirds wingspan $B = 2.213M^{0.528}$, $\rho_{24} = 0.955$; wing area $S = 0.614M^{1.081}$, $\rho_{42} = 0.942$. In both cases, the slopes for hummingbirds are significantly steeper than for other birds and do not differ significantly from constant wing loading similarity; the slopes for other birds are significantly positively allometric compared to geometric (isometric) similarity.

(1982a) and Fig. 5. Lighthill computed how the various constraints on frequency are functions of wing length, with the assumption of geometric similarity of wings and all body components. I have recalculated these constraints as functions of body mass (Fig. 6a and Table 1), assuming geometric similarity. Wingbeat

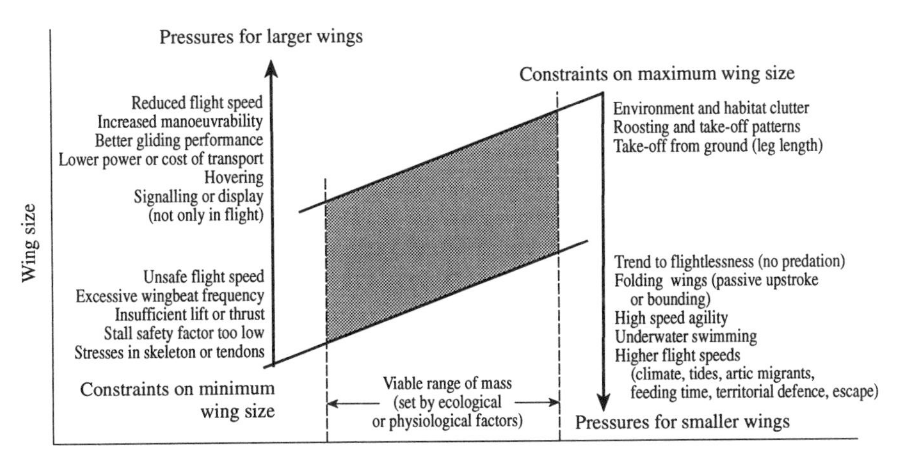

Fig. 4. Schematic diagram indicating the effects of the mosaic of constraints and selective pressures acting on wing size in flying birds. Not all ecological and behavioural factors apply to all birds, but many of the mechanical factors shown are significant to all species. Assuming that body mass is set by independent external ecological and/or physiological factors, and that mechanical constraints are size dependent, measurements of a natural population will fall within the shaded region. This is the origin of the interspecific allometry illustrated in Fig. 3. After Rayner (1988).

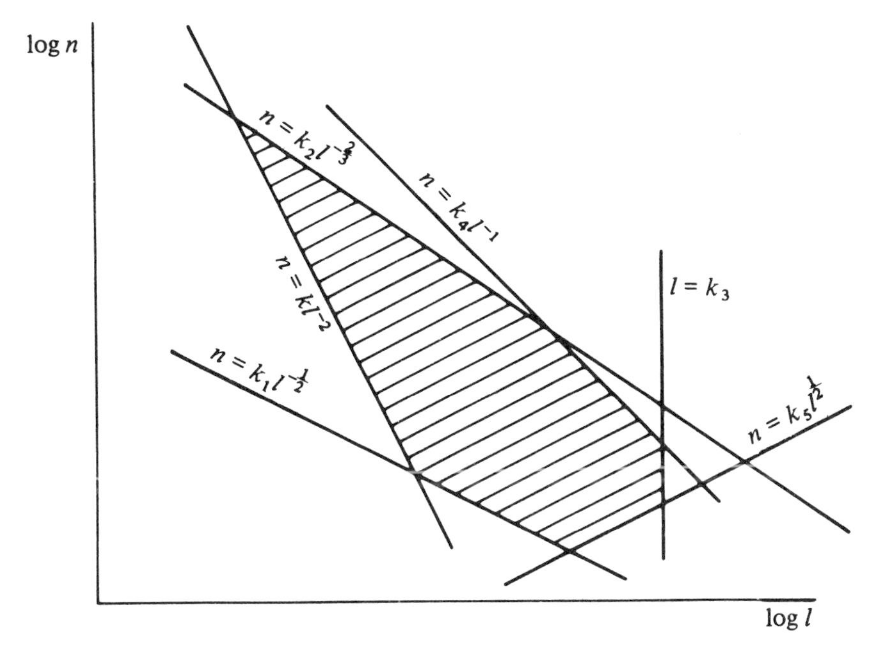

Fig. 5. An example of allometry as the result of adaptation within size-dependent constraints. Frequency n against wing length l in hovering flight, with logarithmic axes. Frequency must fall within the shaded region. The constants of proportionality k_i have not been determined and it has not been confirmed that all constraints are relevant to flying animals. From Alexander (1982a), after Lighthill (1977).

Table 1. Constraints on wingbeat frequency in hovering flight according to different models for scaling of wing morphology.

Constraint	Functional formulation of constraint	Wing morphology scaling model		
		Geometric	Hummingbird	Hybrid
	Scaling of wing dimensions: Scaling of muscle dimensions:	$S \propto M^{2/3}$, $b \propto M^{1/3}$ $A_p \propto M^{2/3}$, $I_p \propto M^{1/3}$	$S \propto M^1$, $b \propto M^{1/2}$ $A_p \propto M^1$, $L_p \propto M^{1/2}$	$S \propto M^1$, $b \propto M^{1/2}$ $A_p \propto M^{2/3}$, $L_p \propto M^{1/3}$
Upper bound on lift coefficient C_L	$f \geqslant (M / Sb^2)^{1/2}$	$f \geqslant M^{-1/6}$	$f \geqslant M^{-1/2}$	
Lower bound on Reynolds' number: Re > 100 for lift generation	$f \geqslant b^{-2}$	$f \geqslant M^{-2/3}$	$f \geqslant M^{-1}$	
Upper bound on profile drag D_{pro} reflecting muscle power	$f \leqslant (M / Sb^3)^{1/3}$	$f \leqslant M^{-2/9}$	$f \leqslant M^{-1/2}$	
Upper bound on skeletal and muscular strength Σ (lift and profile drag)	$f \leqslant (L_p A_p / Sb^3)^{1/2}$	$f \leqslant M^{-1/3}$	$f \leqslant M^{-1/2}$	$f \leqslant M^{-3/4}$
Upper bound on skeletal and muscular strength (induced drag D_{ind})	$f \geqslant M^{3/2} / L_p A_p b$	$f \geqslant M^{1/6}$	$f \geqslant M^{-1/2}$	$f \geqslant M^0$
Upper bound on mechanical (induced) power output P_{ind}	$P_{ind} \propto M^{3/2} / b \leqslant P_{max} \propto M$	$M \leqslant M_{max}$	{No constraint}	
Upper bound on wingbeat frequency set by muscles f_{max}	$f \leqslant f_{max}$	$f \leqslant M^0$	$f \leqslant M^0$	

Formulation of limiting factors follows Lighthill (1977) and Weis-Fogh (1977); these are expressed here in simplified form showing only their dependence on body mass M, wingspan $2b$ and wing area S (and in some cases also on pectoralis cross-sectional area A_p and moment arm about the humeral joint L_p, adopting Lighthill's assumption that the spatial parameters of the wingbeat obey geometrical similarity (i.e. constant stroke amplitude and equal up- and downstroke durations). For simplicity, constants of proportionality are omitted and all quantities are averaged by integration over the wingbeat and wingspan, following Lighthill (1977). The bound on Reynolds' number is not relevant to birds, but is included here for completeness: Re based on wing chord for a hummingbird of 0.002 kg is of the order 10^5.

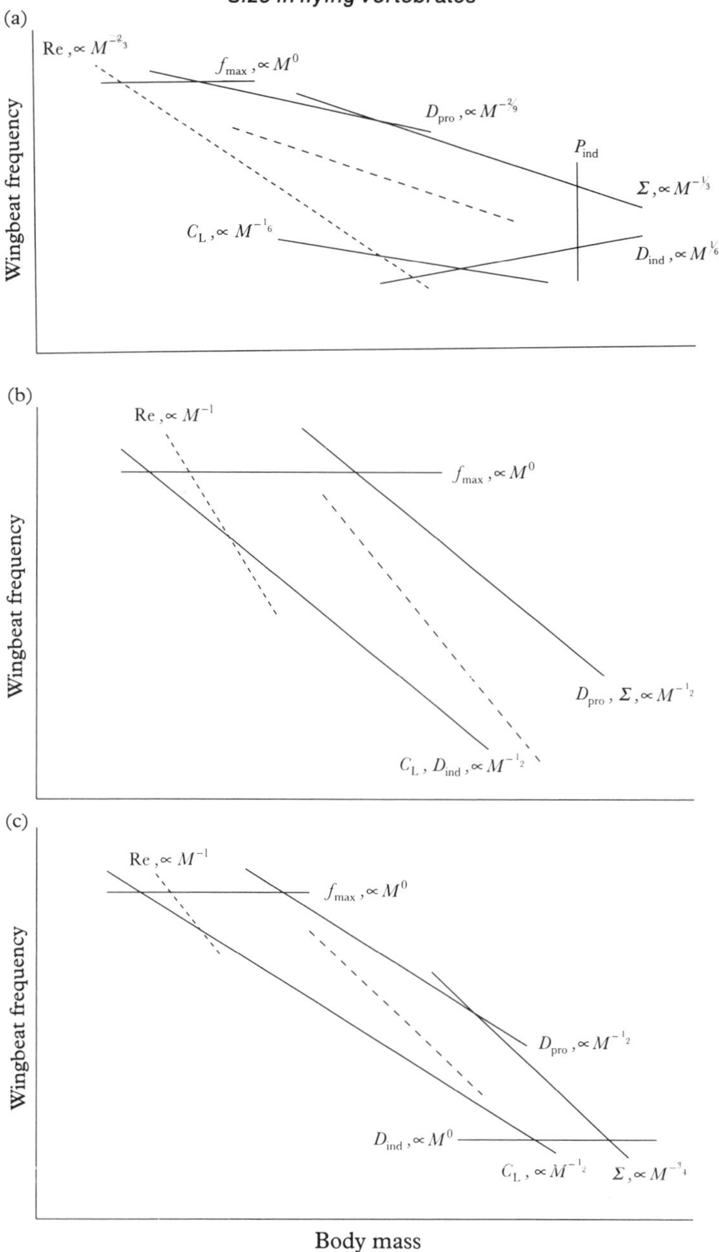

Fig. 6. Theoretical allometry of wingbeat frequency f against body mass M within constraints assuming alternative similarity models, with logarithmic axes. (a) Geometric similarity in forward flight (computation similar to Fig. 5), (b) hummingbird similarity or constant wing loading, for hovering flight, (c) hybrid model, with wings with constant wing loading, but muscle and tendon cross-sections varying geometrically, again for hovering flight. Long dashed lines approximate measured scalings: $f \propto M^{-0.26}$ for forward flight (Fig. 7) and $f \propto M^{-0.612}$ for hovering hummingbirds. Short dashed lines are the scaling of the constraint from Reynolds' number, which probably does not apply to flying birds. See Table 1 for details of notation and derivation. Constants of proportionality are omitted for simplicity; as in Fig. 5 these constants have not been determined, and some of the constraints may not be relevant to birds.

frequency must fall within the constraints, and therefore hovering is possible only over a confined range of body masses. The scaling of wingbeat frequency is a particularly useful example of the interaction of constraints, because frequency is influenced by a range of mechanical, aerodynamic, and physiological factors and, therefore, overlaps a number of separate adaptive domains which share significance for flight. One limitation to using this approach is that it is rarely possible to quantify all the potential constraints: the range of size (wing length) of animals which can hover is limited by energy output, by force generation on the wings, and by strength of the wings, pectoral muscles, and tendons. Given knowledge of the relevant dimensions and material properties these factors may be estimated, within reasonable bounds. But we cannot know *a priori* the safety factors or tolerances which evolution maintains relative to these constraints, although we may surmise that natural selection preserves adequate safety factors in viable individuals.

Similar arguments may be applied to wingbeat frequencies in forward flight (Fig. 7; Rayner 1988, 1995; see also Pennycuick 1986; Norberg 1990). For most of the avian size range only two constraints are relevant: frequency at any mass must be high enough for sufficient thrust to be generated at a realistic wingbeat amplitude, but not so high that there are excessive forces at the wing root. Assuming geometric similarity of wing design the minimum frequency scales as $M^{-1/6}$ (Rayner 1993, 1995). The maximum frequency scales approximately as $M^{-1/3}$. It is therefore not surprising that the measured frequency in various avian clades scales at between $M^{-0.2}$ and $M^{-0.35}$.

In Fig. 7, many birds fall close to or beyond the hypothetical constraints. The diagram displays only frequency and body mass, and does not include other factors such as wingspan and wing area. Moreover, the constraints are predicted on the basis of isometric similarity. The constraining lines are the average for birds as a whole and are only indicative of the constraints on an individual. In most of the extreme cases the apparent anomaly may be explained by departures from isometry. Birds with very high wingbeat frequencies include galliform and anseriform species with very small wings, which may be predicted to have relatively strong tendons and wing bones, and perhaps fly with a lower safety factor in these bones and tendons than is found in some other birds. The potential to vary quantities such as wing design and muscle and skeletal structures is of course central to ecomorphological adaptation of any species. This argument represents an opening for a functional explanation of some aspects of adaptation in engineering terms and leads to the concept that macroevolutionary (i.e. species level or higher) adaptations in locomotory mode evolve to modify or circumvent biomechanical constraints on locomotion imposed directly by structure and locomotor behaviour. The constraints on the radiation of any one species are set, in part at least, by the design and adaptation of that species. This is an alternative expression of the idea that adaptation is limited by '*Bauplan*' constraints, which may be sidestepped only by significant changes in structure.

I have framed these concepts in terms of the pattern of co-variation of frequency and body mass, for which example clear mechanical constraints may be identified,

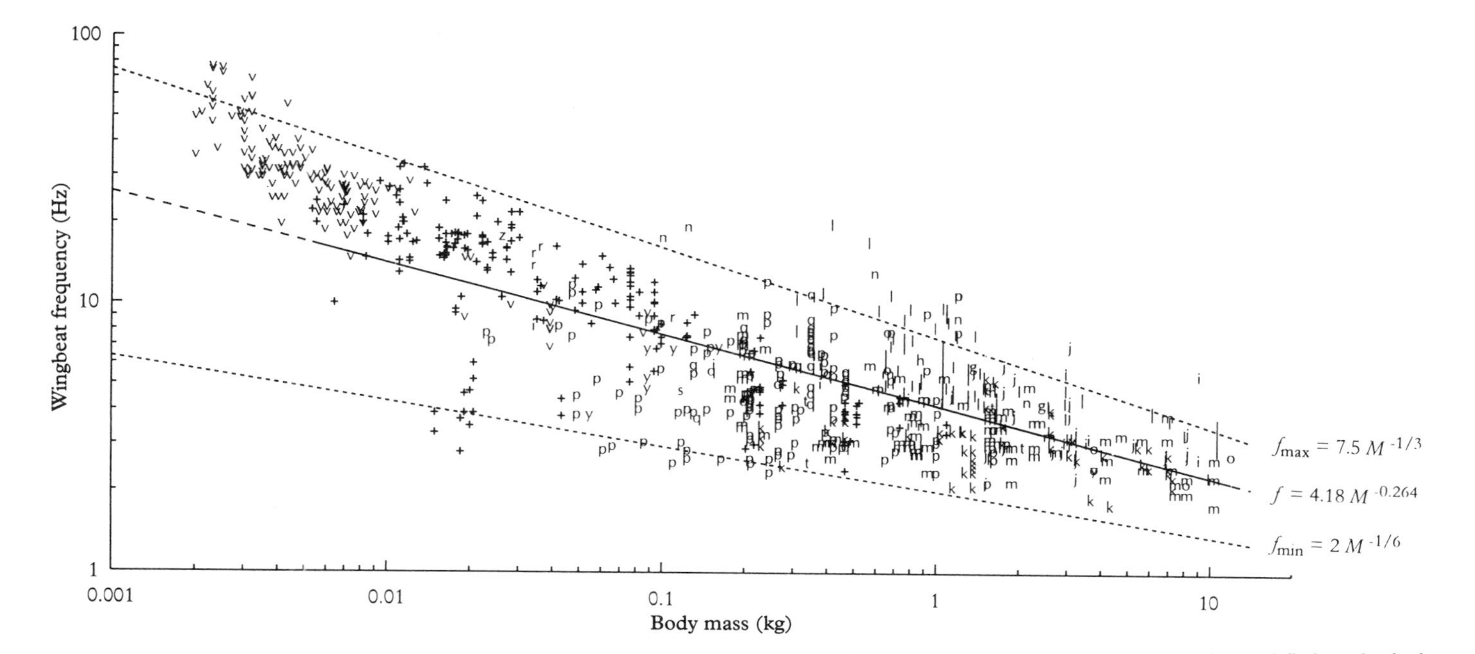

Fig. 7. Measured allometry within constraints. Scaling of measured wingbeat frequency (f, Hz) against body mass (M, kg) in cruising forward flight in birds (but hovering flight in hummingbirds) with real frequency data. From Rayner (1995), after Rayner (1988). Compare the major constraints in Fig. 6a, which also apply to this case. v, hummingbirds; +, passeriforms; other letters refer to other avian orders. Scalings shown are for forward flight only: the slope is steeper in hummingbirds, for which $f = 1.208M^{-0.612}$, $\rho_{132} = -0.796$.

but they may equally be illustrated through variation of—for example—wing design (Rayner 1988), flight speed, or flight energy. In practice, though, we should not expect to separate these co-variates. We might 'explain' extreme wingbeat frequencies by departures from isometry; but, equally, the pattern could be seen from the alternative view that small wings of ducks and pheasants result from adaptation to the environment (Rayner 1988), and in turn *demand* high wingbeat frequencies for effective thrust generation in flight. Every species has evolved the combination of design solutions, trophic ecology, behaviour, and physiology appropriate for reproductive fitness. That some biomechanical constraints on this process are strict is evinced by the close adherence to isometry and by the narrow range of wing designs in flying birds.

Constraints on large size in birds

The largest modern birds able to fly have body masses of little more than 12 kg (Fig. 1). Pennycuick (1975) and Tucker (1977) explained this limit by convergence of limits on mechanical energy consumption, arguing that power available from the flight musculature scales as $M^{2/3}$ or $M^{3/4}$, while power required to fly scales as $M^{7/6}$. This argument has been contradicted by Marden (1994), who argued from measurements of burst performance that peak power output scales as $M^{1.13}$ and therefore that the rate of energy output does not constrain take-off in very large fliers. Pennycuick (1986) argued that the same limit would also be imposed by convergence of aerodynamic and mechanical scaling limits on frequency. However, measured wingbeat frequencies in birds (Fig. 7) suggest that, while the upper mechanical and lower aerodynamic limits on frequency do converge, they do so at a considerably larger mass than that of the largest living flying birds. Part of the problem with these arguments is that model predictions based on bivariate scaling rules against mass or size ignore the important influence of wing design and of other size-independent shape parameters. None the less, both arguments show how the margin between the limits narrows as size increases and this is the reason why the largest birds rely on less energy-intensive modes of flight such as gliding and soaring, and have less flexibility in their choice of flight behaviour. As the margin narrows there is a decline in agility, and the largest species experience extreme difficulty in achieving take-off.

Although these mechanical factors evidently set an absolute upper size for flapping animal flight, it is unreasonable to claim that they set the upper size limit of 12 kg for living flying tetrapods: many pterodactyloid pterosaurs greatly exceeded this limit, as did several bird groups in the Miocene and Quaternary (see p. 87). In practice the factors which set the upper limit probably vary between species, and there is no convincing explanation for the absence of especially large forms in the modern fauna. Similar conceptual problems apply to some groups of Quaternary mammals (Lister, this volume, pp. 277–292), and to gigantism in dinosaurs in the Mesozoic (Benton 1990). It is possible that larger organisms exploited macroadaptations permitting them to circumvent some of the more critical constraints on size: however, if selection favoured these adaptations, their absence in the modern fauna is remarkable. The teratorns were morphologically very similar to their close living relatives, the cathartid vultures, and presumably experienced similar mechanical constraints. Some authors (e.g. Harlé & Harlé 1911) have speculated that past atmospheric conditions were radically different, but there is no evidence that air properties as recent as the Miocene could have been sufficiently different to explain the viability of extremely large fliers; others, with negligible regard for scientific rigour, have proposed that the giant pterosaurs and dinosaurs exploited reduced gravity in the Mesozoic.

It seems more likely that a combination of factors relating to environmental conditions and community structure are responsible for the upper size limit in any taxon. This is consistent with the expectation that animals will maintain a modest safety factor, and will not be designed close to absolute mechanical limits. Selective pressures experienced by large birds or pterosaurs in the past might have been weaker than in contemporary ecosystems, enabling the largest fliers to tolerate lower safety factors than modern fliers do. Or, perhaps more simply, exceptional environmental factors which facilitated flight at large size might have been more reliable than today: Campbell & Tonni (1983) argued that in the Miocene Neotropics trade winds blew continually across the entire continent, guaranteeing that *Argentavis* could become airborne. The absence of such special conditions in contemporary environments explains the absence of exceptionally large flying forms; mechanical factors do not impose an absolute upper limit on the size of extant flying birds.

Bats may represent an alternative example of this process: the largest pteropodids are only one-tenth of the mass of the largest modern birds, and no giant forms are known from the chiropteran fossil record. I argued on pp. 85–86 that the upper size limits in bats are imposed by a combination of trophic niche, sensory constraints, and the difficulty of soaring for a predominantly nocturnal, arboreal flier. There is no obvious mechanical limit associated with flight that would determine the largest size of bats.

Small size in birds

Constraints on small size

Calder (1984) reviewed the factors which may constrain small size in birds. This is a complex problem because—as so often in this discussion—many significant physiological and mechanical factors interact. An ultimate factor, for example, might be the size of the pelvic canal in females in relation to minimum viable egg size. Other contributions to this volume identify a range of non-mechanical factors which probably limit small size in vertebrates.

It has long been known that the cost of transport (energy to carry unit weight through unit distance) decreases with size within different modes of locomotion (Schmidt-Nielsen 1972). This arises in part from the low speeds of small fliers and in part from the negative allometry of mass-specific power in movement. Thus, small animals suffer from increased energy demands, but lower speeds. Greenewalt (1977) argued that this dilemma may be resolved by noting that small animals generally have small home ranges (Alexander 1982b: fig. 7.3). However, I consider it preferable to observe that home-range size is determined by locomotor parameters. Small animals benefit from the greater margin between the energy required for locomotion and the maximum energy they can produce in burst exercise; this gives small birds the advantages of flexible choice of flight pattern, with the ability to take off easily, and perhaps to hover. Flights made by small birds are often short, with frequent manoeuvres and accelerations. This gives them indirect advantages in the ability to avoid predation, and to exploit habitats and food sources denied to larger animals. It is difficult at present to estimate the associated flight energy costs, but they are likely to be enhanced compared to sustained flight, and these flight modes are therefore accessible only to birds of rather small body mass.

I consider it reasonable to suppose that the ultimate lower limit to size in birds is not set by aerodynamic factors associated with the generation of lift and thrust by the flapping wings, because insects comparable in size to the smallest birds or smaller fly well, and wing size appears to be affected by similar constraints in the two groups (Fig. 2). This does not mean that it is possible to ignore factors associated with wing structure, such as those related to the size and cross-section of long wing bones and feathers in very small birds, the design of the avian pectoral girdle, or the biophysics of the pectoral muscles. There are at present very few data relevant to this question, but some indication of the most important factors can be drawn from consideration of the ways in which birds have responded to problems of flight at small size. The smallest birds, the hummingbirds, are an important example, which I consider in more detail below.

Flight efficiency

The rate at which mechanical work must be done in steady flight varies with size, as $M^{7/6}$ if the wings are assumed to obey geometric similarity. The index is slightly lower if the wings are positively allometric (Rayner 1988). On the other hand, the rate of energy input in flight scales approximately as $M^{0.75}$ (Rayner *et al.* in prep.). The ratio of these two quantities is a gross efficiency, which in sustained flapping flight is estimated to increase with body mass, scaling approximately as $M^{0.3}$ (Rayner 1990). No mechanism for this rise in efficiency has yet been proposed, but it is probably related to the contraction dynamics of the flight muscles and to the fact that basal metabolism is larger, relative to the mechanical energy costs of flight, at smaller sizes. Small birds of mass 0.01 kg or less may have to tolerate flight efficiencies as low as 5–10% in steady flight: comparable values have been measured in hovering hummingbirds (Wells 1993). The surplus energy must be radiated as heat, but scalings of physiological energy rates in flight follow those of other physiological processes such as basal or resting metabolism at approximately the three-quarters power of body mass. As a result, energy intake does not have to increase in proportion to body mass in smaller birds.

Although efficiency itself is unlikely to form a meaningful constraint on size, it may be evidence of physiological mechanisms which are important. Low values of efficiency correspond to large quantities of 'wasted' energy, which is likely to have a deleterious effect on reproductive fitness. Birds with certain lifestyles or exploiting some habitats may find that excessive heat loss in continuous flight presents problems, and this factor has often been invoked as an explanation for the use of daily torpor by hummingbirds.

Bounding flight

Many birds need to fly in different ways, among which are the need to vary flight (air) speed and to fly with varying body mass and, in some species, with additional and substantial loads. Since these factors alter the mechanical power needed to fly, but the power source (the muscles) can only change over a long time, birds need gears. Large birds have diverse populations of muscle fibres which are differentially matched to different contraction strain rates and stresses, and hence, by varying wingbeat kinematics, are indirectly matched to different flight patterns, such as gliding, steady flapping or rapid flapping, take-off, and escape (see Norberg (1990) for a review of avian flight muscle biochemistry). By contrast, smaller birds do not do this. It may be that at small sizes the cost of these varied muscles is too great, either because of the weight penalty of the greater flight muscle mass, or because of an enhanced developmental cost. Perhaps, more likely, the range of energy rates with which they would need to cope is too large, owing to the larger power margin in smaller birds. Yet small birds often face severe problems. For instance, body mass may show considerable hypertrophy prior to a long migration flight and in many species mechanical power output at the preferred flight speed may be increased by perhaps 180% compared to flight while lean. To cope with these problems, and to permit a gearing of mean power output, many smaller birds use bounding flight (Rayner 1985b). Brief bursts of wing flaps are interspersed by rest phases during which the wings are folded against the body and the bird follows a ballistic flight path.

Bounding is a flexible flight mode: energy output is modulated by controlling the mean time spent flapping. A bird achieves maximum energy output with sustained continuous flapping, but some species have sufficient muscle capacity in reserve that they bound even while hovering. Kipp (1950) argued that bounding is most important to species which fly vertically upwards, and this is consistent with this model of bounding flight.

A scaling analysis of the energy available from the flight muscles compared with the power requirements of steady flight showed that bounding is feasible only for smaller birds, lighter than approximately 0.1 kg (Rayner 1977). Few species larger than this, nearly all of which are woodpeckers (Piciformes), use bounding, and this mode appears to be confined to birds with relatively low aspect ratio, including most but not all passeriforms (Fig. 8); this apparently reflects both the increased energy cost of flight with low aspect ratio (Rayner 1988) and the difficulty of folding a longer wing.

It is perhaps better to view bounding not as a response to a strict constraint, but as an enabling mechanism which permits birds to achieve a range of flight

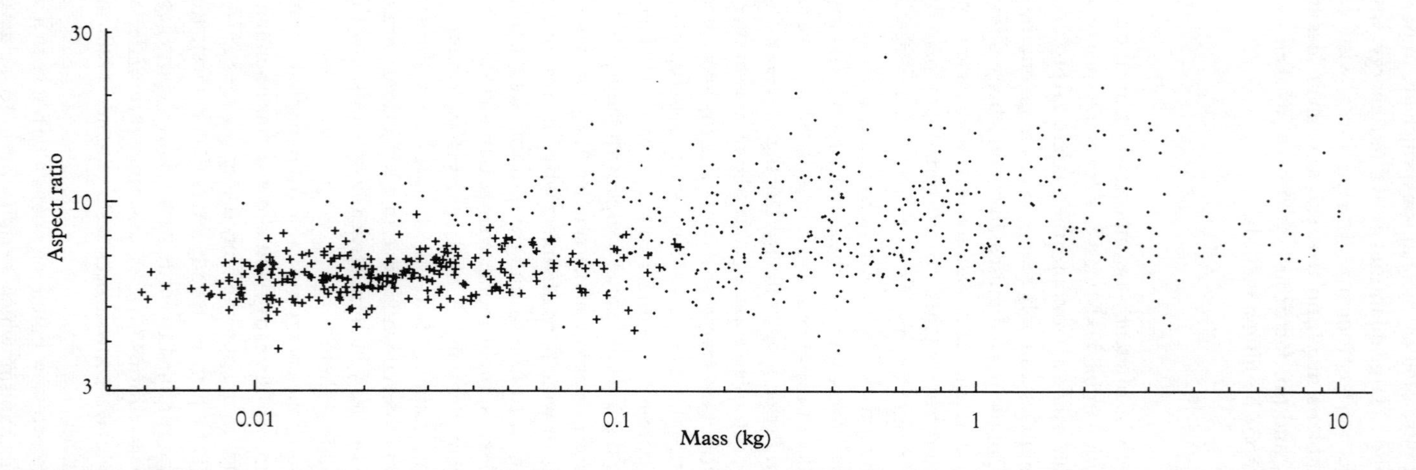

Fig. 8. Aspect ratio of birds known to use bounding flight (+) compared with other species (omitting hummingbirds). Bounding flight is confined to birds of mass less than approximately 0.1 kg and aspect ratio less than approximately 8; few other species fall in this region, and several of these are relatively little-known birds for which flight behaviours have not been described.

performance which they could not otherwise reach. The mechanism is not available at body masses greater than 0.1–0.2 kg, but because of the reduced energy margin at larger sizes there is less need for a broad range of power outputs, and more scope to adopt other means of controlling energy output by altering wingbeat kinematics (Rayner 1995).

Hummingbirds

Hummingbirds are the smallest birds. As discussed on pp. 85–86, they reach the smallest masses of all birds and very few are large enough to overlap in size with other birds. It has long been recognized that hummingbirds form a different adaptive domain, primarily because of the extensive morphological and behavioural adaptations they have evolved in association with their hovering flight and nectarivory (see, for example, Greenewalt 1960a; Scheithauer 1967; Cotton, this volume, pp. 239–258). Although they are closely related to swifts (Apodiformes, Apodidae), with which they share features such as a stiffened wing skeleton (Knoll 1934; Stolpe & Zimmer 1939; Saville 1950), they show none of the adaptations for sustained flight and insect catching of the swifts.

Hovering in hummingbirds is the most energy-demanding activity known in vertebrates, and can be achieved only by carbohydrate, rather than lipid, metabolism (Suarez, Brown, & Hochachka 1986). This has extensive consequences for their behaviour, including the need to feed regularly while they are actively flying, and in many species to defend feeding territories. Their hovering mechanism differs from that of all other birds, since the up- and downstrokes are symmetric and generate lift approximately equally; all other species produce lift predominantly on the downstroke.

Even allowing for their small body mass, hummingbirds have wings of remarkably small area (Rayner 1988) (Figs 3 and 9). This is surprising for birds which rely so much on hovering, for which they would be expected to have relatively long wings. I have proposed that this wing design is an adaptation to fast flight, permitting hummingbirds to feed within a short time period, to move rapidly between inflorescences, and to defend feeding territories effectively (Rayner 1988). The reliably high rates of energy intake from flowers offset the enhanced cost of hovering to feed. Moreover, because of the broader power margin at very small sizes, selective pressures on wing design for flight economy are probably weak.

A potential test of this hypothesis is a comparison of the two hummingbird subfamilies: Phaethornithinae and Trochilinae (Bleiweiss 1990). These differ in behaviour and bill morphology, and phaethornithines do not defend territory, and presumably therefore have less need for high fast flight. If other factors are equal, phaethornithines can be predicted to have relatively longer wings than trochilines. Unfortunately the current sample of wing morphology for phaethornithines is too small to explore this further.

As remarked above, hummingbird wings appear to have evolved within a different set of mechanical constraints than the wings of other birds. The rma lines for wing size for hummingbirds are not simply extrapolations of the lines for other birds, but have an appreciably greater slope (Figs 3 and 9). Hummingbird wings vary so that aspect ratio and wing loading are independent of size.

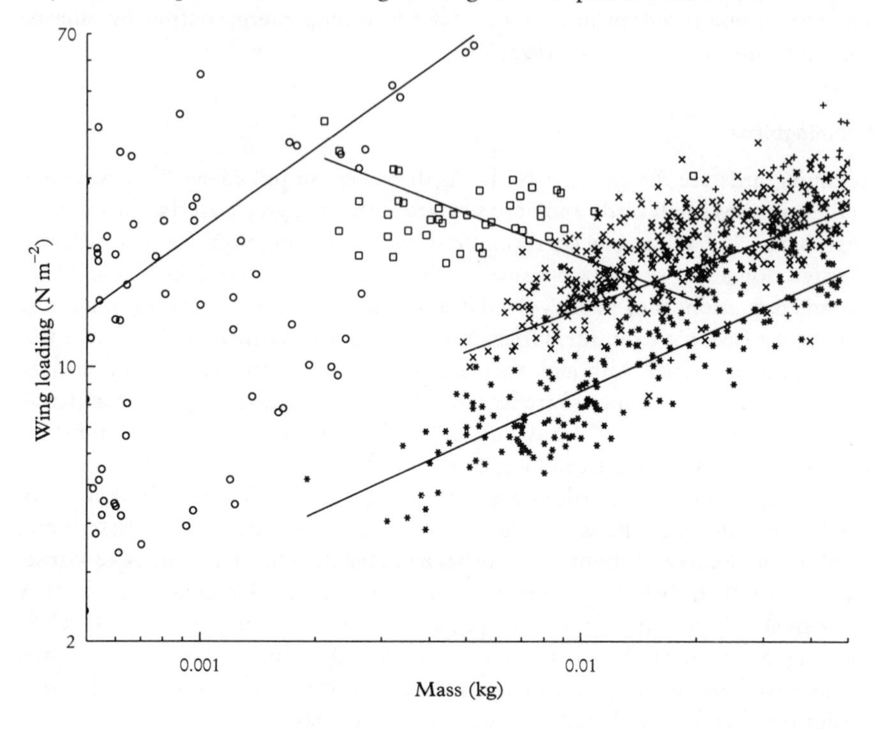

Fig. 9. Wing design in small birds and bats (below 0.05 kg) and large insects (0.005 kg), showing scaling of wing loading $N = Mg/S$ (N m^2) by rma. o, insects (data from Greenewalt 1962); □, hummingbirds (data from Greenewalt 1962, 1975, etc.); ×, passeriform birds (data from Rayner 1988); +, non-passeriform birds (apart from hummingbirds) (data from Rayner 1988); *, bats (data from Norberg & Rayner 1987). Wing loading is an inverse measure of wing size: higher wing loadings imply relatively smaller wings and faster flight speeds. Insect wing size is very scattered; the largest insects shown consist mainly of Coleoptera (larger loadings) and Lepidoptera (smaller loadings), but there are no data for the largest flying coleopterans. Birds and bats have lower loadings than most large flying coleopterans. Birds and bats have lower loadings than most large insects, and bats have appreciably larger wings than birds (see Norberg & Rayner 1987). Hummingbirds have much smaller wings than other small birds; the negative allometry of wing loading with mass is an artefact of the small sample range, and the scaling index does not differ significantly from zero. Interestingly, hummingbirds have smaller wings than the nectarivorous lepidopterans and glossophagine bats with which they are often compared; I argue that this reflects their need to defend feeding territories for some parts of the day. The rma scaling relations are as follows: insects $N = 2655M^{0.693}$, $\rho_{277} = 0.463$; birds (pooled, except hummingbirds) $N = 74.77M^{0.363}$, $\rho_{1100} = 0.841$; hummingbirds $N = 3.459M^{-0.369}$, $\rho_{42} = -0.066$; bats $N = 66.54M^{0.442}$, $\rho_{227} = 0.887$.

The anomalous scaling of wing area in hummingbirds was first reported by Greenewalt (1960a, 1962), albeit with the incorrect index of 4/3 fitted by eye; Greenewalt (1975) later derived the correct index of 1 (Fig. 2). While the steeper scalings are evidently associated with the hovering flight mechanism of these birds, no viable explanation has hitherto been proposed. Greenewalt (1960a, b,

1962) thought it was evidence for a resonant tuning within the bones and muscles of the pectoral girdle, which he invoked—correctly—to explain the scaling of wingbeat frequency in hovering insects as well as hummingbirds. Subsequent studies (Weis-Fogh 1973) have found no mechanism in birds for such a resonance, and moreover wingbeat frequency does not scale inversely with wing length as Greenewalt claimed (Figs 6b and 7).

I propose two linked explanations for the scalings in hummingbirds. The first and simplest is flight speed: if hummingbirds were designed according to geometric similarity, the smallest species would fly unacceptably slowly; flight speed is important to these birds, as is evinced by their disproportionately small wings (above). The second arises from the effect of this wing design model on the scaling constraints on wingbeat frequency. Figures 5 and 6a were derived by assuming geometrical similarity of all design elements. With this model the only lower constraint on body mass in birds is imposed by the maximum wingbeat frequency (probably of the order of 50 Hz), and this constraint can be circumvented by evolving relatively longer and larger wings which can be flapped more slowly (Rayner 1995) as size is reduced. By contrast, the upper limit to the range of mass is set by induced power, which is known to be the dominant mechanical energy demand in hovering. When the constraints are recalculated to account for the hummingbird (constant wing loading) similarity, the diagram changes considerably (Fig. 6b and c, Table 1). (Two versions are shown in the figure; in Fig. 6b I assume that all linear dimensions scale as $M^{1/2}$, while in Fig. 6c I assume that wing dimensions scale as $M^{1/2}$ but that the linear dimensions of muscles and tendons scale geometrically as $M^{1/3}$; both hypotheses are plausible, but there are insufficient data to distinguish between them. In both cases lift generation from the flapping wings becomes more severe as a constraint on mass (since its gradient is steeper); this is to be expected if the wings are larger but are beating more slowly.

It would be false to argue that constraints of this kind have determined the scaling of wing design in hummingbirds. Rather, this phenomenon should be viewed the other way round: the design paradigm which preserves flight speed in small hummingbirds has the effect of modifying the constraints on frequency and on body mass, and of ensuring that hovering in hummingbirds is viable only over a confined range of relatively small masses. The outstanding question remains specification of the genetic and developmental mechanisms which ensure that this adaptation is confined to hummingbirds.

Passeriforms

The other group of birds which has attained the smallest sizes is the Passeriformes. The four species of Regulidae (goldcrest and firecrest of the Palaearctic and kinglets of the Nearctic) have body mass ranging from 0.0045 to 0.008 kg, and some *Phylloscopus* and related warblers have a mean mass of around 0.005 kg; individuals of several other passeriform families approach these small sizes; the smallest passeriform species listed by Dunning (1993) is

Aethopyga shelleyi, a sunbird (Nectariniidae) from the Philippines, in which females may be as small as 0.0031 kg. However, all these birds are significantly larger than the smallest hummingbirds (Fig. 1). European regulids are short-distance migrants, but the migration flights of some *Phylloscopus* are among the longest. Small size does not appear to inhibit migratory behaviour, despite the increased cost of transport (see pp. 99–100), and bounding flight is important in facilitating this behaviour.

It is frustrating not to be able to answer the important question of why no passerines are smaller than this. There is little idea of any individual factor which sets this limit: as I have argued in this paper, the absolute limit does not appear to be mechanical, although a maximum wingbeat frequency set by contraction of vertebrate striated muscle may be significant. Calder (1984) concluded that the limit was probably set by physiological, ecological, digestive, or reproductive factors. However, these concepts alone cannot resolve the question of how hummingbirds can evolve small body masses approximately half those of the smallest passeriforms.

Acknowledgements

I am grateful to Rowan Lockwood for assistance in preparing Fig. 1. My work on vertebrate flight mechanics has been funded by the Royal Society, SERC, NERC, BBSRC, and the Nuffield Foundation.

References

Alexander, R. McN. (1981). Factors of safety in the structure of animals. *Sci. Prog., Lond.* **67**: 109–130.

Alexander, R. McN. (1982a). Size, shape, and structure for running and flight. In *A companion to animal physiology*: 309–324. (Eds Taylor, C. R., Johansen, K., & Bolis, L.). Cambridge University Press, Cambridge.

Alexander, R. McN. (1982b). *Locomotion of animals*. Blackie, Glasgow.

Alexander, R. McN. (1992). Optimization of skeletal structure in vertebrates. *Belg. J. Zool.* **122**: 23–29.

Benton, M. J. (1990). Evolution of large size. In *Palaeobiology: a synthesis*: 147–152. (Eds Briggs, D. E. G. & Crowther, P. R.). Blackwell, Oxford.

Blackburn, T. M. & Gaston, K. J. (1994). The distribution of body sizes of the world's bird species. *Oikos* 70: 127–130.

Bleiweiss, R. (1990). Ecological causes of clade diversity in hummingbirds: a neontological perspective on the generation of diversity. In *Causes of evolution, a palaeontological perspective*: 354–380. (Eds Ross, R. M. & Allmon, W. D.). University of Chicago Press, Chicago.

Calder, W. A. (1984). *Size, function, and life history*. Harvard University Press, Cambridge, MA.

Campbell, K. E. & Marcus, L. (1992). The relationship of hindlimb bone dimensions to body weight in birds. In *Papers in avian paleontology honoring Pierce Brodkorb*: 395–412. (Ed. Campbell, K. E.). (*Sci. Ser. nat. Hist. Mus. Los Angeles Cty* No. 36.)

Campbell, K. E. & Tonni, E. P. (1983). Size and locomotion in teratorns (Aves: Teratornithidae). *Auk* **100**: 390–403.

Charnov, E. L. (1993). *Life history invariants: some explanations of symmetry in evolutionary ecology*. Oxford University Press, Oxford.

Dunning, J. B. (Ed.) (1993). *CRC handbook of avian body masses*. CRC Press, Roca Raton, FL.

Economos, A. C. (1982). On the origin of biological similarity. *J. theor. Biol.* **94**: 25–60.

Eisenberg, J. F. (1981). *The mammalian radiations: an analysis of trends in evolution, adaptation, and behavior*. University of Chicago Press, Chicago.

Fullerton, J. D. (Ed.) (1911). *First report of the bird construction committee*. Aeronautical Society, London.

Gould, S. J. (1971). Geometric similarity in allometric growth: a contribution to the problem of scaling in the evolution of size. *Am. Nat.* **105**: 113–136.

Greenewalt, C. H. (1960a). *Hummingbirds*. Doubleday, New York. (Reprinted Dover, New York, 1991.)

Greenewalt, C. H. (1960b). The wings of insects and birds as mechanical oscillators. *Proc. Am. phil. Soc.* **104**: 605–611.

Greenewalt, C. H. (1962). Dimensional relationships for flying animals. *Smithson. Misc. Collns* No. 144 (2): 1–46.

Greenewalt, C. H. (1975). The flight of birds. *Trans. Am. phil. Soc.* No. 65 (4): 1–67.

Greenewalt, C. H. (1977). The energetics of locomotion—is small size really disadvantageous? *Proc. Am. phil. Soc.* **121**: 100–106.

Günther, B. & Morgado, E. (1982). Theory of biological similarity revisited. *J. theor. Biol.* **96**: 543–559.

Halstead, B. & Middleton, J. (1976). Fossil vertebrates of Nigeria. Part I. *Nigerian Fd* **41**: 55–63.

Harlé, E. & Harlé, A. (1911). Le vol des grands reptiles et insectes disparus semble indiquer une pression atmospherique élevée. *Bull. Soc. géol. Fr.* **4**: 117–121.

Harvey, P. H. & Pagel, M. D. (1991). *The comparative method in evolutionary biology*. Oxford University Press, Oxford.

Hazlehurst, G. & Rayner, J. M. V. (1992). Flight characteristics of Jurassic and Triassic Pterosauria: an appraisal based on wing shape. *Palaeobiology* **18**: 447–463.

Hill, A. V. (1950). The dimensions of animals and their muscular dynamics. *Sci. Prog., Lond.* **38**: 209–230.

Howell, A. B. (1944). *Speed in animals: their specialization for running and leaping*. University of Chicago Press, Chicago.

Kipp, F. (1950). Der wellenförmige Flug der Kleinvögel. *Vogelwarte* **15**: 233–235.

Knoll, W. (1934). Untersuchungen über den Kolibriflug. *Zool. Anz.* **107**: 97–114.

LaBarbera, M. (1986). The evolution and ecology of body size. In *Patterns and processes in the history of life*: 69–98. (Eds Raup, D. M. & Jablonski, D.). Springer Verlag, Berlin. (*Life Sci. Res. Rep.* **36**.)

Langston, W. (1981). Pterosaurs. *Scient. Am.* **244** (2): 92–102.

Lawson, D. A. (1975). Could pterosaurs fly? *Science, N.Y.* **188**: 676–678.

Lighthill, J. (1977). Introduction to the scaling of aerial locomotion. In *Scale effects in animal locomotion*: 365–404. (Ed. Pedley, T. J.). Academic Press, London.

McMahon, T. A. (1973). Size and shape in biology. *Science, N.Y.* **179**: 1201–1204.

McMahon, T. A. & Bonner, J. T. (1983). *On size and life*. Scientific American Library, New York.

Marden, J. H. (1994). From damselflies to pterosaurs—how burst and sustainable flight performance scale with size. *Am. J. Physiol.* **266**: R1077–R1084.

Norberg, U. M. (1990). *Vertebrate flight: mechanics, physiology, morphology, ecology and evolution.* Springer-Verlag, Heidelberg. (*Zoophysiology* **27**: 1–291.)

Norberg, U. M. (1994). Wing design, flight performance, and habitat use in bats. In *Ecological morphology*: 205–239. (Eds Wainwright, P. C. & Reilly, S. M.). University of Chicago Press, Chicago.

Norberg, U. M. & Rayner, J. M. V. (1987). Ecological morphology and flight in bats (Mammalia; Chiroptera): wing adaptations, flight performance, foraging strategy and echolocation. *Phil. Trans. R. Soc. Lond. (B)* **316**: 335–427.

Padian, K. & Rayner, J. M. V. (1993). The wings of pterosaurs. *Am. J. Sci.* **293-A**: 91–166.

Pedley, T. J. (Ed.) (1977). *Scale effects in animal locomotion.* Academic Press, London.

Pennycuick, C. J. (1975). Mechanics of flight. In *Avian biology* **5**: 1–75. (Eds Farner, D. S. & King, J. R.). Academic Press, London.

Pennycuick, C. J. (1986). Mechanical constraints on the evolution of flight. In *The origin of birds and the evolution of flight*: 83–98. (Ed. Padian, K.). (*Mems Calif. Acad. Sci.* No. 8.)

Rayner, J. M. V. (1977). The intermittent flight of birds. In *Scale effects in animal locomotion*: 437–443. (Ed. Pedley, T. J.). Academic Press, London.

Rayner, J. M. V. (1981). Flight adaptations in vertebrates. *Symp. zool. Soc. Lond.* No. 48: 137–172.

Rayner, J. M. V. (1985a). Linear relations in biomechanics: the statistics of scaling functions. *J. Zool., Lond. (A)* **206**: 415–439.

Rayner, J. M. V. (1985b). Bounding and undulating flight in birds. *J. theor. Biol.* **117**: 47–77.

Rayner, J. M. V. (1987). The mechanics of flapping flight in bats. In *Recent advances in the study of bats*: 23–42. (Eds Fenton, M. B., Racey, P. A., & Rayner, J. M. V.). Cambridge University Press, Cambridge.

Rayner, J. M. V. (1988). Form and function in avian flight. *Curr. Orn.* **5**: 1–66.

Rayner, J. M. V. (1990). The mechanics of flight and bird migration performance. In *Bird migration: physiology and ecophysiology*: 283–299. (Ed. Gwinner, E.). Springer Verlag, Berlin, Heidelberg etc.

Rayner, J. M. V. (1993). On aerodynamics and the energetics of vertebrate flapping flight. In *Fluid dynamics in biology*: 351–400. (Eds Cheer, A. Y. & van Dam, C. P.). American Mathematical Society, Providence. (*Contemp. Math.* No. 141.)

Rayner, J. M. V. (1995). Flight mechanics and constraints on flight performance. *Israel J. Zool.* **41**: 321–342.

Rayner, J. M. V., Speakman, J. R., Kunz, T. H., Norberg, U. M., & Jones, G. (In preparation). *Scaling of power and efficiency in flying vertebrates.*

Saville, D. B. O. (1950). The flight mechanism of swifts and hummingbirds. *Auk* **67**: 499–504.

Scheithauer, W. (1967). *Hummingbirds—flying jewels.* Arthur Barker, London.

Schmidt-Nielsen, K. (1972). Locomotion: energy cost of swimming, flying and running. *Science, N.Y.* **177**: 222–228.

Schmidt-Nielsen, K. (1984). *Scaling: why is animal size so important?* Cambridge University Press, Cambridge.

Sibley, C. G. & Ahlquist, J. E. (1990). *Phylogeny and classification of birds: a study in molecular evolution.* Yale University Press, New Haven.

Sokal, R. R. & Rohlf, F. J. (1994). *Biometry.* (3rd edn). W. H. Freeman, New York.

Stolpe, M. & Zimmer, K. (1939). Der Schwirrflug des Kolibri im Zeitlupenfilm. *J. Orn. Lpz.* **87**: 136–155.

Suarez, R. K., Brown, G. S., & Hochachka, P. W. (1986). Metabolic sources of energy for hummingbird flight. *Am. J. Physiol.* **251**: R537–R542.

Thomas, S. P. (1987). The physiology of bat flight. In *Recent advances in the study of bats*: 75–99. (Eds Fenton, M. B., Racey, P. A. & Rayner, J. M. V.). Cambridge University Press, Cambridge.

Tucker, V. A. (1977). Scaling and avian flight. In *Scale effects in animal locomotion*: 497–509. (Ed. Pedley, T. J.). Academic Press, London.

Weis-Fogh, T. (1973). Quick estimates of flight fitness in hovering animals, including novel mechanisms for lift production. *J. exp. Biol.* **59**: 169–230.

Weis-Fogh, T. (1977). Dimensional analysis of hovering flight. In *Scale effects in animal locomotion*: 405–420. (Ed. Pedley, T. J.). Academic Press, London.

Wellnhofer, P. (1991). *The illustrated encyclopedia of pterosaurs*. Salamander Books, London.

Wells, D. J. (1993). Muscle performance in hovering hummingbirds. *J. exp. Biol.* **178**: 39–57.

Winter, W. (1895). *Der Vogelflug. Erklärung der wichtigsten Flugarten der Vögel mit Einschluss des Segelns und Kreisens*. Theodor Ackermann, München.

Symp. zool. Soc. Lond. (1996) No. 69: 111–128

Does echolocation constrain the evolution of body size in bats?

GARETH JONES

School of Biological Sciences
University of Bristol
Woodland Road
Bristol BS8 1UG, UK

Synopsis

Most bats weigh less than 10 g. Megachiropterans are considerably heavier than microchiropterans, possibly because of their different ancestry or because they do not use laryngeal sonar. The predominance of medium-sized to small species of bats is compatible with general theories of extinction and speciation rates in relation to body size. There are few obvious links between diet and body size. It is more likely that biophysical factors associated with flight and/or echolocation constrain body size in bats. Models of flight power predict that flying vertebrates could weigh up to 12 kg, but the largest bat weighs under 1.5 kg. Bats radiated to fill the niche of nocturnal aerial insectivores, and echolocation is an effective way of detecting insects in the dark. However, high frequencies must be used to give strong echoes from small targets. Call frequency scales negatively with mass in five families of bats. Bats must probably be small to produce ultrasound. Because larger echolocating bats call at lower frequencies (hence long wavelengths), it has been proposed that they may not be able to detect small insects effectively. Large bats might therefore have a narrow dietary niche, and the need to rely on large, scarce prey would not favour the evolution of large size. However, large bats in fact seem to have broader dietary niche breadths than smaller species. Flapping and calling are coupled in echolocating bats as an energy-saving mechanism. Wingbeat frequency decreases as body size increases and aerial insectivorous bats generally emit one or fewer pulses per wingbeat. Aerial insectivorous bats may therefore be prevented from evolving large size because they would not be able to produce echolocation calls at a rate high enough to catch sufficient insects to meet the energetic demands associated with being large.

Introduction

Twenty-three per cent, or 977, of extant mammal species belong to the order Chiroptera; a further 42% of all mammal species are rodents (Corbet & Hill 1991). It seems that most mammals are small. Several general hypotheses have been proposed to explain species diversity in relation to body size in animals

ZOOLOGICAL SYMPOSIUM No. 69
ISBN 0–19–857787–7

(Hutchinson & MacArthur 1959; Van Valen 1973; May 1978; Dial & Marzluff 1988; Brown, Marquet & Taper 1993; Blackburn & Gaston 1994; Purvis & Harvey, this volume pp. 159–174). Such general ecological and evolutionary hypotheses may help our understanding of the distribution of body sizes in bats, but at the same time their generality may limit their applicability to bats in particular. Bats, like all other animals, may follow general ecological rules which relate extinction and speciation rates to body size. However, all bats fly and most echolocate, and constraints associated with flight and echolocation may be more important in determining the evolution of body size in these mammals than are more general ecological principles. Advocates of general hypotheses may argue that biophysical factors associated with flight and echolocation constrain maximum and minimum body size in bats, but within these constraints the distribution of size is determined by more global factors. In this paper, I intend to determine whether small bats are more speciose than large bats and to evaluate how any relationships between species diversity and body size fit predictions from hypotheses about the evolution of body size, and then to consider more specific hypotheses which propose that body size in bats may be constrained by biophysical principles associated with flight and echolocation.

Are bats miniature vertebrates?

To investigate the size range of bats, I obtained data on the body masses of 465 species from the literature, approximately half the total species of bats described in Corbet & Hill (1991). Sampling biases may be critical to the interpretation of the relation between size and diversity (Blackburn, Harvey, & Pagel 1990). Since small species of bats are more likely to remain unnoticed and undescribed than are larger species, it is likely that the discovery of any new species of bats will further skew the size distribution towards the smallest classes. Bats range in mass from approximately 2 g (*Craseonycteris thonglongyai*) to approximately 1200 g (*Pteropus giganteus* and *Pteropus vampyrus*). The distribution of body masses shows that over 30% of all bats are < 10 g (Fig. 1). Most bat species are 5–10 g, with relatively few species under 5 g. The smallest size class is therefore not the most speciose.

Members of the suborder Megachiroptera, which contains one family, the Pteropodidae, the Old World fruit bats or flying foxes (162 species), are clearly much larger than microchiropterans (Fig. 2). The largest microchiropteran weighs approximately 160 g (*Cheiromeles torquatus*) and very few weigh more than 50 g (Fig. 1b). The data in Fig. 2 indicate that megachiropterans are on average significantly heavier than all families of microchiropterans where masses are available for more than 20 species ($F_{6,424} = 29.4$, $P < 0.001$; Tukey tests all significant at $P < 0.001$). Pettigrew *et al.* (1989) considered ancestral megachiropterans to be moderately large, but the evolutionary history of the Megachiroptera is at present unclear. Pettigrew *et al.* (1989) present a series of arguments

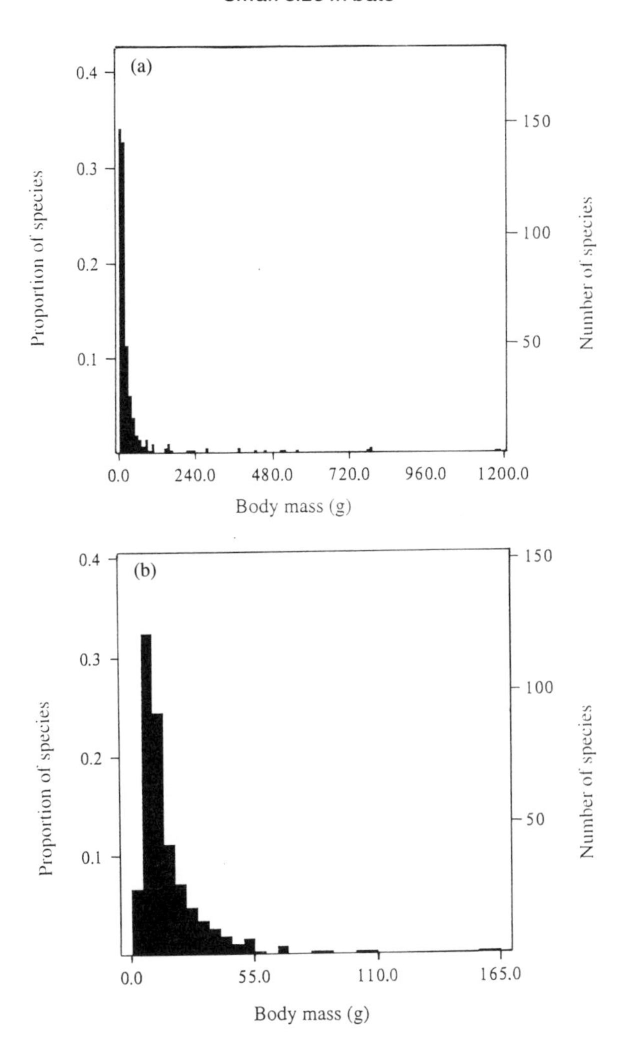

Fig. 1. Distributions of body masses of bats. (a) All bats, $n=465$, mass classes = 10 g; (b) Microchiroptera only, $n=376$, mass class = 5 g. Note differences in scales. Data taken from Jones (1971), Kingdon (1974), Silva Taboada (1979), Strahan (1983), Payne & Francis (1985), Happold, Happold, & Hill (1987), Norberg & Rayner (1987), Eisenberg (1989), Schober & Grimmberger (1989), King (1990), Schmidly (1991), Kalko & Handley (1994), and from G. Jones (unpubl. data) and *Mammalian Species* (American Society of Mammalogists).

linking the Megachiroptera more closely to the Primates than to the Microchiroptera. Others argue that the Chiroptera are monophyletic (Baker, Novacek, & Simmons 1991). Even molecular data, which have been interpreted as supporting monophyly, are likely to be equivocal because the genome of bats is AT-biased (Pettigrew 1994, 1995). I will therefore treat the question of whether the Megachiroptera are bats or flying primates as unanswered at present, and include them in the present discussion.

Pettigrew *et al.* (1989) believed that large size was a derived character in the Microchiroptera. Among the microchiropterans, significant differences occur in mean mass among families ($F_{5,370} = 10.2$, $P < 0.001$), with Tukey tests showing molossids to be significantly heavier than rhinolophids and vespertilionids, and phyllostomids heavier than vespertilionids.

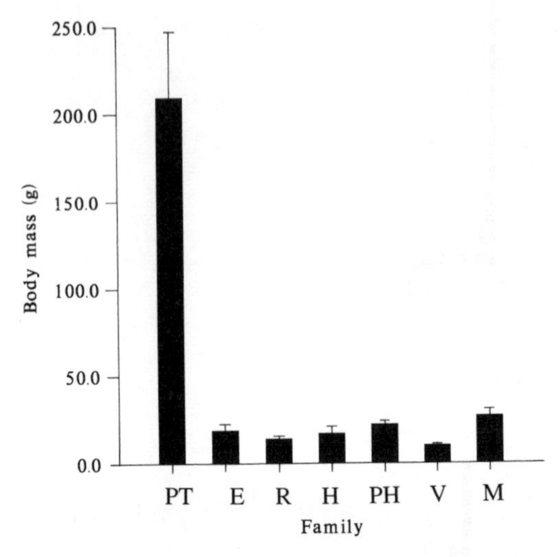

Fig. 2. Body masses (mean ± SE) for families of bats where data are available for >20 species per family. Sample sizes are Pteropodidae (PT), $n=55$; Emballonuridae (E), $n=31$; Rhinolophidae (R), $n=22$; Hipposideridae (H), $n=23$; Phyllostomidae (PH), $n=106$; Vespertilionidae (V), $n=150$; Molossidae (M), $n=44$. See text for ANOVA details.

What do the distributions of body masses of bats tell us? First, it is clear that bats nowhere attain the maximum size predicted from models of power requirements in relation to power available for flight (Pennycuick 1975; see below). Such models predict a maximum size of approximately 12 kg, yet the heaviest bat weighs little over 1 kg.

A second point to note from Figs 1 and 2 is that microchiropterans really are miniature vertebrates. Over 70% of species are under 20 g and almost 40% weigh less than 10 g. The large differences in the mean masses of megachiropterans and microchiropterans could exist because of the different phylogenies of the two groups or could be a consequence of the two suborders using different primary methods for sensory perception. Megachiropterans are primarily visual animals, while microchiropterans echolocate. Bats in only one megachiropteran genus—*Rousettus*—are known to echolocate and their echolocation calls are simple, produced by tongue clicking (Kulzer 1958) rather than by the more sophisticated laryngeal mechanisms employed by microchiropterans.

So why is there a preponderance of small bats? There are broadly two classes of hypotheses about the relations between size and species diversity. Ecological hypotheses may apply to a wide range of taxa, but some design-related hypotheses may be specific to flying vertebrates, or even to echolocating bats. These hypotheses are explored below.

General ecological theories and body size in bats

The niche 'grain-size' and speciation/extinction hypotheses

Several authors have argued that the smaller taxa should be the most speciose (Hutchinson & MacArthur 1959; Van Valen 1973; May 1978). Small animals should be able to divide their environment more finely than large animals can and, given a habitat of a particular size, more small animals than large animals would find niches (MacArthur & Levins 1964). Such niche 'grain-size' hypotheses predict that the smallest animals should be the most speciose, while empirical studies generally show that small to medium-sized animals tend to be more speciose than the smallest size class (May 1978). The niche 'grain-size' hypothesis is therefore unsuitable in explaining the patterns of body size in bats, because of the paucity of species under 5 g. Of course, it can be argued that the interpretation of the distribution of body sizes depends on which size classes are used in analyses: if the data in Fig. 1b were placed into 10 g mass classes, then one would indeed conclude that the smallest bats are the most speciose!

Dial & Marzluff (1988) argued that small to medium-sized animals should be the most speciose. Extinction rates would be highest in extreme-sized animals and the radiation of small-sized species should be most rapid. In combination, these effects would result in an abundance of small to medium-sized animals. This hypothesis is compatible with the observed distribution of body sizes in bats.

Diet, foraging strategy, and energy flow hypotheses

Although animals are generally larger at higher trophic levels, there are many exceptions to this trend (Louw 1993). For bats, there is no obvious influence of diet on body size except perhaps in a few cases. The Old World megachiropterans tend to be larger than New World phyllostomids (Fig. 2), even though both groups eat almost exclusively plant products. Fleming (1993) noted that pteropodid bats were approximately twice the size of plant-visiting phyllostomids in terms of forearm length. Within the microchiropterans, phyllostomids do not differ in mean size from emballonurids, rhinolophids, hipposiderids, or molossids, all of which are exclusively insectivorous (Fig. 2). It is not possible to compare dietary specializations within families because each family is almost exclusively insectivorous, except for the mainly plant-eating phyllostomids. However, from cross-family comparisons there is no evidence to support the idea that plant eaters are larger than insect eaters. Among animal-eating bats, carnivorous species tend to be heavier than insectivorous species, probably because carnivorous bats need to be bigger to handle large prey items (Norberg & Fenton 1988). It is possible that folivory has failed to evolve in bats because the low energy content of leaves requires a large gut capacity and hence a greater body mass. The high energetic costs of flapping flight, especially at high mass, may have constrained the evolution of folivory in bats (Dudley & Vermeij (1992); but see Kunz & Ingalls (1994) for an alternative perspective).

Brown *et al.* (1993) attempted to develop a general model for the effect of body size on fitness. Fitness was redefined as the rate of conversion of energy into offspring. This conversion was assumed to be a two-stage process, involving the rate of acquisition of energy from the environment and the rate of conversion of acquired energy into offspring. The former is assumed to scale as body mass to the 0.75 power, the latter as mass to the -0.25 power. The two processes combined predict an optimum body size for mammals around 100 g and a right log-skewed distribution of masses around this optimum. Although the distribution of body masses of bats is right skewed, few species attain 100 g (Fig. 1).

Barclay & Brigham (1991) argued that bats which catch prey on the wing—aerial insectivores—were smaller than bats that took prey from surfaces—gleaners. However, their analysis did not control for phylogenetic effects. Of the bat families, only the Vespertilionidae has a sufficient number of insectivorous species for a comparison to be made between gleaners and aerial insectivores. Taking data from Barclay & Brigham (1991) and supplementing it with information for European species in Schober & Grimmberger (1989), it is apparent that there is no difference between the mean masses of aerial insectivorous and gleaning vespertilionids (aerial insectivores mean mass 10.38 ± 8.69 (SD) g ($n=62$); gleaners 13.02 ± 7.39 g ($n=13$), $t_{73}=1.02$, N.S.). The smallest bats in the world tend to be aerial insectivores (all bats under 5 g), but aerial insectivores can also be relatively large. Indeed, molossids are normally aerial insectivores, yet the largest microchiropteran is a molossid (*Cheiromeles*) that captures prey by aerial hawking. In the absence of very small gleaning species, hypotheses about small size in bats have concentrated on aerial insectivorous species (Barclay & Brigham 1991; Jones 1994).

General ecological arguments may not help greatly in understanding the size distributions of bats. Bats have two specializations—flight and echolocation—which may impose constraints on body size not felt by animals lacking these abilities. It is therefore now opportune to consider some of these arguments in more detail.

Mechanical constraints on flight performance may limit the body size of bats

Power required for flight and power available from flight muscles

The power margin (power available minus power required) for flight decreases with body mass (Pennycuick 1975). Regression lines of power available and power required cross at approximately 12 kg as the maximum size for horizontal flight in vertebrates. The heaviest birds in several orders weigh close to 12 kg (Pennycuick 1975) and so birds can clearly outweigh bats. Some fossil pterosaurs and subfossil birds may have weighed 40 kg (Norberg 1990). It remains unclear why the largest bat is relatively small for a flying vertebrate, especially since birds and bats have similar flight physiology. However, some microchiropterans may be unable to sustain flight at maximal range speeds (Carpenter 1985), and low endurance of flight by large bats may contribute towards setting a ceiling on body size. Many large birds use gliding or soaring to save energy. Bats rarely soar because of the absence of convective air currents at night (Rayner 1981). They rarely glide because the coupling of echolocation with flapping results in reduced sensory performance during gliding (Jones 1993). The separated primary feathers of birds also assist in gliding (Rayner 1981). Most large birds are aquatic or carnivorous and the radiation of bats into these niches may be severely limited because of the incorporation of limbs into the flight membranes (Rayner 1981).

The lower limit to size in bats may be because the animals are unable to contract muscles at a sufficiently fast rate to achieve the high wingbeat frequencies necessary for sustained flight at low mass (Norberg 1990).

Wing-loading constraints

Because surface area increases as the square of linear dimensions, while volume (hence mass) increases as the cube, larger bats tend to have higher wing loadings. Flight speed increases and manoeuvrability decreases at higher wing loadings (Norberg & Rayner 1987). Hence, larger bats may not be sufficiently manoeuvrable to catch prey within their limits of detection as successfully as smaller species do, a situation favouring small size in bats. Given that detection distances for bats are usually 1–5 m (e.g. Kick 1982; Kalko & Schnitzler 1993), bats will clearly not have much time in which to perform capture manoeuvres, especially since larger species may fly at approximately 10 $m\,s^{-1}$ (Baagøe 1987; Norberg 1987).

Before discussing why echolocation may constrain the evolution of body size in bats it is worth considering (1) why bats echolocate and (2) why usually they use ultrasound for echolocation.

Why bats echolocate and why most bats use ultrasonic calls

Why echolocation?

The vast majority of bats are nocturnal. Insectivorous bats are almost exclusively nocturnal and it is likely that diurnality is strongly selected against in the extant echolocating bats because the limited perception of the world available through echolocation renders bats susceptible to predation by diurnal and crepuscular raptorial birds (Speakman 1990, 1991; Jones & Rydell 1994). Echolocation calls are normally ultrasonic (see below), and attenuate rapidly in air. Bats therefore probably perceive objects only at close range and within a narrow directional cone. This would render them vulnerable to predation from raptorial birds (Speakman 1991).

Although it is easy to accept why extant bats are nocturnal, the question may be asked as to why they became nocturnal in the first place. Major adaptive radiations of birds occurred before bats appeared in the fossil record 50 million years ago (Novacek 1985). Because birds are primarily visual animals, they may have dominated the niches for diurnal aerial insectivores before bats evolved. Echolocating bats presumably evolved from nocturnal insectivorous ancestors (Fenton *et al.* 1995) and therefore showed some pre-adaptations for catching insects by night. Vision may not, however, be an effective means of locating small insects in darkness and bats could have evolved elaborations on a primitive echolocation system already present in their ancestors. Once neural processing became developed for echolocation, it may have been impossible to devote neural resources to a well-developed visual system.

It is unclear whether aerial insect densities are higher in the day or in the night. Many temperate insect groups have maximum densities in daytime (Lewis & Taylor 1965), though one tropical study showed highest biomasses of insects to occur for up to 2 h after sunset (Rautenbach, Kemp, & Scholtz 1988). Bats probably underwent major adaptive radiation in the tropics and the situation there could be more relevant for their evolution. Although it is difficult to speculate about past patterns of diurnal variation in insect abundance, it is clear that there might have been a mostly vacant niche of nocturnal aerial insectivores for early bats to occupy. High-frequency echolocation would be an ideal sensory system for exploiting this niche. Since coupling the production of echolocation calls with flapping the wings allows intense sound pulses to be produced at no extra energetic cost to flapping alone (Speakman & Racey 1991), the combination of flight and echolocation was potentially of great fitness value in terms of the exploitation of nocturnal insects. Plant-eating bats which echolocate, such as many phyllostomids, may have become limited to nocturnality if plant eating was derived from insectivory and the development of echolocation imposed the same problems of predator avoidance as those encountered by insectivorous bats. Megachiropterans may be nocturnal to avoid problems resulting from overheating by daytime flight in the tropics (Speakman, Hays, & Webb 1994).

Why ultrasound?

Most echolocating bats use ultrasonic call frequencies, up to 212 kHz in *Cloeotis percivali* (Fenton & Bell 1981). So why are bat echolocation calls generally ultrasonic? The frequencies of many structures such as air-filled pipes and diaphragms show an inverse relation to their linear dimensions when size is varied isometrically (Pye 1979). Drum membranes and strings also produce lower frequency as linear size increases, so it is clear that animals should be small to produce high-frequency sounds. Frequency scales negatively with body mass in five families of bats for which sufficient data are available (Fig. 3). In all cases the slopes of the double logarithmic plot lie between -0.36 and -0.49 if standard linear regression is used, or -0.48 and -0.82 if the reduced major axes equations are used (Table 1). It is probable that a similar negative relation would hold for the Phyllostomidae if more data were available. The slopes for the different families do not differ statistically (ANCOVA: family × log mass interaction $F_{5,117} = 1.78$, N.S.). However, the majority of bats in the Hipposideridae and Rhinolophidae produce most energy in the second harmonics of their calls, so that, for a given body mass, these bats have higher call frequencies than bats in the Vespertilionidae, Molossidae, and Emballonuridae.

Because many bats eat small insects, they will need to produce high-frequency pulses to detect them in the first instance. This is because high frequencies have shorter wavelengths and wavelength determines the amplitude of echoes in relation to target size. The velocity of sound in air and water equals frequency times wavelength. Since the velocity of sound in air is approximately 340 m s^{-1} at 16°C, it follows that the wavelength of a 10 kHz sound is 3.4 cm and that of 100 kHz is 3.4 mm.

Theory predicts that, in general, objects larger than a wavelength reflect sound, while smaller structures absorb or scatter sound. The predicted backscatter of sound from a sphere is illustrated in Fig. 4. Basically, for small targets whose circumference is less than the wavelength, the intensity of the backscattered echo is inversely proportional to the fourth power of wavelength. For objects of large circumference (10 wavelengths or more), echo intensity is proportional to cross-sectional area. Between these extremes, the intensity of the backscattered echo fluctuates. Theory thus predicts that sounds with wavelengths longer than the circumference of the target will give only weak echoes (Pye 1993). Target strength is defined as the logarithmic ratio of incident energy to the reflected energy, measured at 1 m from the target, and the target strength of insects is too complex to be modelled theoretically (Møhl 1988). Interestingly, some measurements of target strength from mounted insects showed that, contrary to models based on spheres or discs, prey species echo strength was virtually independent of frequencies between 20 and 100 kHz. Nevertheless, as expected, large targets such as moths gave higher values for target strengths than did smaller dipterans (Waters, Rydell, & Jones 1995). Target strength was also considerably lower than that predicted for spheres or discs (Møhl 1988). The biological reality of models based on simple targets such as spheres or discs must therefore be questioned. However, bats must probably use ultrasound, with its associated short wavelengths, in order to receive strong echoes from small targets. At the same time, for physical reasons, bats must be small to produce high-frequency sounds.

How echolocation may constrain body size in bats

Echolocation call frequency may determine dietary niche breadth

Barclay & Brigham (1991) used the predicted relation between call frequency and echo strength in a theory to explain why aerial insectivorous bats are small. As seen in Fig. 3, call frequency decreases as body size increases. Low frequencies also reflect weakly from small targets, at least in theory (Fig. 4). Barclay & Brigham argued that larger echolocating bats cannot detect small insects or, if they could detect them, could not have time to manoeuvre and catch them. Large aerial insectivorous bats should therefore be restricted to eating large prey and, because such prey are relatively scarce, they would need long foraging times.

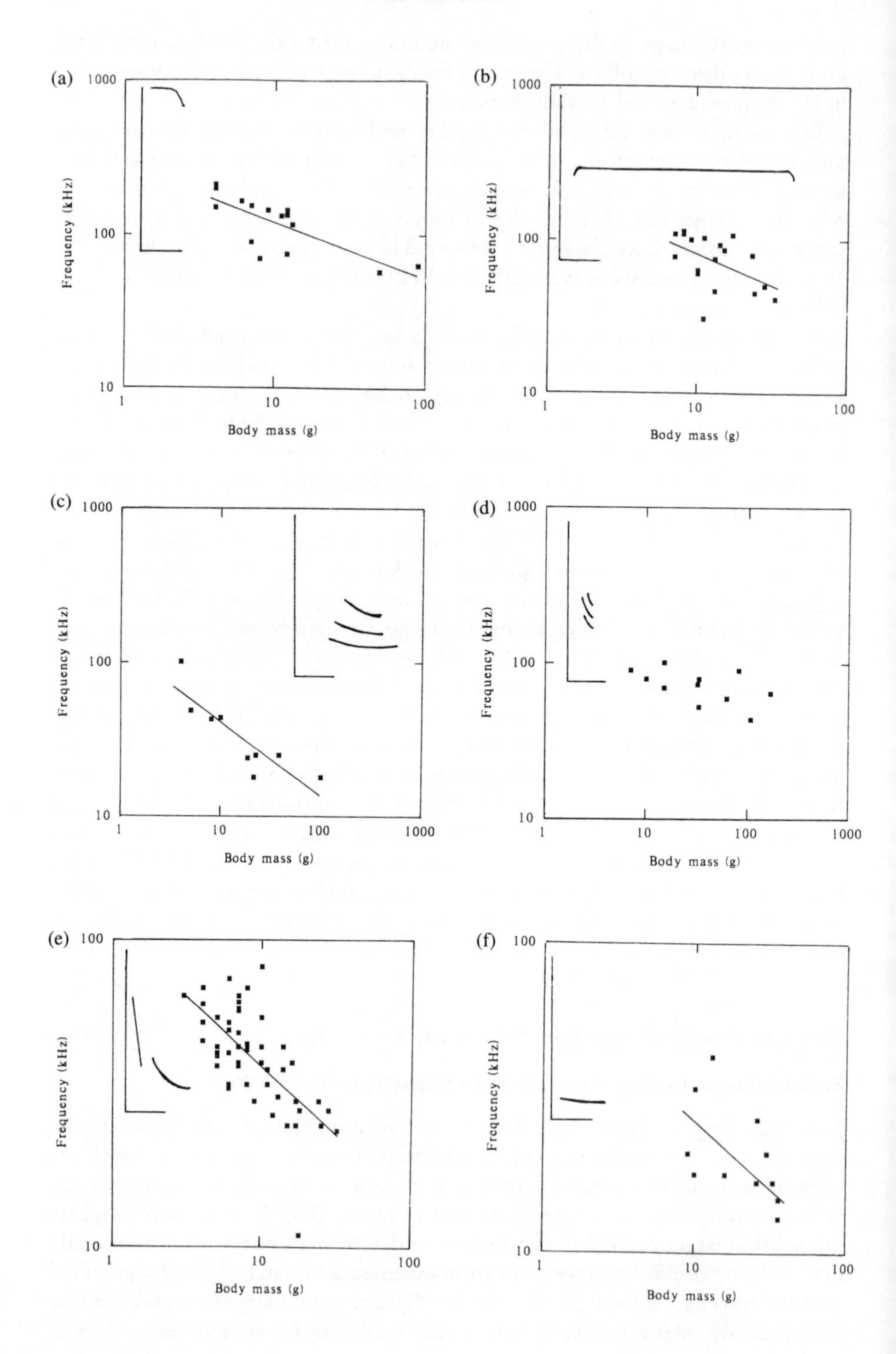
(a)
1000
100
10
1
10
100
Frequency (kHz)
Body mass (g)
(b)
1000
100
10
1
10
100
Frequency (kHz)
Body mass (g)
(c)
1000
100
10
1
10
100
1000
Frequency (kHz)
Body mass (g)
(d)
1000
100
10
1
10
100
1000
Frequency (kHz)
Body mass (g)
(e)
100
10
1
10
100
Frequency (kHz)
Body mass (g)
(f)
100
10
1
10
100
Frequency (kHz)
Body mass (g)

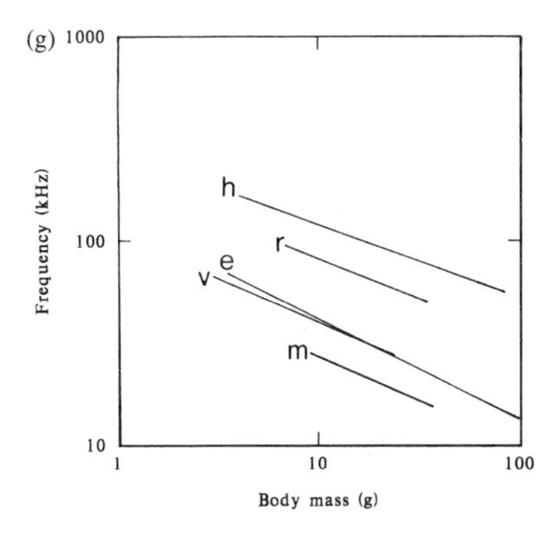

Fig. 3. Relations between frequency of maximal amplitude in echolocation calls and body mass for six bat families: (a) Hipposideridae; (b) Rhinolophidae; (c) Emballonuridae; (d) Phyllostomidae; (e) Vespertilionidae; (f) Molossidae. All lines are standard linear regressions. Note the differences in scales. In (g) regression lines for all families are plotted on the same graph — see text for equations: Hipposideridae (h), Rhinolophidae (r), Emballonuridae (e), Vespertilionidae (v), and Molossidae (m). For each family, characteristic echolocation calls are shown, with the *y*-axis terminating at 150 kHz and the *x*-axis at 10 ms. Species depicted are (a) short, high-frequency CF/FM *Hipposideros fulvus*; (b) long, high-frequency FM/CF/FM *Rhinolophus ferrumequinum*; (c) long, multiharmonic, low-frequency shallow FM *Taphozous kachensis*; (d) short, multiharmonic FM *Phyllostomus discolor*; (e) short FM *Myotis evotis* (left) and FM/CF *Eptesicus fuscus*; (f) long, low-frequency shallow FM *Tadarida aegyptiaca* (CF = constant frequency, FM = frequency modulated). Sources of echolocation call data were Pye (1972, 1980), Barclay (1983, 1986), Aldridge & Rautenbach (1987), Heller (1989), Heller & Helversen (1989), Obrist, Aldridge, & Fenton (1989), Zingg (1990), Barclay & Brigham (1991), Schumm, Krull, & Neuweiler (1991), Jones & Corben (1993), Surlykke *et al.* (1993), and G. Jones (unpubl. data).

Table 1. Scaling of dominant call frequency (kHz) for six families of bats.

Family	Number of species	Intercept (SLR)	Slope (SLR)	Intercept (RMA)	Slope (RMA)	*F*	*P*
Rhinolophidae	18	203.45	–0.401	593.52	–0.820	4.8	*
Hipposideridae	15	273.47	–0.361	356.87	–0.476	17.0	***
Emballonuridae	9	126.27	–0.485	153.32	–0.555	22.0	**
Phyllostomidae	11	114.65	–0.135	169.56	–0.247	3.8	N.S.
Vespertilionidae	52	107.25	–0.443	170.39	–0.661	40.8	***
Molossidae	12	69.53	–0.420	169.71	–0.716	5.3	*

*$P < 0.05$; **$P < 0.01$; *** $P < 0.001$; N.S. – not significant. In the analysis, body mass (g) and dominant call frequency (kHz) were transformed to natural logarithms. Intercepts and slopes are given for standard linear regression (SLR) equations and for reduced major axis (RMA) equations.

Does Barclay & Brigham's (1991) prediction about dietary niche breadth and body size hold? It does not appear to for European bats. Relatively large bats, such as the noctule *Nyctalus noctula* (*c.* 30 g) and Leisler's bat *Nyctalus leisleri* (*c.* 17 g) eat mainly small dipterans such as midges (Chironomidae) and dung-flies (Muscidae) (Sullivan *et al.* 1993; Jones 1995). Paradoxically, the bats

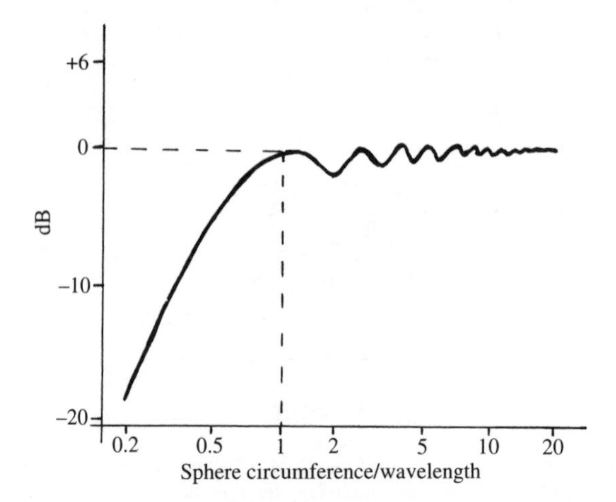

Fig. 4. Backscatter from a sphere for sound waves. Where sphere circumference/wavelength = 1, the curve begins to stabilize (to the right of the intersection of dashed lines). When the wavelength is longer than the sphere circumference, echo intensity falls off rapidly as wavelength increases. The x-axis represents the scale in frequency for a sphere of diameter 10.8 cm. After Pye (1993).

may be able to detect small prey only by using low frequencies. If higher ones were used, the maximum detection distance would fall into the zone of pulse–echo overlap because of excess atmospheric attenuation (Waters *et al.* 1995). Because relatively large bats also eat large prey and small bats eat only small prey (e.g. Swift, Racey & Avery 1985; see Table 2), the dietary niche breadth of large aerial insectivorous bats appears to be broader than that of small species—the converse of Barclay & Brigham's prediction.

Coupling of wingbeat and echolocation limits pulse repetition rate

An alternative hypothesis about how echolocation may constrain body size in bats was proposed by Jones (1994). The coupling of calling with flapping makes the production of intense echolocation calls no more costly than beating the wings and not calling (Speakman & Racey 1991). Calling without flapping could be energetically costly (Speakman, Anderson, & Racey 1989). Calls are produced generally at the end of the upstroke and coincide with exhalation (Suthers, Thomas, & Suthers 1972). Bursts of activity from muscles in the lateral abdominal wall accompany vocalizations and these coincide with activity of the flight muscles (Lancaster, Henson, & Keating 1995).

Aerial insectivorous bats generally produce one (occasionally fewer) pulse per wingbeat when searching for prey, as predicted by the coupling hypothesis (Fig. 5). Gleaning bats sometimes emit more than one pulse per wingbeat, but these pulses are of low intensity. The maximal wingbeat frequency decreases with body size in flying vertebrates because the forces necessary to impart angular acceleration to the wing are inversely proportional to the wing length (Pennycuick 1972). Thus,

Table 2. Diets of some European aerial insectivorous bats.

Species	Mass (g)	Dominant call frequency (kHz)	Diet	Source
Nyctalus noctula	26	21	44% Coleoptera 18% Lepidoptera 22% small Diptera (<5 mm wing length) 3–30 mm body length	Jones (1995) v
Nyctalus leisleri	17	25	43% chironomids 12% dung flies 22% Lepidoptera, Coleoptera, Trichoptera combined 3–25 mm body length	Waters *et al.* (1995) v
Vespertilio murinus	12	25	70% small Diptera 24% Lepidoptera, Coleoptera, Trichoptera combined 3–30 mm body length	Rydell (1992) v
Eptesicus serotinus	20	28	15% small Diptera 63% dung beetles 3–30 mm body length	Catto, Huston, & Racey (1994) f
Eptesicus nilssonii	10	28	47% small Diptera 17% Lepidoptera 18% Coleoptera 3–30 mm body length	Rydell (1986, 1989) v
Pipistrellus pipistrellus	5	46–57	66% small Diptera Only 5% of diet >12 mm body length	Swift *et al.* (1985) f
Myotis mystacinus	5	47	49% small Diptera 35% Lepidoptera 64% of prey <5 mm wing length	Hollyfield (1993) f

Echolocation data from Zingg (1990). Under source, v refers to volume, f to frequency as methods of quantification.

bigger bats flap more slowly and echolocate at lower repetition rates than small bats. Such coupling may constrain maximum body size in bats because very large bats may be unable to echolocate at a sufficiently high rate to detect enough insects to meet the high energetic demands associated with large size. In this way, mechanical constraints associated with flight performance may determine echolocation calling rate, and may contribute to setting an upper limit to body size in echolocating bats which emit intense calls.

Conclusions

Microchiropteran bats can be considered as miniature vertebrates, with most species being between 5 and 10 g. Large size may, however, be a derived condition within the Microchiroptera. The larger mean size of megachiropterans might be the result of different ancestry or a consequence of not echolocating. Even the largest megachiropterans are much smaller than many flying birds. Small size in bats has been investigated in a suite of ecological and mechanical theories, but it seems likely that bats need to be small to produce high-frequency sounds for the echolocation of small insects, and constraints associated with echolocation may explain why there are so many small bats.

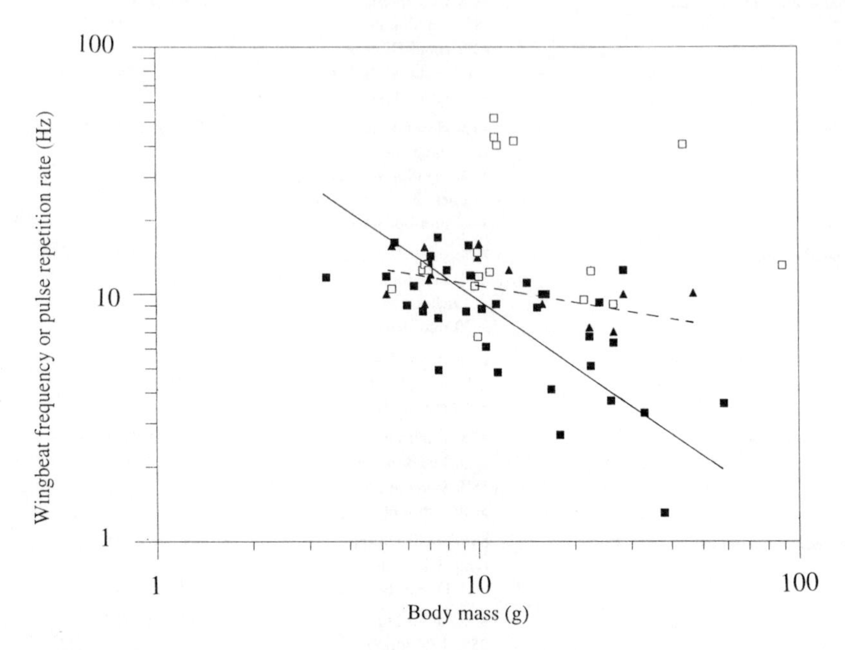

Fig. 5. Scaling of wingbeat frequency and pulse repetition rate in microchiropteran bats. Key: wingbeat frequencies (triangles), pulse repetition rate (PRR) of flutter detectors and whispering bats (open squares), PRR of aerial insectivores (solid squares). Regression lines (reduced major axis) for wingbeat frequencies (dashed line) and PRR (solid line — for aerial insectivores and other species emitting intense echolocation calls only) in relation to body mass are shown. Wingbeat frequency scales as $2.679M^{-0.326}$, $F_{1,18} = 12.0$, $P < 0.01$, and pulse repetition rate as $0.264M^{-0.775}$, $F_{1,42} = 35.0$, $P < 0.001$, where M = body mass (kg). After Jones (1994), reproduced with permission from Blackwell Scientific Publications.

References

Aldridge, H. D. J. N. & Rautenbach, I. L. (1987). Morphology, echolocation and resource partitioning in insectivorous bats. *J. Anim. Ecol.* **56**: 763–778.

Baagøe, H. J. (1987). The Scandinavian bat fauna: adaptive wing morphology and free flight in the field. In *Recent advances in the study of bats*: 57–74. (Eds Fenton, M. B., Racey, P. A. & Rayner, J. M. V.). Cambridge University Press, Cambridge.

Baker, R. J., Novacek, M. J., & Simmons, N. B. (1991). On the monophyly of bats. *Syst. Zool.* **40**: 216–231.

Barclay, R. M. R. (1983). Echolocation calls of emballonurid bats from Panama. *J. comp. Physiol.* **151**: 515–520.

Barclay, R. M. R. (1986). The echolocation calls of hoary (*Lasiurus cinereus*) and silver-haired (*Lasionycteris noctivagans*) bats as adaptations for long- versus short-range foraging strategies and the consequences for prey selection. *Can. J. Zool.* **64**: 2700–2705.
Barclay, R. M. R. & Brigham, R. M. (1991). Prey detection, dietary niche breadth, and body size in bats: why are aerial insectivorous bats so small? *Am. Nat.* **137**: 693–703.
Blackburn, T. M. & Gaston, K. J. (1994). Animal body size distributions: patterns, mechanisms and implications. *Trends Ecol. Evol.* **9**: 471–474.
Blackburn, T. M., Harvey, P. H., & Pagel, M. D. (1990). Species number, population density and body size in natural communities. *J. Anim. Ecol.* **59**: 335–346.
Brown, J. H., Marquet, P. A., & Taper, M. L. (1993). Evolution of body size: consequences of an energetic definition of fitness. *Am. Nat.* **142**: 577–584.
Carpenter, R. E. (1985). Flight physiology of flying foxes, *Pteropus poliocephalus*. *J. exp. Biol.* **114**: 619–647.
Catto, C. M. C., Hutson, A. M. & Racey, P. A. (1994). The diet of *Eptesicus serotinus* in southern England. *Folia zool.* **43**: 307–314.
Corbet, G. B. & Hill, J. E. (1991). *A world list of mammalian species*. (3rd edn). Natural History Museum Publications, London & Oxford University Press, Oxford.
Dial, K. P. & Marzluff, J. M. (1988). Are the smallest organisms the most diverse? *Ecology* **69**: 1620–1624.
Dudley, R. & Vermeij, G. J. (1992). Do the power requirements of flapping flight constrain folivory in flying animals? *Funct. Ecol.* **6**: 101–104.
Eisenberg, J. F. (1989). *Mammals of the Neotropics. The northern Neotropics.* **1.** *Panama, Columbia, Venezuela, Suriname, French Guiana*. University of Chicago Press, Chicago & London.
Fenton, M. B., Audet, D., Obrist, M. K., & Rydell, J. (1995). Signal strength, timing and self-deafening: the evolution of echolocation in bats. *Paleobiology* **21**:229–242.
Fenton, M. B. & Bell, G. P. (1981). Recognition of species of insectivorous bats by their echolocation calls. *J. Mammal.* **62**: 233–243.
Fleming, T. H. (1993). Plant-visiting bats. *Am. Sci.* **81**: 460–467.
Happold, D. C. D., Happold, M., & Hill, J. E. (1987). The bats of Malawi. *Mammalia* **51**: 337–414.
Heller, K.-G. (1989). Echolocation calls of Malaysian bats. *Z. Säugetierk.* **54**: 1–8.
Heller, K.-G. & von Helversen, O. (1989). Resource partitioning of sonar frequency bands in rhinolophoid bats. *Oecologia* **80**: 178–186.
Hollyfield, A. M. (1993). *Diet in relation to prey availability and the directionality and design of echolocation calls in three species of British bats*. PhD thesis: Bristol University.
Hutchinson, G. E. & MacArthur, R. H. (1959). A theoretical ecological model of size distributions among species of animals. *Am. Nat.* **93**: 117–125.
Jones, C. (1971). The bats of Rio Muni, West Africa. *J. Mammal.* **52**: 121–140.
Jones, G. (1993). Flight and echolocation in bats: coupling, and constraints on optimal design. In *Trends in comparative biochemistry and physiology*: 595–606. (Ed. Menon, G.). Council for Scientific Research Integration, Trivandrum.
Jones, G. (1994). Scaling of wingbeat and echolocation pulse emission in bats: why are aerial insectivorous bats so small? *Funct. Ecol.* **8**: 450–457.
Jones, G. (1995). Flight performance, echolocation and foraging behaviour in noctule bats *Nyctalus noctula*. *J. Zool., Lond.* **237**: 303–312.
Jones, G. & Corben, C. (1993). Echolocation calls from six species of microchiropteran bats in south-eastern Queensland. *Aust. Mammal.* **16**: 35–38.

Jones, G. & Rydell, J. (1994). Foraging strategy and predation risk as factors influencing emergence time in echolocating bats. *Phil. Trans. R. Soc. (B)* **346**: 445–455.

Kalko, E. K. V. & Handley, C. O., Jr (1994). Evolution, biogeography, and description of a new species of fruit-eating bat, genus *Artibeus* Leach (1821), from Panama. *Z. Säugetierk.* **59**: 257–273.

Kalko, E. K. V. & Schnitzler, H.-U. (1993). Plasticity of echolocation signals of European pipistrelle bats in search flight: implications for habitat use and prey detection. *Behav. Ecol. Sociobiol.* **33**: 415–428.

Kick, S. A. (1982). Target-detection by the echolocating bat, *Eptesicus fuscus. J. comp. Physiol.* **145**: 431–435.

King, C. M. (Ed.) (1990). *The handbook of New Zealand mammals.* Oxford University Press, Auckland.

Kingdon, J. (1974). *East African mammals. An atlas of evolution in Africa* **2A**. *Insectivores and bats.* Academic Press, London & New York.

Kulzer, E. (1958). Untersuchungen über die Biologie von Flughunden der Gattung *Rousettus* Gray. *Z. Morph. Ökol. Tiere* **47**: 374–402.

Kunz, T. H. & Ingalls, K. A. (1994). Folivory in bats: an adaptation derived from frugivory. *Funct. Ecol.* **8**: 665–668.

Lancaster, W. C., Henson, O. W. Jr, & Keating, A. W. (1995). Respiratory muscle activity in relation to vocalization in flying bats. *J. exp. Biol.* **198**: 175–191.

Lewis, T. & Taylor, L. R. (1965). Diurnal periodicity of flight by insects. *Trans. R. ent. Soc. Lond.* **116**: 393–476.

Louw, G. (1993). *Physiological animal ecology.* Longman Scientific & Technical, Harlow, Essex.

MacArthur, R. H. & Levins, R. (1964). Competition, habitat selection, and character displacement in a patchy environment. *Proc. natn. Acad. Sci. USA* **51**: 1207–1210.

May, R. M. (1978). The dynamics and diversity of insect faunas. In *Diversity of insect faunas*: 188–204. (Eds Mound, L. A. & Waloff, N.). Blackwell Scientific Publications, Oxford.

Møhl, B. (1988). Target detection by echolocating bats. *In Animal sonar: processes and performance*: 435–450. (Eds Nachtigall, P. E. & Moore, P. W. B.). Plenum Press, New York.

Norberg, U. M. (1987). Wing form and flight mode in bats. In *Recent advances in the study of bats*: 43–56. (Eds Fenton, M. B., Racey, P. A., & Rayner, J. M. V.). Cambridge University Press, Cambridge.

Norberg, U. M. (1990). Vertebrate flight. *Zoophysiology* **27**: 1–291.

Norberg, U. M. & Fenton, M. B. (1988). Carnivorous bats? *Biol. J. Linn. Soc.* **33**: 383–394.

Norberg, U. M. & Rayner, J. M. V. (1987). Ecological morphology and flight in bats (Mammalia; Chiroptera): wing adaptations, flight performance, foraging strategy and echolocation. *Phil. Trans. R. Soc. (B)* **316**: 335–427.

Novacek, M. J. (1985). Evidence for echolocation in the oldest known bats. *Nature, Lond.* **315**: 140–141.

Obrist, M., Aldridge, H. D. J. N., & Fenton, M. B. (1989). Roosting and echolocation behavior of the African bat *Chalinolobus variegatus. J. Mammal.* **70**: 828–833.

Payne, J. & Francis, C. M. (1985). *A field guide to the mammals of Borneo.* The Sabah Society/World Wildlife Fund Malaysia, Selangor.

Pennycuick, C. J. (1972). *Animal flight.* Edward Arnold, London. (*Stud. Biol.* No. 33.)

Pennycuick, C. J. (1975). Mechanics of flight. In *Avian biology* **5**: 1–75. (Eds Farner, D. S. & King, J. R.). Academic Press, New York.

Pettigrew, J. D. (1994). Flying DNA. *Curr. Biol.* **4**: 277–280.

Pettigrew, J. D. (1995). Flying primates: crashed, or crashed through? *Symp. zool. Soc. Lond.* No. 67: 3–26.

Pettigrew, J. D., Jamieson, B. G. M., Robson, S. K., Hall, L. S., McAnally, K. I., & Cooper, H. M. (1989). Phylogenetic relations between microbats, megabats and Primates (Mammalia: Chiroptera and Primates). *Phil. Trans. R. Soc. (B)* **325**: 489–559.

Pye, J. D. (1972). Bimodal distribution of constant frequencies in some hipposiderid bats (Mammalia: Hipposideridae). *J. Zool., Lond.* **166**: 323–335.

Pye, J. D. (1979). Why ultrasound? *Endeavour* **3**: 57–62.

Pye, J. D. (1980). Adaptiveness of echolocation signals in bats. Flexibility in behaviour and evolution. *Trends Neurosci.* **3**: 232–235.

Pye, J. D. (1993). Is fidelity futile? The 'true' signal is illusory, especially with ultrasound. *Bioacoustics* **4**: 271–286.

Rautenbach, I. L., Kemp, A. C., & Scholtz, C. H. (1988). Fluctuations in availability of arthropods correlated with microchiropteran and avian predator activities. *Koedoe* No. 31: 77–90.

Rayner, J. M. V. (1981). Flight adaptations in vertebrates. *Symp. zool. Soc. Lond.* No. 48: 137–172.

Rydell, J. (1986). Foraging and diet of the northern bat *Eptesicus nilssoni* in Sweden. *Holarct. Ecol.* **9**: 272–276.

Rydell, J. (1989). Food habits of northern (*Eptesicus nilssoni*) and brown long-eared (*Plecotus auritus*) bats in Sweden. *Holarct. Ecol.* **12**: 16–20.

Rydell, J. (1992). The diet of the parti-coloured bat *Vespertilio murinus* in Sweden. *Ecography* **15**: 195–198.

Schmidly, D. J. (1991). *The bats of Texas.* Texas A&M University Press, College Station, Texas.

Schober, W. & Grimmberger, E. (1989). *A guide to bats of Britain and Europe.* Hamlyn, London.

Schumm, A., Krull, D., & Neuweiler, G. (1991). Echolocation in the notch-eared bat, *Myotis emarginatus. Behav. Ecol. Sociobiol.* **28**: 255–261.

Silva Taboada, G. (1979). *Los murciélagos de Cuba.* Editorial Cientifico-Técnica, Ciudad de la Habana, Habana.

Speakman, J. R. (1990). The function of daylight flying in British bats. *J. Zool., Lond.* **220**: 101–113.

Speakman, J. R. (1991). Why do insectivorous bats in Britain not fly in daylight more frequently? *Funct. Ecol.* **5**: 518–524.

Speakman, J. R., Anderson, M. E., & Racey, P. A. (1989). The energy cost of echolocation in pipistrelle bats (*Pipistrellus pipistrellus*). *J. comp. Physiol. (A)* **165**: 679–685.

Speakman, J. R., Hays, G. C., & Webb, P. I. (1994). Is hyperthermia a constraint on the diurnal activity of bats? *J. theor. Biol.* **171**: 325–341.

Speakman, J. R. & Racey, P. A. (1991). No cost of echolocation for bats in flight. *Nature, Lond.* **350**: 421–423.

Strahan, R. (Ed.) (1983). *The Australian Museum complete book of Australian mammals.* Angus & Robertson, Sydney.

Sullivan, C. M., Shiel, C. B., McAney, C. M., & Fairley, J. S. (1993). Analysis of the diets of Leisler's *Nyctalus leisleri*, Daubenton's *Myotis daubentoni* and pipistrelle *Pipistrellus pipistrellus* bats in Ireland. *J. Zool., Lond.* **231**: 656–663.

Surlykke, A., Miller, L. A., Møhl, B., Andersen, B. B., Christensen-Dalsgaard, J., & Jørgensen, M. B. (1993). Echolocation in two very small bats from Thailand: *Craseonycteris thonglongyai* and *Myotis siligorensis. Behav. Ecol. Sociobiol.* **33**: 1–12.

Suthers, R. A., Thomas, S. P., & Suthers, B. J. (1972). Respiration, wing-beat and ultrasonic pulse emission in an echo-locating bat. *J. exp. Biol.* **56**: 37–48.

Swift, S. M., Racey, P. A., & Avery, M. I. (1985). Feeding ecology of *Pipistrellus pipistrellus* (Chiroptera: Vespertilionidae) during pregnancy and lactation. II. Diet. *J. Anim. Ecol.* **54**: 217–225.

Van Valen, L. (1973). Body size and numbers of plants and animals. *Evolution, Lawrence, Kans.* **27**: 27–35.

Waters, D. A., Rydell, J., & Jones, G. (1995). Echolocation call design and limits on prey size: a case study using the aerial-hawking bat *Nyctalus leisleri*. *Behav. Ecol. Sociobiol.* **37**: 321–328.

Zingg, P. E. (1990). Akustische Artidentifikation von Fledermäusen (Mammalia: Chiroptera) in der Schweiz. *Rev. Suisse Zool.* **97**: 263–294.

Symp. zool. Soc. Lond. (1996) No. 69: 129–142

The genetics and physiology of size reduction in mice

IAN M. HASTINGS

Institute of Cell, Animal and Population Biology
University of Edinburgh
West Mains Road
Edinburgh EH9 3JT, Scotland, UK

Synopsis

Artificial selection for reduced body size in mice has been performed in numerous laboratories. Typically, body weight at age 6 weeks has been reduced from around 20 to 13 g and/or body weight at 10 weeks from around 32 g to 16 g. In our laboratory, we have further reduced body weight in our 'small' line by introducing a gene disrupting growth hormone production (the *little* gene), resulting in mice with a 10-week body weight of less than 10 g. Such mice appear healthy and viable. The genetic and physiological basis of reduced body weight in selected lines is reviewed and the relevance of experiments to natural populations is discussed. The biggest differences are the reduced environmental fluctuations experienced by laboratory stocks and their smaller population sizes compared to natural populations. It is concluded that laboratory stocks must be considered as paradigm populations evolving under very specific conditions. With this caveat, it is concluded that considerable size reduction may be achieved very rapidly in mouse populations and that such size reduction appears not to entail significant alterations in physiology, biochemistry, or behaviour nor to invoke significant reduction in viability as a consequence of scale effects.

Introduction

Growth rate and mature body weight are important economic traits in commercial species of livestock. Their genetic bases have been studied directly in such species as sheep, cattle, and poultry but experiments in these tend to be expensive and to run relatively slowly because of the long generation time of such animals. For this reason mice (and, less frequently, rats) tend to be used as a model species: they are cheaper to maintain and their generation time can be reduced to 9 weeks allowing selection on body weight to be applied faster and on a large number of animals. Such selection experiments usually consist of several replicate lines drawn from the same base population since consistency between the replicate

ZOOLOGICAL SYMPOSIUM No. 69
ISBN 0-19-857787-7

lines is necessary before concluding that observed changes are a direct consequence of selection for reduced body weight rather than a result of the inevitable random genetic changes ('drift') which occur during the selection experiment. Most interest is in lines selected for increased body weight but, fortunately for our purposes, technical considerations often dictate that selection is divergent so that several lines are started from the same base population and are selected for either increased body weight or reduced body weight. Reviews of the genetic value of mouse selection experiments and reference to earlier literature can be found in Falconer (1955), McCarthy (1983), and Eisen (1989), while typical examples of mouse lines selected on body weight are described by Falconer (1973) and Beniwal, Hastings, Thompson, & Hill (1992a, b). More general reviews of the value and interpretation of artificial selection experiments are provided by Hill & Mackay (1989) and Hill & Caballero (1992). Numerous studies have been made of the genetic, physiological, and biochemical changes associated with selection on body weight. They will be briefly reviewed and, where appropriate, access to the often considerable literature will be given.

Experiment and theory

When mice are selected for reduced body weight, the following characteristics of such experiments may be considered.

1. The direct response to selection, giving an idea of the likely evolutionary dynamics of size reduction.
2. Correlated responses in growth, reproductive and behavioural characters: response is rarely restricted to the character under selection, and many genes affecting size also affect other traits which will therefore change as a consequence of selection.
3. Correlated responses in physiological and biochemical traits.
4. The genetic basis of reduced body size, which should provide further clues as to the likely evolutionary dynamics of size reduction.
5. The relevance of these studies to other species.
6. The relevance of these studies to natural populations of mice.

These questions need to be addressed in considering the likely response of a population whose environment has changed so as to favour reduced body size. In particular we need to consider the rate at which a response will occur and the likely side-effects in terms of changes in correlated characters. It is assumed that the reader has no knowledge of quantitative genetics, and some elementary principles will be introduced at the appropriate points.

The direct response to selection for reduced body size

The critical measure in understanding how a population will respond to selection is to calculate how much of the observed phenotypic variation (V_p) is due to 'additive' genetic variance (V_a), i.e. the 'inherent' properties of the genes affecting body weight. The phenotypic variance can be partitioned as

$$V_p = V_a + V_d + V_i + V_e$$

where V_d is variance due to dominance interactions between alleles at the same locus, V_i variance due to interactions between different loci, and V_e variance due to environmental conditions. The critical measure is heritability (h^2) and

$$h^2 = \frac{V_a}{V_p} = \frac{V_a}{V_a + V_d + V_i + V_e} \qquad (1)$$

which allows the response (R) to be calculated as

$$R = ih^2$$

where R is the response to and i is the intensity of selection (see Crow (1986) or Falconer (1989) for more details). Heritability can be regarded as a type of 'signal-to-noise ratio' relating the additive genetic effects, which will contribute to the response, to the 'noise' introduced by other components of variation. There is no reason to suppose that V_a, V_d, or V_i will vary significantly between natural and laboratory populations, but standard laboratory management is explicitly designed to reduce the noise in the system, so that V_e is reduced by means of standard cages, well-defined food, constant temperature, light/dark cycles, and so on. Thus it seems certain that the heritability in natural populations will be well below that measured in laboratory stocks, but to what extent is unclear, as will be discussed later. Heritability is also a property of a specific population (whose genetic constitution determines V_a, V_d, V_i) in a specific environment (which determines V_e). In practice most populations of mice have similar heritabilities for body weight and growth rate of around 0.3–0.5 (Eisen 1989).

A typical response to artificial selection on body weight is shown in Fig. 1. There was a linear initial response which later slowed in the high line and eventually a 'selection plateau' was reached between generations 32 and 42. Interestingly, further response has occurred in this line in subsequent generations. The point at which the rate of response starts to slow and the magnitude of the eventual plateau are both specific to the population being studied and cannot be predicted either from theory or from past experience (see Falconer 1989: chapter 12, or Hill & Caballero 1992). There are, however, several reasons why such a limit may occur. Firstly, the genetic variation in the line may have been lost, either because all the genes affecting body weight have been fixed or as a result of inbreeding. Secondly, there may be natural selection opposing the force of artificial selection. For example, smaller mice may become more prone to thermoregulatory stress. Thirdly, there may be genetic reasons such as overdominance. Interestingly (and unlike selection for increased body weight), there do not seem to be limits for reduced size; the apparent reduction in response shown in Fig. 1 is largely a scale effect and the response is linear (at least up to generation 45) when plotted on a logarithmic scale (data not shown). In our laboratory we have reduced mean body weight at age 10 weeks from 32 g to 16 g over 50 generations of selection without reaching a selection limit (see Fig. 1) and without significantly reducing the viability of such mice. Crossing the *little* dwarfing gene into this line further reduces body weight by 50% (see below), again without apparently reducing the viability of such mice whose mature weight now averages less than 10 g.

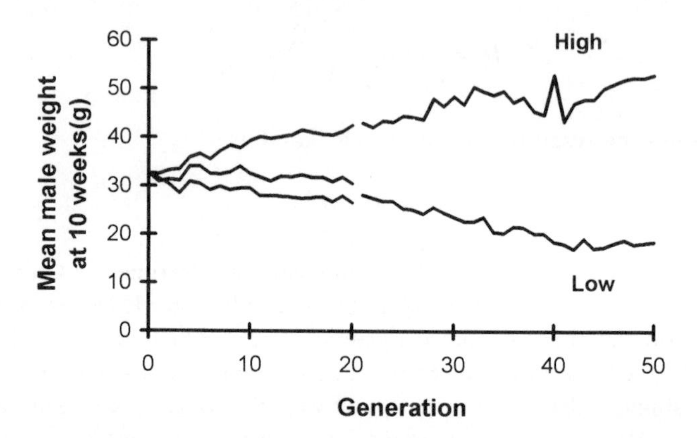

Fig. 1. Changes in the average weight of adult mice over 50 generations of selection. Initially selected for 20 generations on male lean mass at 10 weeks of age (which is closely correlated with body weight; $r_a = 0.94$) and subsequently on body weight in both sexes at 10 weeks of age; further details in Sharp, Hill & Robertson (1984) and Beniwal *et al.* (1992a).

Correlated responses in growth, reproductive, and behavioural characters

In a perfect world, each trait of an animal would be optimized for the environment in which it lives. However, the same genes may influence several traits and, as the frequency of these genes changes in response to selection pressures, its effects are exerted on other traits which consequently change in magnitude. Such changes are called 'correlated responses' and are a near-universal observation in laboratory-based selection experiments. The correlated change in litter size which arose as a consequence of the selection illustrated in Fig. 1 is shown in Fig. 2. The degree to which the two traits (the one being selected and the one exhibiting a correlated response) are determined by the same set of genes can be measured by the genetic correlation r_a between the traits. Typical values of r_a observed between body weight and other physical characteristics are given in Table 1. There is a large positive correlation with litter size, apparently due to differences in ovulation rate (Land 1970; Brien 1986). At first sight, this reduction in litter size may constitute a selective force opposing the evolution of small size (in contrast with the observation that reduced viability was not associated with small size). However, these were analyses of the first litter size and a more detailed study (Roberts 1961) showed that mice from lines selected for reduced weight have more litters and that their overall lifetime reproductive output was greater than that of mice from the lines selected for increased weight. Interestingly, there did not appear to be any change in metabolic rate, which continued to scale to lean mass raised to the power 0.75 (analyses performed at generation 16; Hill & Bishop 1986).

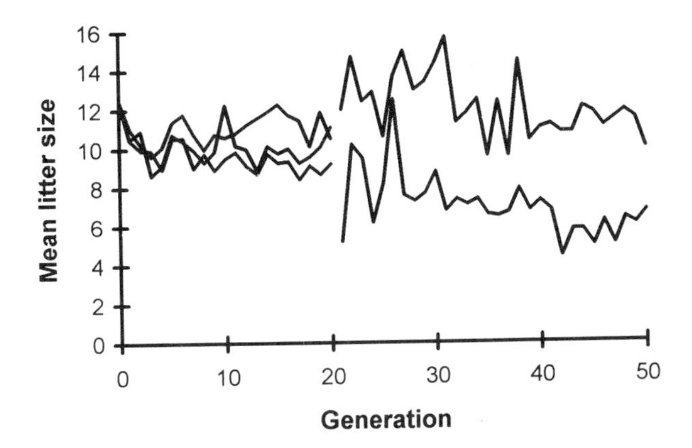

Fig. 2. Changes in a correlated character (mean primiparous litter size) as a consequence of the selection experiment illustrated in Fig. 1.

Table 1. Magnitudes of genetic correlations (r_a) obtained from the experiment illustrated in Fig. 1.

Trait	r_a	Reference
Fat content (%)	+0.5	Beniwal *et al.* (1992b)
Litter size	+0.3	Beniwal *et al.* (1992b)
Behavioural traits		
Food intake	+0.5	Holmes & Hastings (1995)
Ultrasound calling rate	$< \pm 0.1$	Holmes & Hastings (1995)
Response to novel stimulus	$< \pm 0.1$	Holmes & Hastings (1995)
Response in open field	$< \pm 0.1$	Holmes & Hastings (1995)

Recently we have undertaken a study designed to detect behavioural changes which may have occurred as a consequence of selection on body weight (Holmes & Hastings 1995). The behavioural traits measured were feeding activity, open field behaviour, response to a novel object, and pup ultrasound calling rate. In all cases, the magnitude of estimated r_a was less than 10% and not significantly different from zero (Table 1 and unpublished). This suggests that changes consequent on physiological or hormonal alterations in small mice had not greatly affected their behaviour. This is noteworthy because, for example, male hormones in mice are known to affect both body weight and behaviour (particularly aggression: Van Oortmerssen & Bakker 1981; Deruiter *et al.* 1992).

Correlated responses in physiology and biochemistry

The correlated physical and behavioural responses (or lack thereof) considered above provide little information about how a particular change in phenotype is achieved and reveal nothing about the genetic basis of response. The physiological changes which have occurred as a correlated response to selection have been measured in numerous experiments. In lines of mice selected on body weight, interest has focused, not surprisingly, on putative changes in growth hormone (GH) metabolism and its mediators such as insulin-like growth factors 1 & 2 (Igf1, Igf2) (McKnight & Goddard 1989; Siddiqui *et al.* 1989; Bootland, Hill, & Sinnett-Smith 1991; Medrano *et al.* 1991). As might be expected, lines selected for increased weight have increased levels of Igf1, and vice versa. Similarly, lines divergently selected on Igf1 levels show a positive correlated response in body weight (Blair *et al.* 1989). However, counter-intuitive results are noted for levels of circulating GH where lines selected for increased size have lower levels of circulating GH and vice versa. This illustrates a major problem with physiological measurements, that it is difficult to disentangle cause from effect. It is unlikely that reduced levels of GH have *caused* increased growth and more probable that increased growth has stimulated feedback regulatory systems designed to decrease GH production. To overcome this problem more direct experimental approaches have been taken. Lines divergently selected on body weight do not appear to have altered sensitivity to exogenous GH (Bootland *et al.* 1991; Hastings, Bootland & Hill 1993), and changes in the GH axis do not seem to have played a disproportionate role in the response to selection (Pidduck & Falconer 1978; Hastings *et al.* 1993). These experiments of Pidduck & Falconer and Hastings *et al.* are extremely interesting for our purposes as they introduced dwarfing genes (*dwarf* and *little*, respectively) into lines already selected for reduced body weight. Introduction of these genes resulted in body weight at 10 weeks decreasing from 18 to 6 g and body weight at 7 weeks from 17 to 8 g, respectively. Importantly, dwarfs from both lines were fully viable (but with reduced fertility) despite their very small size, implying that fitness reduction due to scaling effects did not operate.

An analogous experiment was performed in our laboratory to investigate the role of testosterone in the response to selection on body weight (Hastings & Hill in prep.). The injection of exogenous testosterone increased body weight by 10%, demonstrating that it is an area of metabolism which may be expected to be differentially changed by selection. However, both the high and low line responded to the same degree thereby suggesting that sensitivity to the hormone had not been altered by selection.

Falconer, Gauld, & Roberts (1981) took embryos from their high and low selected lines and made embryonic chimeras containing cells from both selection lines. The resulting animals differed both in their overall proportion of cells from each parent, and in the proportion of cells within individual tissues. Body weight was affected by the overall proportion of cells but not by the proportion of cells within 'key' metabolic tissues such as the liver, pituitary, or testis. They concluded that growth regulation was largely systemic and not controlled by any one of the individual tissue types they examined.

These experiments examining key areas of metabolism and tissues are not conclusive, but are consistent with the hypothesis that the response to selection on body weight is determined by many genes affecting many areas of metabolism. Genes with large effects on body weight tend to have deleterious pleiotropic effects (see later). Conversely, the physiological experiments described above show that large changes in body weight involve *systematic* physiological changes without resulting in obvious declines in viability; the inference is that selection response has not involved genes with major physiological effects.

The genetic basis of response

There is considerable interest at present in identifying the physiological basis of the response to selection as this may allow animal breeds to be 'improved' without the necessity of long-term selection. For example, direct genetic manipulation of GH metabolism may achieve the same result in a much shorter time. One of the earliest and best known among transgenic experiments consisted of inserting additional growth hormone genes into mice which subsequently grew faster (Palmiter *et al.* 1982). Since then, various groups have inserted additional growth hormone genes into species such as pigs (Polge *et al.* 1989; Pursel *et al.* 1989). A problem frequently encountered in such experiments is severe deleterious pleiotropic effects. The same effects are noted in spontaneously arising mutations either decreasing growth rate (Charlton 1984) or increasing growth rate (Berger, Calvert, & Bradford 1989). These results suggest that if genes with very large effects on body weight contribute to response, then decline in viability is more likely than if response is due to change at a large number of loci with smaller individual effects on the phenotype.

Several groups have examined 'candidate loci' which may be expected to contribute to the response of lines selected on body weight. Such candidates have usually been associated with growth hormone metabolism. Salmon *et al.* (1988) identified a restriction fragment length polymorphism (RFLP) in the growth hormone gene which affected body weight and Winkelman & Hodgetts (1992) identified RFLPs in the GH and Igf2 loci which affected body weight. In the latter experiment the growth hormone polymorphism present in the large selected lines was actually associated with reduced body weight in an F2 cross which the authors interpreted as an interaction with the genetic background. Gray & Tait (1993) suggested that an RFLP in the ornithine decarboxylase gene had been selected in our lines selected on body weight. It may be difficult to reconcile these results with those of Pidduck & Falconer (1978), Hastings *et al.* (1993), and Hastings & Hill (in prep.) which suggest that key areas of metabolism had not been differentially affected by selection (see above). It is possible that the candidate loci were in linkage disequilibrium with a chromosomal region which had a significant effect on body weight and that the candidate loci served as markers for this region rather than exerting an independent effect.

This type of candidate locus study aims to elucidate the genetic basis of response which remains obscure. In essence we need to ascertain whether response is due to a very large number of genes, each with a very small effect ('polygenes'), or a small number with relatively large effects ('major' genes). The results are inconclusive at present. We have investigated our lines selected on body weight and shown that the 'infinitesimal' polygenic model (which assumes that response is due to an extremely large number of genes each with an extremely small effect) is violated (Beniwal *et al.* 1992a, b) but analyses of crosses of these lines did not detect any genes of large effect (although there was a large sex-linked effect: Hastings & Veerkamp 1993; Veerkamp *et al.* 1993). The genetic basis presumably lies somewhere between the two extremes. The relevance of the genetic basis is that major genes tend to have deleterious pleiotropic effects (see above) and as natural populations are larger they are more likely to contain genes with large effects on body weight. Currently, there is great interest in using microsatellite markers to map genes which affect body weight and to measure the magnitude of their effects as, in principle, they ultimately have the power to resolve the debate about the number of genes (or chromosome regions) which contribute to the response from selection.

The most significant limitation to these investigations into the genetic basis of the response may be lack of any information about the developmental genes underlying the trait. This limitation has been widely recognized but remains intractable. We have evidence that the differences in body weight in our lines may appear as early as day 14 *in utero* (G. Guneren, unpubl. data) but the genes controlling this early embryonic growth are unknown. Falconer *et al.* (1978) examined the role of changes in cell number and cell mass in lines of mice divergently selected on body weight. They found that both factors contributed to the response and concluded that 'the main effect of selection for body weight had been to speed up or slow down the normal processes of cellular growth'. It can only be hoped that advances in molecular biology will ultimately allow the genes controlling such traits to be identified.

The relevance of these laboratory studies to other species

The heritability of body size and growth rate seems to be comparable across species (Falconer 1989: table 10.1), lying in the region 0.3–0.6. In addition, the physiology of mammalian metabolism is conserved (Prosser 1973) and the same correlated responses appear in many species. For example, the changes in the activity of lipogenic enzymes noted in mouse lines selected on fat content (Asante, Hill, & Bulfield 1989; Hastings & Hill 1990) were also noted in pigs (Muller 1986) and chickens (Bannister *et al.* 1984) selected on the same trait. The phenotypic relationship between growth and reproductive traits seems relatively conserved in mammals (Brien 1986). This apparent consistency across species is one of the justifications for using mouse selection lines as models for commercial species (McCarthy 1982) and therefore it seems plausible that the results of laboratory studies on mice will also be relevant to other mammalian species.

The relevance of these laboratory studies to natural populations

The most obvious and significant difference between laboratory and natural populations is that the former live in highly controlled environments, typically at a constant temperature with continuous access to food. The environmental variance in nature is likely to be higher than in laboratories so V_e is increased and heritability will presumably be significantly lower than estimated under laboratory conditions (equation 1). The forces of natural selection will also differ between the two populations, and may be reasonably supposed to be more intense in nature. The magnitude of these discrepancies is difficult to estimate as it is extremely difficult to ascertain either heritability or the strength of selective forces in natural populations (Shaw 1987; Mitchell-Olds & Shaw 1987). Selection on a trait may also be opposed by selection forces which operate in the wild but not in the laboratory. An obvious example is that mice selected for small body size have a reduced surface-to-volume ratio so may suffer in periods of cold or reduced food availability which occur in nature but not in thermoregulated animal units where food is constantly available. Again, there is no way of estimating the magnitude of such forces.

The patterns of correlated response may also differ in natural populations. It is important to note the difference between genetic correlation and phenotypic correlation: their magnitudes and even their sign may differ. Correlated responses result from changes in the frequency of genes affecting both traits (i.e. the one under direct selection, in this case body weight, and the one exhibiting a correlate response) over the course of selection. The extent of this correlation is measured by the genetic correlation r_a between the traits which may be positive (when the selected genes increase the value of the correlated trait) or negative (when the genes decrease the magnitude of the correlated trait). The magnitude and even the direction of r_a depends on the genes contributing to the response so the presence or absence of correlated responses in the laboratory may be an imperfect guide to processes acting in the real world. For example, the common observation that mice selected for increased body weight also become fatter (r_a of approximately 0.5, Table 1) may be explained by their access to unlimited food in the laboratory situation: they eat more, so extra energy is available for both increased growth and fat deposition. In natural populations, food is often a limiting factor and increased growth may only be possible at the expense of reduced fat deposition. We might therefore expect larger mice to become leaner, and smaller mice to deposit as fat the energy saved on growth (Fig. 3). Estimating r_a in natural populations is extremely difficult. The extent to which the two characters co-vary between individuals is given by

$$cov_p = cov_a + cov_e$$

where cov_p is the phenotypic co-variance between the traits, cov_a is the additive genetic co-variance between the traits, and cov_e is the environmental co-variance between the traits (Hartl & Clark 1989: 490).

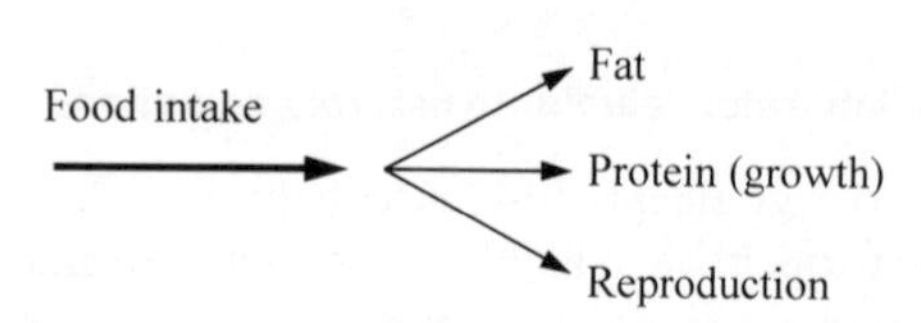

Fig. 3. Correlated responses in laboratory and natural populations. Diagrammatic representation of food intake and its partition in the mouse. In laboratory stocks, selection for reduced size may involve genes which reduce food intake; protein and fat deposition may both be reduced, resulting in a positive genetic correlation. In natural populations, food may be a limiting resource and selection for reduced weight may decrease protein deposition while the extra energy is deposited as fat, resulting in a negative genetic correlation.

The observed phenotypic covariance in a natural population indicates neither the magnitude nor even the sign of the genetic co-variance (and by extension the magnitude and sign of r_a). To continue the above example, mice in natural populations may have a negative genetic co-variance between body weight and fat content but a positive environmental co-variance which dominates the genetic co-variance and which makes the overall phenotypic co-variance positive. A plausible large positive environmental co-variance may come from mouse populations where individuals sampled from beneficial environments such as domestic houses are both large and fat, while those sampled from marginal environments are small and thin. The two co-variances may be disentangled by investigating the resemblance between relatives but, as in estimating heritabilities, this method is plagued by technical difficulties and has not been attempted.

The other major difference (besides environment) between laboratory and natural populations is in population size. Laboratory experiments usually consist of eight to 16 pairs per generation while wild populations are presumably much larger, although they may live in small subpopulations with an indeterminate amount of migration between populations (Berry & Bronson 1992). Small populations lose genetic variation because of random processes ('drift') and the eventual limit to selection (in the absence of recurrent mutation) is proportional to effective population size (Hill & Caballero 1992). Larger populations can also sustain a larger selection response attributable to the input of new mutations (Hill 1982), which is directly proportional to effective population size. However, the exact effects of population size depend on the assumptions made about gene action (recessive, additive, or dominant) and the distribution of mutational effects. Smaller populations are more sensitive to the effects of random change in gene frequency. If an additive allele has a selective advantage $1+s$ (compared to another allele of selective advantage unity) in a population of effective population size N, then its fate will be determined by natural selection when $4Ns > 10$, by random genetic drift when $4Ns < 0.1$, and by both processes when $0.1 < 4Ns < 10$ (Kimura 1983). Smaller populations are therefore more sensitive to the random genetic processes of mutation and drift and may be regarded as less genetically 'stable' than larger populations. This may partly explain why island populations (which may be small or derived from a small 'founder' population) may show such rapid alterations in size.

Conclusion

Many selection experiments have been performed to reduce body weight in laboratory stocks of mice. The problems arise when trying to extend these results to natural populations where natural selection is probably more intense, the magnitude of heritability lower, and populations larger. It is probably best to consider laboratory populations as merely a natural population evolving in a rather benign environment and to try to deduce robust conclusions from laboratory studies. The first conclusion is that populations are *potentially* able to change body size extremely rapidly. The second conclusion is that even rapid reduction in body weight appears not to result in large-scale alterations in critical physiological systems such as growth hormone metabolism. The third conclusion is that reduction in body weight need not result from the fixation of genes with large effects on weight; such genes would be expected to have deleterious pleiotropic effects. The result of the last two observations is that relatively large reductions in body weight may be achieved without *necessarily* rendering smaller individuals less viable.

Acknowledgements

I thank Professors W. G. Hill and D. S. Falconer for helpful comments on the text. This work was supported by the Biotechnology and Biological Sciences Research Council and by the Medical Research Council.

References

Asante, E. A., Hill, W. G., & Bulfield, G. (1989). Analysis of lines of mice selected for fat content. 1. Correlated responses in the activities of NADPH-generating enzymes. *Genet. Res.* **54**: 155–160.

Bannister, D. W., Lee, A., Whitehead, C. C. & Griffin, H. D. (1984). Lipogenic enzyme activity and fructose 2, 6-biophosphate concentration in livers of two lines of domestic fowl selected for different body fat content. *Int. J. Biochem.* **17**: 1301–1305.

Beniwal, B. K., Hastings, I. M., Thompson, R., & Hill, W. G. (1992a). Estimation of changes in genetic parameters in selected lines of mice using REML with an animal model. I. Lean mass. *Heredity* **69**: 352–360.

Beniwal, B. K., Hastings, I. M., Thompson, R., & Hill, W. G. (1992b). Estimation of changes in genetic parameters in selected lines of mice using REML with an animal model. II. Body weight, body composition and litter size. *Heredity* **69**: 361–371.

Berger, T., Calvert, C. C., & Bradford, G. E. (1989). Reduced male reproductive capacity in mice with high genetic potential for post-weaning growth. *J. Reprod. Fert.* **87**: 33–38.

Berry, R. J. & Bronson, F. H. (1992). Life history and bioeconomy of the house mouse. *Biol. Rev.* **67**: 519–550.

Blair, H. T., McCutcheon, S. C., Mackenzie, D. D. S., Gluckman, P. D., Ormsby, J. E. &

Breier, B. H. (1989). Responses to divergent selection for plasma concentrations on insulin-like growth factor-I in mice. *Genet. Res.* **53**: 187–191.

Bootland, L. H., Hill, W. G., & Sinnett-Smith, P. A. (1991). Effects of exogenous growth hormone on growth and composition in genetically selected mice. *J. Endocr.* **131**: 19–24.

Brien, F. D. (1986). A review of the genetic and physiological relationships between growth and reproduction in mammals. *Anim. Breed. Abstr.* **54**: 975–996.

Charlton, H. M. (1984). Mouse mutants as models in endocrine research. *Q. Jl exp. Physiol.* **69**: 655–676.

Crow, J. F. (1986). *Basic concepts in population, quantitative, and evolutionary genetics.* W. H. Freeman, New York.

Deruiter, A. J. H., Koolhaas, J. M., Keijser, J. N., Van Oortmerssen, G. A., & Bohus, B. (1992). Differential testosterone secretory capacity of the testes of aggressive and non-aggressive house mice during ontogeny. *Aggressive Behav.* **18**: 149–157.

Eisen, E. J. (1989). Selection experiments for body composition in mice and rats: a review. *Livestk Prod. Sci.* **23**: 17–32.

Falconer, D. S. (1955). Patterns of response in selection experiments with mice. *Cold Spring Harbor Symp. quant. Biol.* No. 20: 178–196.

Falconer, D. S. (1973). Replicated selection for body weight in mice. *Genet. Res.* **22**: 291–321.

Falconer, D. S. (1989). *An introduction to quantitative genetics.* (3rd edn). Longman, Harlow.

Falconer, D. S., Gauld, I. K., & Roberts, R. C. (1978). Cell numbers and cell sizes in organs of mice selected for large and small body size. *Genet. Res.* **31**: 287–301.

Falconer, D. S., Gauld, I. K., & Roberts, R. C. (1981). The control of body size in mouse chimaeras. *Genet. Res.* **38**: 25–46.

Gray, A. & Tait, A. (1993). Identification of ornithine decarboxylase as a trait gene for growth in replicated mouse lines divergently selected for lean body mass. *Genet. Res.* **62**: 31–37.

Hartl, D. L. & Clark, A. G. (1989). *Principles of population genetics.* (2nd edn.) Sinauer Associates, Sunderland, Ma, USA.

Hastings, I. M., Bootland, L. H., & Hill, W. G. (1993). The role of growth hormone in lines of mice divergently selected on body weight. *Genet. Res.* **61**: 101–106.

Hastings, I. M. & Hill, W. G. (1990). Analysis of lines of mice selected for fat content. 2. Correlated responses in the activities of enzymes involved in lipogenesis. *Genet. Res.* **55**: 55–61.

Hastings, I. M. & Hill, W. G. (In preparation). *The contribution of male hormones to the response of mouse lines divergently selected for body weight or fat content.*

Hastings, I. M. & Veerkamp, R. F. (1993). The genetic basis of response in mouse lines divergently selected for body weight or fat content. 1. The relative contributions of autosomal and sex-linked genes. *Genet. Res.* **62**: 169–175.

Hill, W. G. (1982). Rates of change in quantitative traits from fixation of new mutations. *Proc. natn. Acad. Sci. USA* **79**: 142–145.

Hill, W. G. & Bishop, S. C. (1986). Genetic control of growth, carcass composition and food utilization in laboratory animals. In *Proceedings of the 3rd world congress on genetics applied to livestock production* **11**: 355–366. (Eds Dickerson, G. E. & Johnson, R. K.). Agricultural Communications, University of Nebraska, Nebraska.

Hill, W. G. & Caballero, A. (1992). Artificial selection experiments. *A. Rev. Ecol. Syst.* **23**: 287–310.

Hill, W. G. & Mackay, T. F. C. (Eds) (1989). *Evolution and animal breeding*. CAB International, Oxford.

Holmes, I. S. & Hastings, I. M. (1995). Behavioural changes as a correlated response to selection. *Genet. Res.* **66**: 27–33.

Kimura, M. (1983). *The natural theory of molecular evolution*. Cambridge University Press, Cambridge, UK.

Land, R. B. (1970). Genetic and phenotypic relationships between ovulation rate and body weight in the mouse. *Genet. Res.* **15**: 171–182.

McCarthy, J. C. (1982). The laboratory mouse as a model for animal breeding: a review of selection for increasing body weight and litter size. In *Proceedings of the 2nd world congress on genetics applied to livestock production* **5**: 66–83. Servicio de Publicaciones Agraria, Madrid.

McKnight, B. J. & Goddard, C. (1989). The effect of food restriction on circulating insulin-like growth factor-I in mice divergently selected for high or low protein or fat to body mass ratios. *Comp. Biochem. Physiol. (A)* **92**: 565–569.

Medrano, J. F., Pomp, D., Sharrow, L., Bradford, G. E., Downs, T. R., & Frohman, L. A. (1991). Growth hormone and insulin-like growth factor 1 measurement in high growth (hg) mice. *Genet. Res.* **58**: 67–74.

Mitchell-Olds, T. & Shaw, R. G. (1987). Regression analysis of natural selection: statistical inference and biological interpretation. *Evolution* **41**: 1149–1161.

Muller, E. (1986). Physiological and biochemical indicators of growth and composition. In *Exploiting new technologies in animal breeding. Genetic developments*: 132–139. (Eds Smith, C., King, J. W. B. & McKay, J. C.). Oxford Science Publications, Oxford.

Palmiter, R. D., Brinster, R. L., Hammer, R. E., Trumbauer, M. E., Rosenfeld, M. G., Birnberg, N. C. & Evans, R. M. (1982). Dramatic growth of mice that develop from eggs microinjected with metallothionein-growth hormone fusion genes. *Nature, Lond.* **300**: 611–615.

Pidduck, H. G. & Falconer, D. S. (1978). Growth hormone function in strains of mice selected for large and small size. *Genet. Res.* **32**: 195–206.

Polge, E. J. C., Barton, S. C., Surani, M. A. H., Miller, J. R., Wagner, T., Rottman, F., Camper, S. A., Elsome, K., Davis, A. J., Goode, J. A., Foxcroft, G. R., & Heap, R. B. (1989). Induced expression of a bovine growth hormone construct in transgenic pigs. In *Biotechnology in growth regulation*: 189–199. (Eds Heap, R. B., Prosser, C. G., & Lamming, G. E.). Butterworths, London.

Prosser, C. L. (Ed.) (1973). *Comparative animal physiology*. (3rd edn). W. G. Saunders, Philadelphia.

Pursel, V. G., Miller, K. F., Bolt, D. J., Pinkert, C. A., Hammer, R. E., Palmiter, R. D., & Brinster, R. L. (1989). Insertion of growth hormone genes into pig embryos. In *Biotechnology in growth regulation*: 181–188. (Eds Heap, R. B., Prosser, C. G., & Lamming, G. E.). Butterworths, London.

Roberts, R. C. (1961). The lifetime growth and reproduction of selected strains of mice. *Heredity* **16**: 369–381.

Salmon, R. K., Berg, R. T., Yeh, F. C., & Hodgetts, R. B. (1988). Identification of a variant growth hormone haplotype in mice selected for high body weight. *Genet. Res.* **52**: 7–15.

Sharp, G. L., Hill, W. G., & Robertson, A. (1984). Effects of selection on growth, body composition and food intake in mice. I. Responses in selected traits. *Genet. Res.* **43**: 75–92.

Shaw, R. G. (1987). Maximum likelihood approaches applied to quantitative genetics of natural populations. *Evolution* **41**: 812–826.

Siddiqui, R. A., McCutcheon, S. N., MacKenzie, D. D. S., Blair, H. T., Ormsby, J. E., Gluckman, P. D. & Breier, B. H. (1989). Role of testosterone in regulating the growth of mice selected for low vs. high plasma insulin-like growth factor-I concentrations. *Acta endocr., Copenh.* **121**: 686–690.

Van Oortmerssen, G. A. & Bakker, T. C. M. (1981). Artificial selection for short and long attack latencies in wild *Mus musculus domesticus*. *Behav. Genet.* **11**: 115–126.

Veerkamp, R. F., Haley, C. S., Knott, S., & Hastings, I. M. (1993). The genetic basis of response in mouse lines divergently selected for body weight or fat content. 2. The contribution of genes with large effect. *Genet. Res.* **62**: 177–182.

Winkelmann, D. C. & Hodgetts, R. B. (1992). RFLPs for somatotropic genes identify quantitative trait loci for growth in mice. *Genetics* **131**: 929–937.

Symp. zool. Soc. Lond. (1996) No. 69: 143–156

Speciation in small mammals

J. B. SEARLE

Department of Biology
University of York
PO Box 373
York YO1 5YW, UK

Synopsis

There are more species of small mammal (less than 1 kg) than species of large mammal, with the rodents, bats, and insectivores particularly speciose. Small mammals show greater differentiation into species at both the local and the regional level. Species of small mammal tend to occupy smaller geographic ranges than large species so that, within a geographic region the size of North America, it is common to find examples of several to many parapatric species of small mammal (often congeneric), each with essentially the same ecological needs. When species of small mammal do have large ranges, they are often subdivided into many parapatric subspecies. However, while these subspecies (like parapatric species) all utilize similar resources, they presumably each possess particular adaptations relevant to the physical and biological characteristics of the area in which they occur. It would seem reasonable to suggest that speciation in small mammals may often involve a smooth evolutionary progression within a continuous species range from locally adapted forms to subspecies to species, i.e. parapatric speciation. In fact, there are theoretical difficulties with this process and speciation in small mammals may usually require geographic isolation, in accordance with Mayr's allopatric model. The Pleistocene glaciations may have generated many such isolation events in small mammals. Speciation may occur particularly rapidly in very small isolated populations (the peripatric version of the allopatric model); there is much evidence for substantial genetic change under these circumstances in small mammals such as the house mouse *Mus musculus domesticus*. A fourth mode of species formation, sympatric speciation, has been suggested for bats.

Introduction

Among ecologists, ethologists, and population geneticists, 'small mammals' are a commonly used, though polyphyletic, taxonomic unit (e.g. Golley, Petrusewicz & Ryszkowski 1975; Stoddart 1979). To these scientists as well as to the lay public, there is clearly a greater similarity among mice, shrews (shrew-mice), and bats (flying-mice), than between such species and rhinoceroses, tigers, and whales. The difficulties come when it is necessary to define some threshold between small and

ZOOLOGICAL SYMPOSIUM No. 69
ISBN 0–19–857787–7

large mammals. Inevitably there will be some large rat-like creature that will exceed a particular value and some miniature deer that will be smaller than it! I shall take 1 kg as my arbitrary threshold. Despite having defined a threshold, I will make little use of it; I will make most mention of distinctly small mammals and compare them with distinctly large mammals.

In this brief review I will firstly, in a very qualitative way, highlight the ecological possibilities there are for species of small mammal. Next, I will consider the extent to which species of small mammal actually exploit these possibilities. Thirdly, I will discuss modes of speciation in small mammals. Finally, I will debate the relative importance of the rate of species formation and the availability of ecological niches in determining how many species of small mammal there are in existence. The comparison of small mammals with their large counterparts will run throughout the review. Inevitably, because of the excellent coverage of the north temperate fauna (for example, by Corbet 1978 and Hall 1981), attention will be concentrated on the mammals of the Nearctic and Palaearctic.

In this article, Mayr's 'biological' definition of species will be applied, so that 'speciation' is the splitting of one species to produce two, with the two daughter forms reproductively isolated (or, more strictly, genetically isolated) from each other (e.g. Mayr 1970). Taxa are said to be reproductively isolated if they would fail to mate, their gametes would fail to unite, or they would produce only inviable or sterile hybrids, given any opportunity to interbreed in nature. Genetic isolation is equivalent to reproductive isolation except in one important situation, that of 'ring species' (see Endler 1977 and below). Thus, ring species are reproductively isolated but genes could potentially flow between them via intermediate forms. Truly genetically isolated taxa are incapable of exchanging genes in nature. The crucial importance of the inability of biological species to exchange genes is that such forms can be considered to be on separate phylogenetic trajectories with the potential to evolve in completely different ways.

Ecological possibilities

Small mammals are absent from the major part of the Earth's surface. They do not occur in open water or in the high Arctic and Antarctic (except in association with humans), even though large mammals (whales, seals, and polar bears) are found in these places. It must be presumed that these particular environmental conditions are unfavourable to small mammals and there are obvious physiological grounds for this presumption.

Therefore, it is the ecological possibilities for small mammals on the major land masses that need to be considered. Clearly, smallness opens up some possibilities unavailable to large mammals. For example, aerial food sources are essentially the preserve of small mammals and only small mammals seem to be able to adopt a completely fossorial lifestyle (Bourlière 1975). However, even if small mammals are able to use more components of the terrestrial habitat than large mammals, it does not automatically follow that more species of small mammal are to be expected. First, those food supplies that can be used by both large and small species could be solely utilized by large species, given the dominance of large species over small. Second, those species of small mammal that are able to survive could conceivably be matched in number by large carnivorous species, each specializing on a different type of small mammal. Even those small mammals that use an underground or aerial environment could be dug up or caught at the roost.

Clearly, large mammals are not so dominant as may be theoretically possible; there are some very species-rich communities of small mammal. Given that small mammals are so speciose, there are energetic constraints on how closely large mammals (especially large carnivorous mammals) can match small mammals in terms of number of species (van Valen 1973; Colinvaux 1980). Individual large mammals need more energy than individual small mammals and therefore species of small mammal have the opportunity to achieve higher densities. If large mammals were to be as speciose as small mammals in a particular community, they would have to occupy much larger geographical ranges to achieve viable population sizes (Nunney & Campbell 1993). Conversely, the individual densities of large mammal species could only approach those of small mammal species in a particular community if they constituted only a very few, generalist species. There is, of course, the third possibility, that large mammals will not be able to match small mammals either in terms of local density or in terms of species number in a community.

The number of species of small mammal

There are considerably more species of small mammal than large mammal in the world (van Valen 1973). There are some notably speciose families which consist entirely or almost entirely of small species (among rodents, Muridae (1138 species) and Sciuridae (260); among insectivores, Soricidae (289); among bats, Pteropodidae (173), Phyllostomidae (148) and Vespertilionidae (355) out of a total of 4444 species of mammal: Nowak 1991). The numerical superiority of small mammals is probably underestimated because the new species that are being described are almost entirely of the small category (compare recent and old editions of *Walker's mammals of the world*).

Small mammals include more species not only globally, but also regionally and locally. Clearly, these data are consistent with the energetic arguments presented above. It should be noted that the numerical superiority of small mammal species over large does not reflect a bias associated with my 'naturalistic' definition of smallness. Terrestrial mammals range in size between 10^{-3} and 10^4 kg, so, on a logarithmic scale, the 10^0 kg threshold biases slightly 'in favour' of large mammals.

Locally, species of small mammal are associated with a much wider variety of ecological niches than are observed for large mammals, translating into a higher number of species. Thus, among the native terrestrial mammals of Britain, for example, there are 30 small species and 11 large species (Corbet & Harris 1991). The density of each small species also tends to be higher, as exemplified by mammals both in Britain (Harris, Morris, Wray & Yalden 1995) and elsewhere (Damuth 1981; but see Lawton 1990).

At the regional level, species of small mammal tend to have smaller geographical ranges than species of large mammal (van Valen 1973). Thus, within a geographic region such as Europe or North America, there are often several species of small mammal with neighbouring, non-overlapping distributions (i.e. parapatric ranges). Each species, although occupying essentially the same niche (living in the same part of the habitat, eating the same food types) and showing great morphological similarity, is presumably better adapted to the precise environmental circumstances where it lives. An example is provided by the sibling species of shrew, *Sorex araneus* and *Sorex coronatus*, which make contact in the Low Countries, Germany, and Switzerland, with *araneus* found to the north and east and *coronatus* to the south and west (Hausser, Catzeflis, Meylan, & Vogel 1985). On the basis of geographic ranges of the two species and the habitats they occupy in the contact zone, it can be argued that *araneus* is better adapted to areas with a continental climate and can occur on either wet or dry soils while *coronatus* is competitively superior where there is an Atlantic climate and balanced soil humidity (Brünner & Neet 1991).

Although species of small mammal tend to have smaller geographical ranges than those of large mammals, there are still some very widespread small species. These are often subdivided into numerous parapatric subspecies. Obviously, these subspecies tend to be even more similar than parapatric species in terms of morphology and ecological needs, yet they are distinct and, like parapatric species, presumably well adapted to the physical and biological characteristics of the area in which they occur. Subspecies have been named in many species of mammal, both large and small, but the most impressive mosaics of described subspecies are found among North American rodents (Hall 1981). The pocket gophers (*Geomys* and *Thomomys*) are subdivided into particularly large numbers of parapatric subspecies and species. There are, for instance, 215 geographic forms within the *Thomomys bottae* complex (Patton & Smith 1989). Although there is a danger that the criteria used to define 'subspecies' could differ between groups of mammals, it is worth noting that, based on Hall's (1981) treatise on the mammals of North America, widespread small mammals tend to be subdivided into more subspecies than widespread large mammals.

In summary, the relative numbers and geographical ranges of small and large mammals are consistent with the energetic arguments presented earlier. However, at both a local and a regional level there are considerably fewer species of large and small mammals than there might be. First, there are some extremely abundant and widespread species that potentially could be subdivided into many daughter forms, but are not. For example, if the above-mentioned *S. araneus* occurs at the same density over its large northern Palaearctic range (Corbet 1978) as it occurs in Britain (Harris *et al.* 1995), the species has a total population size of approximately 3×10^9, many more individuals than necessary for its viability. Secondly, it is common to find that particular 'mammal' niches are not occupied, as a consequence of failure of colonization or lack of speciation. For example, as was realized by Darwin (1872), bats are essentially the only mammals that have colonized oceanic islands by natural means. Bats are not likely to evolve into ground-dwelling forms, so many small mammal niches remained unoccupied on oceanic islands until humans brought rats and mice (e.g. Leader-Williams & Walton 1989). In contrast, the pigmy shrew (*Sorex minutus*), as the only shrew that colonized Ireland (Yalden 1982), would seem a reasonable progenitor of new larger species. However, no such species have arisen in the period that the pigmy shrew has been isolated on Ireland. Clearly, this represents a 'failure' of speciation but not a surprising one, given that Irish pigmy shrews have been separated for only 13 000 years at maximum (Yalden 1982). Speciation, of course, is the primary topic of this article and now needs to be addressed more generally.

Modes of speciation

As early as the nineteenth century there was a debate about whether subspecific forms had to be geographically isolated before they could become full species (Darwin 1872). In this article, the different modes of speciation to be considered are the different geographical contexts in which species formation could occur. I will follow Bush (1975) for the different categories, except that Bush's second version of allopatric speciation will be named 'peripatric speciation' after Mayr (1982). The four different modes of speciation are therefore as follows (see Fig. 1).

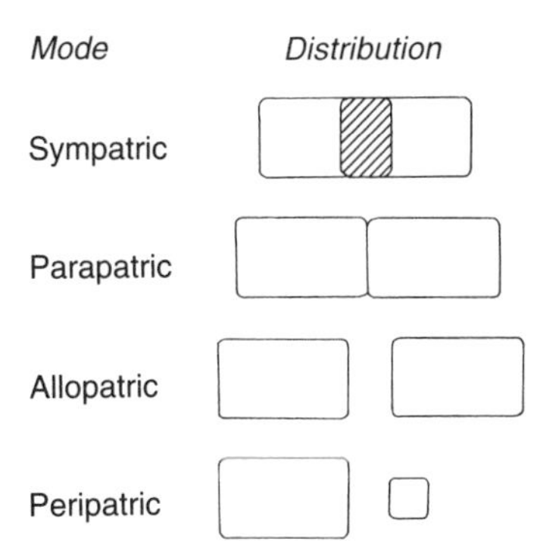

Fig. 1. The relative distribution of two subspecific forms attaining species status according to the different modes of speciation.

1. *Sympatric speciation*, where subspecific forms speciate while showing substantial geographic overlap.
2. *Parapatric speciation*, where there is speciation of subspecific forms which have ranges that are in contact but not overlapping.

3. *Allopatric speciation*, where subspecific forms become reproductively isolated while geographically separate.
4. *Peripatric speciation*, a special case of allopatric speciation, where a population in a small geographic isolate undergoes rapid genetic change, including development of reproductive isolation.

It should be noted that the adjectives 'sympatric', 'parapatric', etc., can be used to describe the relative position of species as well as modes of species formation. Thus, 'parapatric species' need not necessary have arisen by 'parapatric speciation' (see below).

Sympatric speciation

The possibility of sympatric speciation has been analysed in most detail for phytophagous insects (Bush 1993). Often, the larvae of a particular species of phytophagous insect feed on only a single plant species and adult courtship occurs on the surface of the same plant. If certain members of the insect species start to specialize on a new food plant, then reproductive isolation may arise as a correlated response (Rice 1987). In mammals, courtship is far less rigidly tied to particular food-related microhabitats and sympatric speciation appears an unlikely process. However, if there were to be sympatric speciation in the Mammalia, it would be more likely to occur in small species than in large (see Bush 1993). For example, in a small ground-dwelling tropical mammal species which feeds on fruit fallen from the forest canopy, individuals might spend their whole lifetime on the forest floor. However, it is not difficult to envisage that some members of the species might become specialized to exploit the fruit that still hangs in the canopy and thus (given the huge distance between forest floor and the tree-tops from the viewpoint of a small species) would need to spend their whole lives in the tree-tops.

While sympatric speciation is unlikely to be an important process for most groups of mammal, it may occur in bats. Jones & van Parijs (1993) discovered that the widespread European bat *Pipistrellus pipistrellus* is in fact composed of two sibling species which differ in the frequency of their echolocation call. Jones & van Parijs suggested that these two species could have evolved in sympatry as a consequence of members of the ancestral species specializing on different prey types (large and small) by using calls of different frequency (low and high). If the frequency of courtship calls is the same as that of feeding calls and if bats choose mates that have similar calls to themselves, then once again reproductive isolation could occur as a correlated response. Clearly, much work needs to be done to follow up this interesting idea. If sympatric speciation is shown to be a likely possibility for the two forms of *P. pipistrellus*, then it is noteworthy that there are three other *Pipistrellus* species in Europe, all morphologically very similar and with overlapping ranges (Macdonald & Barrett 1993). As a generality, bats, to a much greater extent than other mammals, consist of complexes of sibling species with overlapping ranges. Some of these cases could represent sympatric speciation, although given the mobility of bats, a sympatric distribution of a new species should certainly not be considered a guarantee of a sympatric origin.

Parapatric speciation

From the many examples of small mammals subdivided into parapatric subspecies, this would, at first sight, seem the most likely form of speciation. Thus, it is not difficult to envisage a smooth evolutionary progression within a continuous species distribution from locally adapted forms of subspecies to species; Endler (1977) developed a detailed genetic model of this process. Even for subspecies which have arisen in geographical isolation, if contact is made subsequently ('secondary contact' as opposed to the 'primary contact' shown by subspecies that arise *in situ*) then, potentially, reproductive isolation could be attained in parapatry. In fact, there is little clear evidence of a parapatric speciation process in small mammals or any other taxa.

The first requirement of parapatric speciation is, nevertheless, easily met. Parapatric speciation will occur only if the hybrids are unfit relative to either subspecies. Hybrid unfitness is commonly observed in the areas of hybridization between subspecies ('hybrid zones'), including examples involving small mammals (Barton & Hewitt 1985; Harrison 1990; Searle 1993). For reproductive isolation to evolve when subspecies are in contact, either the hybrids must continue to become more unfit so that they become completely inviable or completely sterile, or the subspecies must become homogamous (i.e. develop completely assortative mating or completely assortative fertilization). There has been particular interest in the possibility that selection against unfit hybrids in hybrid zones may lead to evolution of completely assortative mating ('reinforcement'). With reinforcement, subspecies could be able to develop reproductive isolation more quickly in parapatry than in allopatry. However, there is considerable scepticism as to whether reinforcement is a frequent process (Butlin 1987; Howard 1993). For example, in most situations, it is difficult to see how assortative mating genes and unfitness genes could stay consistently linked in individuals within the hybrid zone. Also, interactions with individuals from outside the hybrid zone might disrupt the process, as they would be under different types of selective constraint from individuals within the zone. The only real candidate for a reinforcement event among small mammals involved two chromosomal races of house mouse which had only a single type of unfit hybrid and which were found together in a small isolate (Capanna & Corti 1982; Hauffe & Searle 1992). Similarly suitable circumstances may be rare.

There are two further reasons for scepticism about parapatric speciation. Firstly, a premise of the parapatric speciation process is that hybrid zones should evolve, but in fact many may be stable for thousands of years (Mayr 1970). Secondly, as well as the possibilities that, with time, hybrids become more unfit or that homogamy evolves, there is also a possibility that natural selection may promote greater fitness in hybrids (Harrison 1990; Barton 1993). Among small mammals, there are several examples of hybrid zones between chromosomal races which appear to have 'ameliorated' in this fashion (Searle 1993).

Whatever the criticism of the parapatric speciation process, it is a possible mode of species formation. As such, given the tendency for species of small mammal to be subdivided into parapatric subspecies, there is an expectation that parapatric speciation could occur more often in small mammals than in large mammals.

Allopatric speciation

The allopatric speciation model of Mayr (1970) is the most universally accepted mode of speciation for sexually reproducing organisms. If a species is separated into two geographic isolates, then the isolated populations over time are bound to develop genetic differences including reproductive isolation.

Allopatric speciation is most easily envisaged in association with major vicariance events. On the drifting apart of continents many species of mammals would have been subdivided into two, with the separated populations later becoming reproductively isolated. This would have happened in the Cretaceous when South America and Africa parted. A more recent example, which unfortunately does not involve mammals, is the separation of marine faunas when the Panama isthmus was formed 3 million years ago; this was a major biogeographical event which led to numerous instances of allopatric speciation (Vermeij 1993).

Allopatric speciation has probably been an important process in small mammals in connection with the alternating glacial (cold) and interglacial (warm) periods that have occurred over the last 2 million years. These climatic changes have resulted in dramatic fluctuations in faunal and floral distributions, sometimes causing a once-continuous large species range to split into a number of small isolates within 'refugia' of suitable habitat. This process occurred in many regions of the world, including perhaps tropical areas (Mayr & O'Hara 1986). To take an example nearer home, in Europe at the height of the last glaciation (20 000 years ago) many species of temperate small mammals would have been isolated in separate refugia in Iberia, Italy, the Balkans, and areas further east (Mayr 1970; Searle 1984; Thorpe 1984; Hewitt 1993; D. T. Bilton, P. M. Mirol, & J. B. Searle unpubl.). Under these circumstances, the populations in the separate refugia could have evolved along distinct pathways and thereby attained subspecific or specific status. At the end of the glaciation (10 000 years ago) such subspecies or species of temperate small mammal would have migrated from each refugium into central and northern Europe, forming secondary contacts with related forms from other refugia. Thus, in the example of sibling species of shrew given above, *S. coronatus* emanated from an Iberian refugium and *S. araneus* from a refugium (or refugia) further east and so the species now make secondary contact in central Europe (Searle 1984).

In association with the ice-age refugia in North America (Hoffman 1981) there is clear evidence that speciation occurred more frequently in small mammals than in large mammals (Rogers, Rogers, Hoffman, & Martin 1991). As small mammals tend to have a larger population size and a shorter generation time than large mammals, new selectively advantageous mutations may have arisen more often in the refugial populations of small mammals. In addition, the small mammals, as more specialized forms, may have had to adapt more to fit the local circumstances of the refugium. In these ways the small mammals may have evolved faster and speciated more readily than the large mammals.

Although the fluctuating climate of the Pleistocene may have contributed to species formation in small mammals, it also led to the extinction of many subspecies and species. But even this process could have enhanced speciation. Thus, in the case of a 'ring species' (Endler 1977), the end-members of a sequence of parapatric subspecies may show reproductive isolation from one another, but they are not genetically isolated, because genes can flow through the intermediate subspecies. Clearly, the extinction of these intermediate forms may promote the genetic isolation of the end-members, allowing them to be considered separate biological species. Given that small mammals often consist of long chains, rings, or mosaics of subspecies, this process of extinction of subspecies could certainly have occurred on occasion.

There are undoubtedly a variety of possible ways in which small mammals can speciate allopatrically and it is reasonable to propose that this has been an important mode of speciation for these animals—except perhaps for bats, which are, of course, far more mobile than other small mammals and therefore less influenced by geographical barriers.

Peripatric speciation

It is not always easy to distinguish allopatric and peripatric speciation and it is probably best to consider peripatric speciation an extreme example of the allopatric process. However, a small isolated population may have distinctly different properties from a large isolated population. In particular, there may be rapid changes in gene and genotype frequencies as a result of genetic drift in the small population. Alleles which are normally rare within a particular species may become common or even fixed in a small isolate. Also, genic heterozygosity may reach extremely low levels. Among small mammals, the genetic effect of small population size has been particularly well studied in the house mouse. Fixation of rare alleles has been observed in small populations on offshore islands (Berry & Peters 1977) and in a small geographically isolated population in mainland Italy (S. Fraguedakis, H. C. Hauffe, & J. B. Searle unpubl.). An extraordinarily low heterozygosity has been recorded in mice from the tiny Isle of May near Edinburgh (Berry *et al.* 1990). Results of this kind have led some to believe that genetic drift is a critical aspect of the rapid speciation that does frequently occur when animals and plants are isolated on small oceanic islands (Giddings, Kaneshiro & Anderson 1989). However, there are also very different selection pressures associated with life on a small island or in a small mainland isolate which could be the cause of more important genetic change from the aspect of reproductive isolation than genetic drift (Williamson 1981; Barton & Charlesworth 1984).

There are features of small mammals that may predispose them to form small isolated populations, thereby increasing the opportunity for peripatric speciation. Most species of small mammal (except bats) show poor powers of active dispersal, so that a few individuals in a habitat fragment or on an island may be completely isolated from centres of population. Small mammals could come to be present in a habitat fragment if such a fragment represents a remnant of a previously larger area and small mammals may colonize islands by passive transport (rafting or transport with humans). Even though they are good dispersers, bats may become isolated on mid-oceanic islands. Large mammals, because of their greater abilities to disperse actively, may be less likely than small mammals to be confined to habitat fragments and, with a few exceptions, they are unable to colonize oceanic islands because neither flight nor passive transport is possible for mammals of large size.

Another feature of small mammals that may help their persistence in isolated populations is their ability to expand quickly from a situation close to extinction, because of short generation times and large litter sizes. Populations of large mammals are much slower to expand from low numbers and are more likely to succumb to disease or some other catastrophe as a result. For a detailed treatment, Lawton (1990) discusses this and other aspects of the population dynamics and dispersal abilities of small and large animal species.

There are several data which support the occurrence of peripatric speciation in small mammals. For example, Godfrey (1958) attempted to cross bank voles (*Clethrionomys glareolus*) from different small islands off the coast of Britain (where they have been isolated probably since the last glaciation) and found assortative mating and reduced hybrid viability and fertility. Similar results were obtained when he crossed the mainland and island forms. Another example is the marked tendency for novel chromosomal rearrangements to occur in isolated populations of some species of small mammals, such as the western pocket gopher *Thomomys talpoides* (Thaeler 1985). These chromosomal rearrangements are apparently more easily fixed in small populations, as a result of genetic drift or local selection. Once fixed these mutants can contribute to reproductive isolation because of the high frequency of meiotic error in chromosomal heterozygotes (King 1993).

Conclusions

It is of interest to establish the pattern of speciation in small mammals, not only because they are an obvious and important part of the fauna, but also because all extant mammals (including *Homo sapiens*) are derived ultimately from small forms (Stanley 1973; Lillegraven, Thompson, McNab, & Patton 1987).

Considering the present day, there are more species of small mammal than large mammal, consistent with fewer ecological opportunities for large species. Also, I believe that all modes of speciation are more likely in small mammals. There are no clearly argued cases of sympatric or parapatric speciation involving large mammals. At first sight, allopatric and peripatric speciation might be expected to be rather common processes in large mammals. Over their sizeable ranges there would appear to be plentiful opportunities for subdivision into isolates where new genetic forms (and, ultimately, species) might develop. In fact, in North America there tend to be fewer subspecies in large mammals than in small mammals (Hall 1981), reflecting less differentiation either *in situ* or in refugia during glacial maxima. With regard to the possibility of differentiation *in situ*, many geographical barriers are of less consequence to large mammals than small because large mammals are more mobile (for example, see Ayres & Clutton-Brock 1992). In this way, the ranges of large mammals may tend to be more truly continuous than the ranges of small mammals (except bats). Both large and small isolated populations of large mammals can occur, of course, but, as outlined in the sections above on allopatric speciation and peripatric speciation, speciation may occur rather rarely within them.

Returning to small mammals, it is now of interest to establish the relative importance of the ecological possibilities for species and the rate of species formation, in determining the number of species of small mammal that are observed. Clearly, the ecological possibilities put an upper limit on the number of species that can exist; each species must be able to maintain a viable population size by uniquely exploiting a particular niche over a large enough geographical area. However, for the Palaearctic and Nearctic at least, the number of species of small mammal at the regional level is well below this upper limit; the net rate of production of species (absolute production rate minus the extinction rate) has clearly been insufficient. This failure of species number to match the ecological possibilities at a regional level may be considered to be the result of the rarity of the parapatric speciation process in small mammals. There is little evidence that small local differences in climate and geology promote the splitting of single widespread species, even though viable daughter forms could theoretically be produced in this way. If parapatric speciation events do occur in small mammals they presumably involve either (1) cases of dramatic variation in habitat over short distances (e.g. dry savannah abutting rain forest in tropical areas) sufficient to create large selective pressures on a widespread species such that it may speciate *in situ* or (2) examples where forms develop substantial genetic differences in allopatry but actually speciate after resuming contact.

Thus, for the Palaearctic and Nearctic, the number of species observed regionally is probably largely determined by the rate of allopatric and peripatric speciation. The number of forms within any particular set of closely related species tends to correlate well with the distribution of present-day geographical barriers and incidence of range subdivision during the Pleistocene.

It is clear, therefore, that there could be substantially more species of small mammal at the regional level, if forms tended to have less extensive ranges. Likewise, there could be a greater number of species locally, if small mammals tended to occupy narrower niches. The fact that the numbers of small mammal species observed locally do not match the ecological possibilities can be viewed as the result of insufficient sympatric speciation in small mammals (in a similar way as the lack of small mammals regionally can be ascribed to a failure of parapatric speciation). If sympatric speciation does occur in small mammals, it is presumably only when there is a very clearly exploitable 'empty niche' and when courtship is associated with utilization of this niche (see the section on sympatric speciation, p. 148).

So, the number of species of small mammal observed locally is, again, probably determined by the rate of allopatric and peripatric speciation. If, during isolation, ecological differences develop between the speciating pair, then an overlapping distribution may develop with minimal morphological change to either of the species. However, in other cases, species formed by an allopatric or peripatric process may have to undergo character displacement (Brown & Wilson 1956; Schluter & McPhail 1993) on becoming sympatric.

In conclusion, allopatric and peripatric speciation are probably the most common modes of species formation in small mammals, although parapatric and sympatric speciation may occur under certain circumstances. Given the ease with which small mammals can be studied, there is every hope that there will be further elucidation of the factors determining the numbers and distributions of this group of organisms in the future.

Acknowledgements

I am very grateful to Drs Angela Douglas and Peter Miller for comments on the article and to Dr David Bilton, Dr Stella Fraguedakis, Dr Heidi Hauffe, and Patricia Mirol for unpublished results. I also wish to thank the Natural Environment Research Council and the European Commission for providing grants for my studies of within-species variation and speciation in small mammals (GR3/07569 and CHRXCT930192).

References

Ayres, J. M. & Clutton-Brock, T. H. (1992). River boundaries and species range size in Amazonian primates. *Am. Nat.* **140**: 531–537.

Barton, N. H. (1993). Why species and subspecies? *Curr. Biol.* **3**: 797–799.

Barton, N. H. & Charlesworth, B. (1984). Genetic revolutions, founder effects, and speciation. *A. Rev. Ecol. Syst.* **15**: 133–164.

Barton, N. H. & Hewitt, G. M. (1985). Analysis of hybrid zones. *A. Rev. Ecol. Syst.* **16**: 113–148.

Berry, R. J. & Peters, J. (1977). Heterogeneous heterozygosities in *Mus musculus* populations. *Proc. R. Soc. (B)* **197**: 485–503.

Berry, R. J., Triggs, G. S., Bauchau, V., Jones, C. S., & Scriven, P. (1990). Gene flow and hybridization following introduction of *Mus domesticus* into an established population. *Biol. J. Linn. Soc.* **41**: 279–283.

Bourlière, F. (1975). Mammals, small and large: the ecological implications of size. In *Small mammals: their productivity and population dynamics*: 1–8. (Eds Golley, F. B., Petrusewicz, K., & Ryszkowski, L.). Cambridge University Press, Cambridge.

Brown, W. L. & Wilson, E. O. (1956). Character displacement. *Syst. Zool.* **5**: 49–64.

Brünner, H. & Neet, C. R. (1991). A parapatric scenery: the distribution and ecology of *Sorex araneus* and *S. coronatus* (Insectivora, Soricidae) in southwestern Germany. *Z. Säugetierk.* **56**: 1–9.

Bush, G. L. (1975). Modes of animal speciation. *A. Rev. Ecol. Syst.* **6**: 339–364.

Bush, G. L. (1993). A reaffirmation of Santa Rosalia, or why are there so many kinds of small animals? In *Evolutionary patterns and processes*: 229–249. (Eds Lees, D. R. & Edwards, D.). Academic Press, New York. (*Linn. Soc. Symp. Ser.* **14**).

Butlin, R. (1987). Speciation by reinforcement. *Trends Ecol. Evol.* **2**: 8–13.

Capanna, E. & Corti, M. (1982). Reproductive isolation between two chromosomal races of *Mus musculus* in the Rhaetian Alps (northern Italy). *Mammalia* **46**: 107–109.

Colinvaux, P. (1980). *Why big fierce animals are rare.* George Allen & Unwin, London.

Corbet, G. B. (1978). The mammals of the Palaearctic region: a taxonomic review. *Publs Br. Mus. nat. Hist.* No. 788: 1–314.

Corbet, G. B. & Harris, S. (Eds) (1991). *The handbook of British mammals.* (3rd edn). Blackwell, Oxford.

Damuth, J. (1981). Population density and body size in mammals. *Nature, Lond.* **290**: 699–700.

Darwin, C. (1872). *The origin of species by means of natural selection.* (6th edn). John Murray, London.

Endler, J. A. (1977). *Geographic variation, speciation, and clines.* Princeton University Press, Princeton (*Monogr. Popul. Biol.* No. 10: 1–246).

Giddings, L. V., Kaneshiro, K. Y., & Anderson, W. W. (Eds) (1989). *Genetics, speciation, and the founder principle.* Oxford University Press, Oxford.

Godfrey, J. (1958). The origin of sexual isolation between bank voles. *Proc. R. phys. Soc. Edinb.* **27**: 47–55.

Golley, F. B., Petrusewicz, K., & Ryszkowski, L. (Eds) (1975). *Small mammals: their productivity and population dynamics.* Cambridge University Press, Cambridge.

Hall, E. R. (1981). *The mammals of North America.* (2nd edn). John Wiley, New York.

Harris, S., Morris, P., Wray, S., & Yalden, D. (1995). *The status of British mammals.* Joint Nature Conservancy Council, Peterborough.

Harrison, R. G. (1990). Hybrid zones: window on evolutionary process. *Oxford Surv. evol. Biol.* **7**: 69–128.

Hauffe, H. C. & Searle, J. B. (1992). A disappearing speciation event? *Nature, Lond.* **357**: 26.

Hausser, J., Catzeflis, F., Meylan, A., & Vogel, P. (1985). Speciation in the *Sorex araneus* complex (Mammalia: Insectivora). *Acta zool. Fenn.* No. 170: 125–130.

Hewitt, G. M. (1993). Postglacial distribution and species substructure: lessons from pollen, insects and hybrid zones. In *Evolutionary patterns and processes*: 97–123. (Eds Lees, D. R. & Edwards, D.). Academic Press, New York. (*Linn. Soc. Symp. Ser.* **14**).

Hoffmann, R. S. (1981). Different voles in different holes: environmental restrictions on refugial survival of mammals. In *Evolution today*: 24–45. (Eds Scudder, G. E. & Reveal, J. L.). Hunt Institute for Botanical Documentation, Carnegie-Mellon University, Pittsburgh.

Howard, D. J. (1993). Reinforcement: origin, dynamics, and fate of an evolutionary hypothesis. In *Hybrid zones and the evolutionary process*: 46–69. (Ed. Harrison, R. G.). Oxford University Press, New York.

Jones, G. & van Parijs, S. M. (1993). Bimodal echolocation in pipistrelle bats: are cryptic species present? *Proc. R. Soc. (B)* **251**: 119–125.

King, M. (1993). *Species evolution: the role of chromosome change.* Cambridge University Press, Cambridge.

Lawton, J. H. (1990). Species richness and population dynamics of animal assemblages. Patterns in body size: abundance space. *Phil. Trans. R. Soc. (B)* **330**: 283–291.

Leader-Williams, N. & Walton, D. (1989). The isle and the pussycat. *New Scient.* No. 1651: 48–51.

Lillegraven, J. A., Thompson, S. D., McNab, B. K., & Patton, J. L. (1987). The origin of eutherian mammals. *Biol. J. Linn. Soc.* **32**: 281–336.

Macdonald, D. W. & Barrett, P. (1993). *Collins field guide: mammals of Britain and Europe.* Harper Collins, London.

Mayr, E. (1970). *Populations, species, and evolution.* Harvard University Press, Cambridge, Ma.

Mayr, E. (1982). Processes of speciation of animals. In *Mechanisms of speciation*: 1–19. (Ed. Barigozzi, C.). A. R. Liss, New York. (*Progr. clin. biol. Res.* **96**: 1–19.)

Mayr, E. & O'Hara, R. J. (1986). The biogeographic evidence supporting the Pleistocene forest refuge hypothesis. *Evolution* **40**: 55–67.

Nowak, R. M. (1991). *Walker's mammals of the world.* (5th edn). John Hopkins University Press, Baltimore.

Nunney, L. & Campbell, K. A. (1993). Assessing minimum viable population size: demography meets population genetics. *Trends Ecol. Evol.* **8**: 234–239.

Patton, J. L. & Smith, M. F. (1989). Population structure and the genetic and morphologic divergence among pocket gopher species (genus *Thomomys*). In *Speciation and its consequences*: 284–304. (Eds Otte, D. & Endler, J. A.). Sinauer, Sunderland, Ma.

Rice, W. R. (1987). Speciation via habitat specialization: the evolution of reproductive isolation as a correlated character. *Evol. Ecol.* **1**: 301–314.

Rogers, R. A., Rogers, L. A., Hoffman, R. S., & Martin, L. D. (1991). Native American biological diversity and the biogeographic influence of ice age refugia. *J. Biogeog.* **18**: 623–630.

Schluter, D. & McPhail, J. D. (1993). Character displacement and replicate adaptive radiation. *Trends Ecol. Evol.* **8**: 197–200.

Searle, J. B. (1984). Three new karyotypic races of the common shrew *Sorex araneus* (Mammalia: Insectivora) and a phylogeny. *Syst. Zool.* **33**: 184–194.

Searle, J. B. (1993). Chromosomal hybrid zones in eutherian mammals. In *Hybrid zones and the evolutionary process*: 309–353. (Ed. Harrison, R. G.) Oxford University Press, New York.

Stanley, S. M. (1973). An explanation for Cope's Rule. *Evolution* **27**: 1–26.

Stoddart, D. M. (Ed.) (1979). *Ecology of small mammals.* Chapman & Hall, London.

Thaeler, C. S. (1985). Chromosome variation in the *Thomomys talpoides* complex. *Acta zool. fenn.* No. 170: 15–18.

Thorpe, R. S. (1984). Primary and secondary transition zones in speciation and population differentiation: a phylogenetic analysis of range expansion. *Evolution* **38**: 233–243.

van Valen, L. (1973). Body size and number of plants and animals. *Evolution* **27**: 27–35.

Vermeij, G. J. (1993). The biological history of a seaway. *Science* **260**: 1603–1604.

Williamson, M. (1981). *Island populations.* Oxford University Press, Oxford.

Yalden, D. W. (1982). When did the mammal fauna of the British Isles arrive? *Mammal Rev.* **12**: 1–57.

Ecology

Symp. zool. Soc. Lond. (1996) No. 69: 159–174

Miniature mammals: life-history strategies and macroevolution

ANDY PURVIS[1]
and PAUL H. HARVEY

Department of Zoology
University of Oxford
South Parks Road
Oxford OX1 3PS, UK

Synopsis

Most mammal species are smaller than 60 g. The smallest species are mainly shrews, myomorph rodents, and insectivorous bats. A bias in data availability has led most previous work on patterns of mammal life-history evolution to focus on unusually large species. We review the major findings of such comparative studies, beginning with allometries of key variables and moving on to patterns of correlations among residuals. Every date, weight, and rate scales tightly with adult size across mammal species: small mammals tend to live faster and die younger than large species. This fast–slow continuum is also apparent when body size is factored out. We discuss two very different optimality models, one aiming to understand the driving forces behind life-history variation, the other trying to explain the macroevolutionary pattern of body size among species. Our comparative analysis finds that, among small mammals, most of the variation in life histories is independent of body size, but that correlations among residuals are broadly similar to those in mammals as a whole: the slow–fast continuum is still present. The size independence of the variation shows that quite radical life-history changes are possible without any change in body size being necessary: mammals are not tightly constrained by allometry. Interestingly, mammals that are smaller than their relatives have relatively small litters of large offspring. Our findings lend at best qualified support to the models we discuss.

Introduction: miniature mammals

Mammal species show extreme diversity in body weight, spanning approximately eight orders of magnitude from the smallest shrews, bats, and rodents (around 2 g) to the blue whale (nearly 2×10^8 g). As Fig. 1 shows for the Palaearctic region, the distribution of weights among species is continuous: there is no dichotomy, no obvious dividing line that can be used to delineate miniature species. There is a positive skew to the distribution, even with the body size axis expressed in logarithms: it is not

[1]Present address: Department of Biology, Imperial College, Silwood Park, Ascot SL5 7PY, UK.

ZOOLOGICAL SYMPOSIUM No. 69
ISBN 0-19-857787-7

lognormal. The median body weight is approximately 60 g with the quartiles at approximately 23 and 440 g. A similar skew is found among mammal species in North America (Brown, Marquet, & Taper 1993) and world-wide (Eisenberg 1981) and in other taxa (May 1986); Brown *et al.* and May report higher modes for mammals, presumably because they exclude bats. Extrapolating to the world as a whole, Fig. 1 suggests that approximately a quarter of mammal species are lighter than an ounce, approximately half weigh between an ounce and a pound, and the rest weigh over a pound. The smallest quarter of species might reasonably be called miniature.

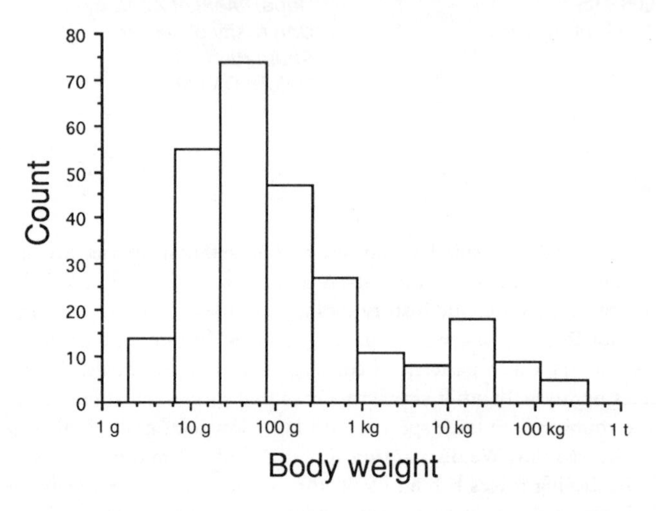

Fig. 1. Distribution of body weights of 268 species of non-aquatic mammal living in the Palaearctic region. Data from Letcher (1994).

These thousand or so species are nearly all shrews, myomorph rodents, and insectivorous bats. Table 1 outlines some broad life-history characters of these groups. As well as small size, most of these species share another feature that is relevant here: very little is known of their life history in the wild. They are predominantly nocturnal and secretive and many do not breed well in the laboratory (Bronson 1989). The bias in our knowledge of life history is striking. For instance, the most comprehensive survey of mammal reproductive data (Hayssen, van Tienhoven, & van Tienhoven 1993) contains more than four times more information about Carnivora than about Insectivora, although the latter group has nearly 100 more species. The carnivore entry is also longer than that for bats, which have nearly four times as many species, and families of small rodents are also generally poorly known. Nearly half the information listed for placental mammals comes from just three orders—primates, carnivores, and artiodactyls—which make up fewer than one-sixth of the species, and which are generally much larger than most of the other five-sixths. The picture worsens when we consider natural mortality data, which are in worryingly short supply for the class as a whole anyway. A recent compilation (Purvis & Harvey 1995) could find valid life tables from natural populations of just 64 species, only 10 of them below the median mammal weight. Whenever we apply comparative findings to small mammals, we are therefore running the risks associated with extrapolating beyond the range of the data (discussed by Schmidt-Nielsen 1984).

Table 1. Some life-history characteristics of the three main groups of very small mammals.

Characteristic	Shrews	Microchiroptera	Myomorpha
Adult weight (g)	2–35	1.5–200	6–2000
Gestation length (days)	13–24	40–240*	15–50
Litter size (*n*)	2–10	1–3	2–10
Neonatal weight (g)	0.1–2.5	1–12	1–28
Litters per year (*n*)	1–10	1–3	1–10
Youngest birth (days)	64	280	39
Lifespan (years)	Up to 1–1.5	4–5 (average); 30 (max.)	Up to 2 (average)

Data from Nowak (1991) and Hayssen *et al.* (1993).
* 240 days includes delayed implantation.

As well as very small mammals, species that are smaller than their relatives can be viewed as miniature (see Lister, this volume, pp. 277–292). The smallest few species in each high-level taxon (e.g. family) are likely to have undergone evolutionary size reduction although, in practice, it is difficult to identify phyletic dwarfs more rigorously, because of the problems associated with recognizing ancestors and comparing fossils and living species. Do such relative miniatures have unusual life histories? If they are the 'wrong size' for their adaptations (Stanley 1973), this could show up in deviations from general life-history trends. Another category of miniatures is species, such as the pygmy hippopotamus, that are markedly smaller than their sister species; such pairwise comparisons are limited in number, however, and we have not considered them here.

We take a comparative approach and focus exclusively on placental mammals. The next section reviews the patterns of life-history variation in (mostly large) mammals, leading to a discussion of Charnov's (1991) and Brown *et al.*'s (1993) ambitious and wide-ranging models of life-history evolution. We then ask whether the patterns of life-history variation among small mammals are consistent with those in larger mammals and whether they fit the predictions of the models. We also test whether relatively small-bodied taxa deviate from allometric trends in a consistent way.

Allometry of life-history variables

Practically all aspects of mammal biology show variation that correlates with body size and a great many of these relationships fit the basic allometric power function:

$$y = a\,x^b \tag{1a}$$

or, in its logarithmic form

$$\log y = \log a + b \log x \tag{1b}$$

A whole host of dates, weights and rates scale allometrically with adult body weight (e.g. Peters 1983; Calder 1984; Reiss 1989). Life-history variables are no exception: Table 2 shows the allometries of some selected life-history variables (from Purvis & Harvey 1995; see also, for example, Millar & Zammuto 1983). Small species generally mature earlier and smaller, give birth to larger litters of smaller neonates, and die younger. There is a slow–fast continuum (Harvey, Read, & Promislow 1989; Read & Harvey 1989), with small mammals living their lives faster than large mammals do. These differences in speed of life are enormous, as can be seen by comparing the life tables for female bank voles (*Clethrionomys glareolus*; adult weight around 17 g) and African elephants (*Loxodonta africana*; adult weight around 2.75 t). A vole population could pass through 17 generations by the time a new-born elephant is weaned and over 50 by the time that it first gives birth. If each vole lived for 18 months, a single pregnant female could give rise to nearly 10^{27} descendants in this time, with a total weight equal to approximately 10 000 billion billion elephants (life-history parameters from Laws 1966; Pucek, Ryszowski, & Zejda 1969; Bobek 1973; Purvis & Harvey 1995).

Table 2. The allometric coefficients of some key life-history variables, assessed from life-table data for natural populations.

Trait	Allometric coefficient ± SE	r
Age at maturity	$+0.236 \pm 0.04$	$+0.75$
Annual fecundity	-0.241 ± 0.06	-0.63
Adult mortality rate	-0.241 ± 0.05	-0.74
Juvenile mortality rate	-0.321 ± 0.05	-0.83

From Purvis & Harvey (1995).

Beyond allometry

Allometric regressions of life-history variables often have impressively high r^2 values, with body weight seeming to 'explain' at least 90% of the variance. The apparent tightness of these relationships led to speculation that body size alone may be the target of selection, with other characteristics being pulled along by allometric 'constraints', presumably mediated by genes with pleiotropic effects (see e.g. Lewontin 1979; Linstedt & Calder 1981; Western & Ssemakula 1982). However, although body size must limit the available life-history strategy set in some ways, estimating allometric slopes does not explain why a particular strategy is adopted by species of a particular size: it may suggest some hypotheses rather than others, but cannot then provide a test (Harvey, Read, & Promislow 1989). Furthermore, the high r^2 in such allometric regressions can be an artefact of the wide range of body sizes included in the data: r^2 increases as the range of x increases. In fact, a significant amount of life-history variation is independent of body size. For instance, adult greater horseshoe bats (*Rhinolophus ferrumequinum*) weigh only a little more than the bank vole (22 versus 17 g), yet live their lives much more slowly: a bat born at the same time as the elephant would have fewer than 100 descendants by the time the elephant gave birth (life-history parameters from Corbet & Harris 1991; Purvis & Harvey 1995).

Intriguingly, the size-independent variation in one parameter was often found to be correlated with size-independent variation in others: the fast–slow continuum still applies when body-size effects are controlled for (Read & Harvey 1989). Clades that reproduce early for their body size have relatively large litters of relatively small neonates, and die relatively young (Table 3). Residual analyses highlighted trade-offs between components of fitness—e.g. longevity and fecundity, or litter number and neonatal weight—that could not be ascribed to influences of or selection for body weight. Other variables—metabolic rate, brain size, and mortality schedules—have been suggested as the drummer to which life-history patterns march; the comparative evidence has firmly rejected the first two while giving support to the third (Sutherland, Grafen, & Harvey 1986; Harvey & Read 1988; Read & Harvey 1989; Promislow & Harvey 1990). Such analyses view allometric lines as lines of functional equivalence rather than of constraint (Gould 1975).

Table 3. Significant interrelationships among some key life-history variables when body size is held constant.

Trait pair	Sign of correlation
Age at maturity and adult mortality rate	–
Age at maturity and juvenile mortality rate	–
Adult mortality rate and juvenile mortality rate	+
Age at maturity and annual fecundity	–
Annual fecundity and adult mortality	+
Annual fecundity and juvenile mortality	+
Gestation period and duration of lactation	+
Gestation period and age at maturity	+
Duration of lactation and age at maturity	+
Neonatal weight and litter size	–
Neonatal weight and gestation period	+

After Read & Harvey (1989) and Purvis & Harvey (1995).

Charnov's model of life-history evolution

Charnov (1991, 1993) has developed the first model that integrates optimality theory with comparative life-history data. It is more ambitious and comprehensive than any other model so far proposed (see Roff (1992) and Stearns (1992) for recent reviews of other models). He takes as his starting point the known allometric relationships of certain key life-history traits (age at maturity, annual fecundity, juvenile and adult mortality rates) and the interrelationships among them when body size is held constant. His model (Fig. 2) is as simple as can be, in that it allows natural selection to shape only one trade-off. He assumes that mammals grow according to a simple growth law until they mature, when they divert the energy that could fuel further growth into reproduction instead. Adult weight (*W*) and, hence, age at maturity, determines reproductive effort (proportional to $W^{0.75}$). Age at maturity is set by a trade-off between reproductive effort (late-maturing animals are larger so can invest more) and the risk of dying before reproducing (early-maturing animals are more likely to survive to reproduce). Charnov assumes that adult mortality rates are set by the environment and are not influenced by allocation decisions; natural selection shapes the age at maturity to maximize lifetime reproductive success. Over the long run, natural populations are numerically stable, so death rates must balance birth rates. Numerical stability is achieved in the model through density-dependent juvenile mortality.

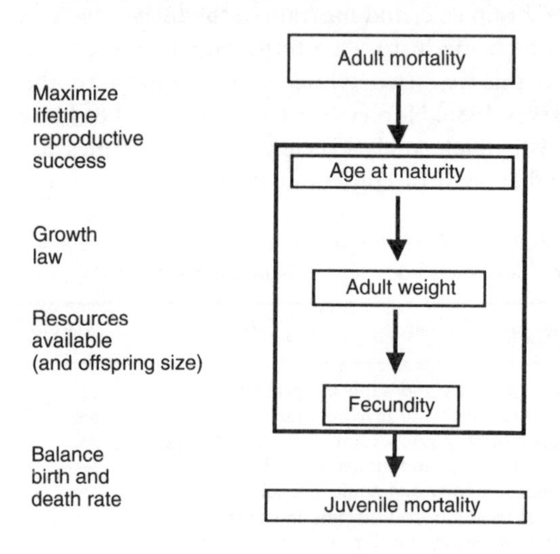

Fig. 2. A summary of Charnov's model of mammal life-history evolution. From Harvey & Nee (1991).

This model correctly recovers the known scalings of age at maturity, annual fecundity, and adult and juvenile mortality rates. The correlations between residuals are also expected if the growth law relating size to age is made slightly more flexible. If different species have different values of *A* in the growth equation $dW/dt = A\ W^{0.75}$, then species with higher values of *A* are predicted to reproduce earlier and die quicker than other species of the same adult weight (Charnov 1991).

Charnov's model also makes a number of novel predictions that can be tested by comparative evidence. He predicts that certain invariants—dimensionless products, ratios, or functions of the characters discussed above—will *not* scale with adult body size, as follows.

1. Product of age at maturity and adult mortality rate. Since the adult mortality rate, as Charnov defines it, is the reciprocal of average life expectancy at maturity and age at maturity is measured from independence from the mother, the interpretation is that all mammals spend the same fraction of

their independent lives as immatures. Comparative evidence bears this prediction out: the average value of this invariant across 64 species for which life tables are known is 0.73 (Purvis & Harvey 1995), suggesting that, at least in large mammals, animals that survive immaturity reach maturity approximately 42% of the way through their lifespan.

2. Survivorship to maturity. Charnov predicts that the fraction of neonates that survives to maturity is independent of body weight, because the scalings of age at maturity and juvenile mortality rate cancel out (the product of these latter two characteristics is also an invariant). Again, the prediction received empirical support. On average, the fraction surviving is approximately 0.37, though there are differences among major clades: more ungulates survive than carnivores or rodents (Purvis & Harvey 1995).
3. Product of age at maturity and annual fecundity. The reciprocal of annual fecundity is the mean spacing of births (female births only are considered here), so this invariant can be viewed as the ratio of the time spent reaching maturity to the time between births. Indeed, this product is invariant with body size. The mean value is 1.05, though there are again differences among orders (Purvis & Harvey 1995).

Other predictions of Charnov's model are also borne out by comparative data, mainly from larger species with known life tables (Berrigan *et al.* 1993; Charnov 1993; Purvis & Harvey 1995). Some predictions, concerning the ratio of weaning weight to adult weight (which is assumed to be another invariant) receive no support, however (Purvis & Harvey 1995); indeed, the data indicate that this ratio is not invariant, but decreases with increasing body weight. The paucity of mortality data for the smallest species limits our ability to test whether the first two invariants are indeed invariant at small body sizes: all we can do is see whether their values are unusual in the smallest species. The invariance of the third is testable, however, as is the assumption that weaning weight is a constant fraction of adult weight. If Charnov's model is right, body size evolves in response to the extrinsic mortality rate around maturity, and growth rate differences are responsible for the size-independent variation in life history. We return to comparative evidence from small mammals after considering another life-history model, which aims to explain the distribution of body sizes among species (Fig. 1).

Brown, Marquet, and Taper's model

Whereas Charnov's model optimizes lifetime reproductive success, Brown *et al.* (1993) assume that selection operates to maximize reproductive power—the rate at which materials from the environment are converted into useful work for reproduction (as opposed to maintenance or growth). Whereas Charnov explicitly excluded costs of reproduction from his model, Brown *et al.* exclude from theirs the trade-off between offspring size and offspring number.

Brown *et al.* (1993) model the conversion of environmental resources into offspring as a two-step process, the intermediate product being energy stores that are surplus to requirements for growth and maintenance:

$$\text{Environmental resource} \rightarrow \text{Surplus} \rightarrow \text{Reproductive work}$$

We denote the rate of the first step as A (for acquisition) and the rate of the second step as C (for conversion). The reproductive power of an individual is the product of C and the amount of surplus it has acquired (which is dependent upon A). At steady state, the reproductive power is given by

$$\text{Power} = \frac{A\ C}{A + C} \qquad (2)$$

Note that power is limited by the rate of resource acquisition as C becomes large and by the rate of conversion when A is large. The twist Brown *et al.* now add is to express A and C as allometric functions of body size. They assume

$$A = k_1\ W^{0.75} \qquad (3a)$$

because 0.75 is the exponent for metabolic rate, growth rate, and productivity (e.g. Schmidt-Nielsen 1984). Because mass-specific metabolism scales with exponent -0.25 (e.g., Schmidt-Nielsen 1984), they use

$$C = k_2\ W^{-0.25} \qquad (3b)$$

If k_1 and k_2 are constant, C will be largest when W is smallest: small mammals are limited by the rate at which they can acquire resources from their environment. On the other hand, A gets larger with increasing body size: big mammals are limited by the rate at which they can convert their energy stores into offspring. The trade-off between A and C implies the existence of an optimal size for mammals. Brown *et al.* (1993) derive an expression for this optimum in terms of k_1 and k_2 and, using estimates of these constants from the intercepts of cross-species allometries, estimate this optimal size to be approximately 100 g. They give three lines of evidence that support the notion of an optimal size at around 100 g. Firstly, the modal size class among North American terrestrial mammals contains 100 g. Secondly, small species on islands tend to become larger, whereas larger species become smaller: the threshold is somewhere between 100 g and 1 kg (Lomolino 1985). Finally, very small land masses only support species of around this size: bigger or smaller species can maintain populations only on larger land masses.

Brown *et al.*'s (1993) model leads immediately to testable comparative predictions. If it is correct, then scaling relationships to do with reproductive power should change sign and slope at the optimum body size. Above this size, reproductive power should scale as $W^{-0.25}$; we have already seen that annual fecundity shows just this pattern for the predominantly large species for which life tables are available. Below the threshold size, species are hypothesized to be limited by acquisition rates, so reproductive power should scale as $W^{0.75}$. The most miniature species should therefore mature later, have smaller litters spaced more widely and live longer than species nearer to the optimum size (Brown 1995). If Brown's model is correct, then miniature mammals could also be defined as those whose reproductive power is limited by their rates of acquisition of surplus energy.

Comparative methodology

We have used three sets of comparative data. The main data set comprises over 400 species smaller than 100 g (the weight at which Brown *et al.* (1993) predict scalings should change). Body weight data came predominantly from Letcher (1994) and J. Prothero (unpubl.). Life-history data were mainly from Hayssen *et al.* (1993), supplemented from Millar (1977, 1981); median values were taken for all variables except age at first reproduction (for which minima were used) and gestation length (we tried to exclude periods of delayed implantation). This data set was used to test Brown *et al.*'s (1993) predictions for small species, and also to ask more generally which scaling relationships and residual correlations that are found in the class as a whole can be demonstrated when only small species are considered—an important exercise, given that such relationships are currently tacitly assumed. Secondly, we used Purvis & Harvey's (1995) compilation of life tables and related data, covering 64 species from nine placental orders. Because only 10 of the species weigh under 60 g, these data can provide only very weak tests. However, they provide the only direct test of predictions involving mortality rates. Thirdly, we used Read and Promislow's (Read & Harvey 1989; Promislow & Harvey 1990) data set, covering over 700 eutherian species of all sizes, to ask whether species that are smaller than their relatives show unusual patterns.

Closely related species are often similar because of shared inheritance from a common ancestor, rather than because of convergent or parallel evolution (Ridley 1983; Harvey & Pagel 1991; Miles & Dunham 1993). Repeated convergence towards a particular association of traits, such as late maturity and low fecundity, suggests that the features may be fundamentally linked, but a single origin of a pattern (however many species inherit it) provides no basis for statistical hypothesis testing. Our tests are, therefore, wherever possible based on comparisons within each of many taxa. The main method we use is Pagel's (1992) modification of Felsenstein's (1985) procedure of independent comparisons (as implemented by Purvis & Rambaut 1995): each node in a presumed phylogeny contributes only one piece of information to the test, no matter how many lineages are descended from it. We have used Corbet & Hill's (1991) taxonomy of the mammals in lieu of a phylogeny, on the assumption that the groups listed there are monophyletic. Branches in the phylogeny were set to equal lengths and the contrasts calculated at nodes were not standardized: the justification for these choices is empirical (based on model criticism during regression analysis—Garland, Harvey & Ives 1992), not theoretical. Hypotheses were generally tested by least-squares regression through the origin (Grafen 1989; Garland *et al.* 1992) or by sign tests. Given the limitations of data and phylogeny, our analysis should be viewed as exploratory, and our conclusions preliminary.

Comparative results

As Table 4 shows, most life-history variables do not scale strongly with adult body size within our set of small mammals. Neonatal weight and weaning weight are exceptions, both scaling strongly with exponents around 0.75. The exponent for weaning weight is significantly less than unity (although least-squares regression tends to underestimate true slopes when, as here, x contains error: Harvey & Pagel 1991), implying that the ratio of weaning weight to adult weight declines with adult size. This result is in line with some previous studies (Millar 1977; Purvis & Harvey 1995), but not with Charnov's (1991) model—he assumes that weaning weight is a constant fraction of adult weight. The lack of any correlation of the other variables with body size neither supports nor refutes Brown *et al.*'s (1993) prediction that their scaling with body size should be reversed in small mammals. When the smallest members of the life-table data set are considered, the relationships between life history and body size tend, if anything, to be in the same direction as in larger mammals (again contrary to Brown *et al.*'s suggestion), though the sample size is far too small to permit rigorous testing.

Table 4. Scaling of life-history variables with body size in the small mammal data set.

Variable	Species	Contrasts	Slope (SE)	P_1	+,−	P_2
Neonatal mass	185	65	0.72 (0.044)	<0.0001	55, 6	<0.0001
Litter size	397	109	0.00 (0.02)	0.84	40, 43	0.83
Weaning age	161	58	0.02 (0.067)	0.77	24, 28	0.68
Weaning mass	95	35	0.78 (0.054)	<0.0001	31, 3	<0.0001
Age at maturity	156	57	0.00 (0.133)	0.97	23, 32	0.28
Gestation	202	73	0.01 (0.036)	0.76	32, 35	0.81
Annual fecundity	170	61	0.00 (0.066)	0.95	25, 28	0.78

Slope is from least-squares regression through the origin; P_1 = corresponding P-value; '+,−' indicates the number of taxa within which the relationship was positive and negative, respectively (with zeros excluded); P_2 is the P-value from a sign test on these numbers.

Many of the correlations found among residuals for large mammals also occur in this data set (Table 5). The trade-off between offspring number and offspring size (at both birth and weaning) is very strong, and species with large litters wean them early. The durations of pre- and post-natal care are correlated positively. Late-maturing species are less fecund—clearly the fast–slow continuum applies even within small mammals. Non-significant associations among other variables are generally in the same direction as found in larger mammals (Read & Harvey 1989), perhaps suggesting that those patterns too would achieve significance if more or better data were available.

Table 5. Interrelationships of life-history variables after body size is factored out.

	Litter size	Weaning age	Weaning mass	Age at maturity	Gestation length	Annual fecundity
Neonatal mass	−0.68** (65)	0.01 (50)	0.59* (35)	0.00 (45)	0.11 (58)	−0.22 (44)
Annual fecundity	(Auto-correlated)	−0.35* (41)	−0.28 (24)	−0.47* (42)	−0.16 (49)	
Gestation length	−0.20 (70)	0.75*** ($53^{\dagger 2}$)	0.26 (33)	0.14 (52)		
Age at maturity	−0.09 (55)	0.17 (42)	−0.11 (27)			
Weaning mass	−0.47** (35)	0.45* ($32^{\dagger 1}$)				
Weaning age	−0.17* (58)					

The first number is the partial correlation coefficient (* $P<0.05$; ** $P<0.01$; *** $P<0.001$); the number of independent contrasts is given in parentheses. † Model criticism led to the exclusion of the succeeding number of extreme outliers (standardized residuals >4). Note that $P<0.10$ for litter size versus gestation length.

As Charnov (1991, 1993) predicts, the product of age at maturity and annual production of female offspring does not scale with body size (least-squares slope = 0.103; $t_{41} = 0.74$, $P = 0.45$). The distribution of this product in the small mammal data set was positively skewed (mean = 1.05, SE = 0.106; median = 0.73; $n = 73$ species). Interestingly, this invariant differs markedly among orders (one-way ANOVA on species values: $F_{4,68} = 5.3$, $P = 0.0009$). Bats have very low values (mean = 0.55; SE = 0.064; $n = 23$), less than half those of rodents (mean = 1.20; SE = 0.151; $n = 42$) and insectivores (mean = 1.58; SE = 0.35; $n = 6$). Bats are known to be at the slow end of the slow–fast continuum once body size has been factored out, while insectivores are at the other extreme (Read & Harvey 1989). The interrelatedness of terms in Charnov's model implies that, since bats have low values of this invariant, they should also mature early in relation to their lifespan, and should have higher than usual survivorship to maturity. Adult lifespans in bats are certainly much longer than for most other mammals of the same size, but we do not know of comparable data about their juvenile survivorship.

The life-table data set indicates that small species (those weighing less than 100 g) do not have unusual values of Charnov's invariants (age at maturity times adult mortality rate: $t_{49} = -0.05$, $P = 0.96$; survivorship to maturity: $t_{48} = 0.21$, $P = 0.84$; age at maturity times annual fecundity: $t_{54} = 1.14$, $P = 0.26$), with the exception of the ratio of weaning weight to adult weight (small species wean relatively heavy young: $t_{43} = 2.83$, $P = 0.007$). Again, this ratio appears to be the weak link in Charnov's model (see also Purvis & Harvey 1995).

What about relative miniatures? We used Read and Promislow's data set (Read & Harvey 1989; Promislow & Harvey 1990) to assess whether the smallest members of families fitted general trends of life-history variation. We classed the lightest quartile of each family as miniature. Then we regressed, across all species, life-history characteristics on adult body size and noted, for each family, whether miniature species had more or less positive residuals than the remainder. Families were then used as units of comparison in sign tests to assess significance. The results are given in Table 6. Relative miniatures tend to have smaller litters of large neonates for their body size, after long gestations. A possible artefactual explanation for this pattern is that the allometric relationships are significantly non-linear across mammal species: with neonatal weight, for instance, there appears to be a kink at approximately 1–10 kg body weight, with a steeper slope at heavier weights than at lighter ones. The line fitted across the whole data is therefore too steep for the species below the kink, so small species have more positive residuals. Above the kink, the pattern would be reversed, but there are fewer species and families there. We tested the adequacy of this explanation by splitting the data into many body-size categories (each 0.1 log units wide)—so avoiding problems associated with choosing a suitable regression model—and using a general linear model to test whether, within size classes, relative miniatures had significantly more larger neonates than did non-miniatures: they did ($t_{390} = 2.78$; $P = 0.02$), but gestation length did not differ significantly ($P = 0.13$). Although there are problems of non-independence with these analyses, they suggest that secondarily small species are indeed unusual, having relatively large neonates. A corollary is that if two mammals are of the same size, the one with larger relatives tends to have fewer but larger offspring in a litter.

Table 6. The position of relative miniatures compared to the allometric lines.

Variable	Families	Above relatives	Below relatives	*P*-value
Neonatal weight	35	28	7	0.0006
Annual fecundity	28	10	18	0.18
Age at maturity	27	10	17	0.25
Litter size	42	13	29	0.02
Gestation	41	31	10	0.001
Lifespan	29	18	11	0.26
Age at weaning	32	18	14	0.60

Families = number of families within which a comparison was possible; *P*-values are computed by using sign test (two-tailed). See text for explanation.

Discussion

Perhaps the most remarkable feature of life-history variation in small mammals is how much of it is independent of adult size, which correlates significantly only with neonatal weight and weaning weight. The fast–slow continuum found in larger species (Read & Harvey 1989) is strongly present, however. Insectivorous bats lead slower lives than do rodents and insectivores of similar body sizes. The comparative analyses presented in Table 5 show that the continuum can be detected even *within* these major clades.

Our analysis confirms and strengthens previous tests of Charnov's (1991, 1993) model of life-history evolution. Because so many of his predictions are borne out, there are grounds to hope that his approach and the causal factors (mortality around maturity and growth rates) he suggests are correct, but his assumption that weaning weight is a constant fraction of adult weight is wrong. Moving to Brown *et al.*'s (1993) model, we find no evidence that 'usual' allometries go into reverse below 100 g as predicted, but nor is there much evidence that they do not: the lack of the usual relationships could be a type II error or may indicate that Brown *et al.* are on the right track. A more telling test must await more mortality and fecundity data on very small mammals, especially non-bats.

The amount of size-independent variation demonstrates that species are not tightly constrained to evolve along allometric lines. Brown *et al.*'s (1993) derivation of the optimum body size assumed that the slope and intercept of allometries were constant across the whole class. Both are probably subject to change. Indeed, the finding that relative miniatures tend to have large neonates and small litters for their body size implies that slopes and/or intercepts change commonly; it also implies that there may be obstacles to further size reduction. Further analyses of relative miniatures may help to identify these constraints, if such they are. If each mammal group has its own constraints and optimum size, then the distribution of body sizes across species will have to be viewed as a set of superimposed distributions. A reappraisal of the allometries of acquisition and conversion processes in different mammal taxa is now in order, to see whether the distribution of body sizes within such groups can successfully be predicted from Brown *et al.*'s (1993) model.

Acknowledgements

This work was funded by the NERC (GR3/8515). We thank Peter Cotgreave, Ruth Levy, and Rod Page for discussion, and Jim Brown for sight of his in-press manuscript.

References

Berrigan, D., Purvis, A., Harvey, P. H., & Charnov, E. L. (1993). Phylogenetic contrasts and the evolution of mammalian life histories. *Evol. Ecol.* 7: 270–278.

Bobek, B. (1973). Net production of small rodents in a deciduous forest. *Acta theriol.* **18**: 403–434.

Bronson, F. H. (1989). *Mammalian reproductive biology.* University of Chicago Press, Chicago.

Brown, J. H. (1995). *Macroecology.* University of Chicago Press, Chicago.

Brown, J. H., Marquet, P. A., & Taper, M. L. (1993). Evolution of body size: consequences of an energetic definition of fitness. *Am. Nat.* **142**: 573–584.

Calder, W. A. (1984). *Size, function and life history.* Harvard University Press, Cambridge, Ma.

Charnov, E. L. (1991). Evolution of life history variation among female mammals. *Proc. natn. Acad. Sci. USA* **88**: 1134–1137.

Charnov, E. L. (1993). *Life history invariants: some explorations of symmetry in evolutionary ecology.* Oxford University Press, Oxford.

Corbet, G. B. & Harris, S. (Eds) (1991). *The handbook of British mammals.* (3rd edn). Blackwells, Oxford.

Corbet, G. B. & Hill, J. E. (1991). *A world list of mammalian species.* (3rd edn). British Museum (Natural History), London & Oxford University Press, Oxford.

Eisenberg, J. F. (1981). *The mammalian radiations: an analysis of trends in evolution, adaptation and behaviour.* University of Chicago Press, Chicago.

Felsenstein, J. (1985). Phylogenies and the comparative method. *Am. Nat.* **125**: 1–15.

Garland, T., Harvey, P. H., & Ives, A. R. (1992). Procedures for the analysis of comparative data using independent contrasts. *Syst. Biol.* **41**: 18–32.

Gould, S. J. (1975). Allometry in primates, with emphasis on scaling and the evolution of the brain. In *Approaches to primate paleobiology*: 244–292. (Ed. Szalay, F.). Karger, Basel.

Grafen, A. (1989). The phylogenetic regression. *Phil. Trans. R. Soc. (B)* **326**: 119–157.

Harvey, P. H. & Nee, S. (1991). How to live like a mammal. *Nature, Lond.* **350**: 23–24.

Harvey, P. H. & Pagel, M. D. (1991). *The comparative method in evolutionary biology.* Oxford University Press, Oxford.

Harvey, P. H. & Read, A. F. (1988). How and why do mammalian life histories vary? In *Evolution of life histories of mammals: theory and pattern*: 213–232. (Ed. Boyce, M.S.). Yale University Press, New Haven.

Harvey, P. H., Read, A. F., & Promislow, D. E. L. (1989). Life history variation in placental mammals: unifying the data with theory. *Oxf. Surv. evol. Biol.* **6**: 13–31.

Hayssen, V., van Tienhoven, A., & van Tienhoven, A. (1993). *Asdell's patterns of mammalian reproduction*. Cornell University Press, Ithaca, New York.

Laws, R. M. (1966). Age criteria for the African elephant, *Loxodonta a. africana*. *E. Afr. Wildl. J.* **4**: 1–37.

Letcher, A. J. (1994). *Studies in the biogeography of birds and mammals*. D.Phil. thesis: University of Oxford, UK.

Lewontin, R. C. (1979). Sociobiology as an adaptationist program. *Behav. Sci.* **24**: 5–14.

Linstedt, S. L. & Calder, W. A. (1981). Body size, physiological time, and longevity of homeothermic animals. *Q. Rev. Biol.* **56**: 1–16.

Lomolino, M. V. (1985). Body sizes of mammals on islands: the island rule re-examined. *Am. Nat.* **125**: 310–316.

May, R. M. (1986). The search for patterns in the balance of nature: advances and retreats. *Ecology* **67**: 1115–1126.

Miles, D. B. & Dunham, A. E. (1993). Historical perspectives in ecology and evolutionary biology: the use of phylogenetic comparative analyses. *A. Rev. Ecol. Syst.* **24**: 587–619.

Millar, J. S. (1977). Adaptive features of mammalian reproduction. *Evolution* **31**: 370–386.

Millar, J. S. (1981). Pre-partum reproductive characteristics of eutherian mammals. *Evolution* **35**: 1149–1163.

Millar, J. S. & Zammuto, R. M. (1983). Life histories of mammals: an analysis of life tables. *Ecology* **64**: 631–635.

Nowak, R. M. (1991). *Walker's mammals of the world*. (5th edn). Johns Hopkins Press, Baltimore.

Pagel, M. D. (1992). A method for analysis of comparative data. *J. theor. Biol.* **156**: 431–442.

Peters, R. H. (1983). *The ecological implications of body size*. Cambridge University Press, Cambridge.

Promislow, D. E. L. & Harvey, P. H. (1990). Living fast and dying young: a comparative investigation of life-history variation among mammals. *J. Zool., Lond.* **220**: 417–437.

Pucek, Z., Ryskowski, L., & Zejda, J. (1969). Estimation of the average length of life in bank voles, *Clethrionomys glareolus* (Schreber, 1780). In *Energy flow through small mammal populations*: 187–201. (Eds Petrusewicz, K. & Ryskowski, L.). PWN, Warsaw.

Purvis, A. & Harvey, P. H. (1995). Mammal life history evolution: a comparative test of Charnov's model. *J. Zool., Lond.* **237**: 259–283.

Purvis, A. & Rambaut, A. E. (1995). Comparative analysis by independent contrasts (CAIC): an Apple Macintosh application for analysing comparative data. *Comput. Appl. Biosci.* **11**: 247–251.

Read, A. F. & Harvey, P. H. (1989). Life history differences among the eutherian radiations. *J. Zool., Lond.* **219**: 329–353.

Reiss, M. J. (1989). *The allometry of growth and reproduction*. Cambridge University Press, Cambridge.

Ridley, M. (1983). *The explanation of organic diversity: the comparative method and adaptations for mating*. Oxford University Press, Oxford.

Roff, D. A. (1992). *The evolution of life histories: theory and analysis*. Chapman & Hall, New York.

Schmidt-Nielsen, K. (1984). *Scaling: why is animal size so important?* Cambridge University Press, Cambridge.

Stanley, S. M. (1973). An explanation for Cope's Rule. *Evolution* 27: 1–26.

Stearns, S. C. (1992). *The evolution of life histories*. Oxford University Press, Oxford.

Sutherland, W. J., Grafen, A. & Harvey, P. H. (1986). Life history correlations and demography. *Nature, Lond.* 320: 88.

Western, D. & Ssemakula, J. (1982). Life history patterns in birds and mammals and their evolutionary interpretation. *Oecologia* 54: 281–290.

Symp. zool. Soc. Lond. (1996) No. 69: 175–199

The functional ecology of small fish: some opportunities and consequences

P. J. MILLER

School of Biological Sciences
The University
Bristol BS8 1UG, UK

Synopsis

The ecology of small teleosts (< 100 mm) is discussed in terms of the essential whole-organism functions (feeding, surviving, growing, and reproducing). Small size makes available new habitat opportunities for teleosts, ranging from nektonic to cryptobenthic and differing along several ecotopic dimensions. The advantages and disadvantages of small size for teleosts are discussed with reference to whole-organism functions, involving both intrinsic size-related factors and directions of selection within habitat. Appropriate responses in life-history patterns relating to survival and reproduction emphasize the significance of contrasting mortality patterns over potential lifespan in producing extremes of reproductive allocation. Compensation for size-related constraints on offspring size (bestowal) and fecundity is discussed with reference to repeat spawning and the occurrence of viviparity, especially the role of superfoetation in combining frequency of reproduction with high bestowal. Functional interactions of small fish with mankind can be adverse or beneficial, with human interest in aquaria providing a new niche for at least some small fishes.

Introduction

The concept of functional ecology has been reviewed by Calow (1987). The present author sees the ecology of small fish as the study of interaction with ecosystem, interactions taking place through function, which is in turn defined as the transduction of energy into work or biosynthesis by the fish system. A hierarchy of function culminates in the integrated functional complexes of the whole animal. Whole-organism functions serve four major requirements for fitness in a particular ecotope. These are feeding, survival, growth, and reproduction, as, for example, quantified by Wootton (1994) in that most famous of small fish, *Gasterosteus aculeatus*. The temporal pattern of energy and material partitioning between these functions over lifespan—in essence life history—and its adaptiveness for ecotope is the subject of functional ecology. Size reduction entails progressive change in function or ecology (Hanken & Wake 1993) and here

ZOOLOGICAL SYMPOSIUM No. 69
ISBN 0–19–857787–7

the functional ecology of small fish is examined in terms of habitat scope and the various advantages and disadvantages of small size relative to whole-organism functions.

This paper deals with teleosts whose asymptotic or observed maximum length is not more than approximately 10 cm, about 15 g in a fish of typically fusiform shape (Miller 1979b). This category corresponds with the lowest subdivision of size distribution in fishes reviewed by Lindsey (1966) and covers at least 2000 species, around 10% of the 23 500 or more teleosts enumerated by Nelson (1994). In teleost phylogeny, a small limiting size must be regarded as a derived condition, with functional constraints on size diminution noted by Miller (1979b) and Harrison (this volume). In this context, the paedomorphic features seen in many small fish (Weitzmann & Vari 1988) can be viewed as dedicated specializations, rather than simply the consequences of gonadal heterochrony (Buckup 1993). Size variation within some teleost clades may be phylogenetically constrained (Warburton 1989), but reduction has occurred in many separate lines, especially in the Otophysi, Atheriniformes, and percomorphs (Nelson 1994), as a major contribution to the diversity of modern teleosts. Towards the lower limit of size, from South America Weitzmann & Vari (1988) list 85 nominal species of teleosts not exceeding 26 mm in standard length and Kottelat & Vidthayanon (1993) 47 freshwater species from southern and south-east Asia in the same category. The record smallest fish at maturity, and also the smallest vertebrates extant, are, in length, the goby *Trimmatom nanus*, with females ripening at 8.0 mm (Winterbottom & Emery 1981) and, in weight, *Schindleria praematura*, at not more than 8 mg (Bruun 1940), now shown also to be a gobioid (Johnson & Brothers 1993). Figure 1 indicates how minute such species appear against a familiar small fish (and a thumbnail).

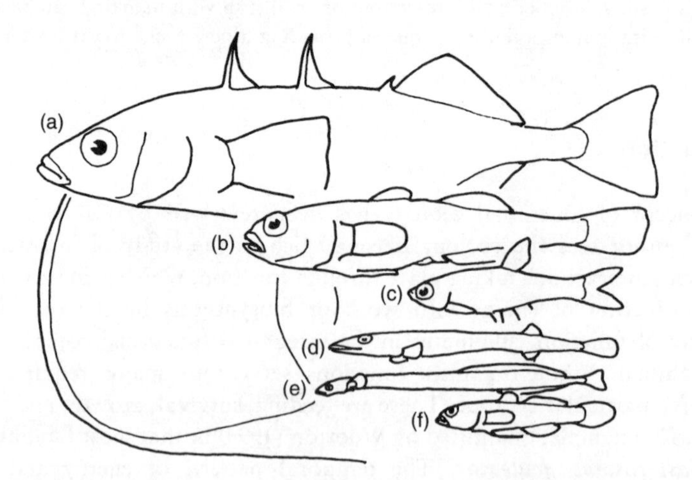

Fig. 1. Minute teleosts at maximum length in comparison with (a) *Gasterosteus aculeatus*, 40 mm standard length (average size at maturity); (b) *Economidichthys trichonis* (Gobiidae: smallest European freshwater species); (c) *Boraras micros* (Cyprinidae: freshwater, Thailand); (d) *Opeatogenys gracilis* (Gobiesocidae: smallest European marine species); (e) *Schindleria praematura* (Gobiidae); (f) *Trimmatom nanus* (Gobiidae: smallest vertebrate). In background: author's thumbnail. Sources noted in text.

Habitats for small teleosts

Diminution in size is accompanied by a resolution of habitat space and structural opportunities not accessible at a larger scale (Levin 1992). The only general environmental constraint for the very smallest freshwater fishes is thought to be flow rate, with perhaps restriction to lentic or only slowly flowing waters (Weitzmann & Vari 1988; Kottelat & Vidthayanon 1993). The expected relationship between smaller size and spatially restricted habitats has been demonstrated (Warburton 1989) and exemplified by surviving populations of desert pupfish species in minuscule water bodies (Pister 1990).

With habitat defined as the feeding area explored by trivial movements rather than migration (Southwood 1981), three major, if somewhat overlapping, categories can be recognized for small fish: nektonic, epibenthic, and cryptobenthic. These are described by Miller (1979b). According to ecological theory (White & Pickett 1985), the nektonic habitat may be regarded as physically continuous and coarse grained, where any patchiness is mobile (such as schools or feeding aggregations); population numbers are potentially high (Damuth 1987) but still often short of capacity and the risk to survival from such an obvious factor as predation is potentially great. In contrast, cryptobenthic living takes place in circumscribed and typically fixed patches, limited in food supply and supporting small populations but with these at capacity and at low risk from nektonic or epibenthic predators. Between the two, epibenthic habitats may be seen to span the interface between water column and substrate, and are subject to a variable mix of selective forces depending on the balance between extreme exposure and crypsis in the lifestyle of a particular species.

If reduced size opens new possibilities for niche acquisition, it is not surprising that this has been associated with resource partitioning in the highly diverse and notionally stable ecosystems of the tropical and warm temperate areas of the world, where the majority of small teleost species, both marine and freshwater, occur (Miller 1979b: fig. 1). Classic examples are the fish communities of hermatypic coral reefs and analogous cichlid species flocks of East African lakes, although many coral species are probably more flexible in habitat requirements than formerly supposed (Sale 1980, 1988).

For less diverse ecosystems, numerous accounts deal with ostensible partitioning between small fish species, such as the microhabitats of resident demersal small fish along a Louisiana marsh edge (Baltz, Rakocinski, & Fleeger 1993). Partitioning is most easily assessed in terms of diet but fine-scale spatial separation is also possible for diminutive fishes. The stream-dwelling rainbow darter, *Etheostoma caeruleum*, forages over flat surfaces while the sympatric fantail darter, *Etheostoma flabellare*, occurs in crevices between rocks (Schlosser & Toth 1984). Similarly, but in the sea, species of the marine blennioid *Acanthemblemaria* segregate on either vertical or horizontal surfaces (Greenfield & Johnson 1990).

Higher numbers and metabolic scaling result in greater use of resources by small than by larger species in communities (Griffiths 1992). In the ecosystem, obvious trophic cascade effects of small fish on nektonic prey have been noted for gambusias and pupfish (Gerking 1994). The very abundant epibenthic common goby, *Pomatoschistus microps*, contributes to the top-down influence of other small predators on meiofauna (Berge & Hesthagen 1981) but to a debatable extent.

Functional benefits of small size

Feeding

Downward scaling in body size permits cost-effective exploitation of potentially vast resources of small food organisms. For a predator, body size and feeding behaviour should relate to the size range of food organisms so as to maximize net energy gain and to minimize the likelihood of mortality while feeding (Townsend & Winfield 1985). Size reduction is thus an appropriate adaptation if such resources are exploited by individual picking. This is energetically less demanding than sieving (Gibson & Ezzi 1992) although selective sieving of meiofauna is feasible for some small fish such as the darter goby, *Gobionellus boleosoma* (Carle & Hastings 1982).

In nektonic habitats, plankton is a major dietary component for small species; in rainforest waters, drift terrestrial arthropods may be of comparable importance. For small fish such as the stickleback, zooplankton is a better source of energy than benthic food (Wootton 1994), despite the higher exposure for a predated species in its acquisition. For epibenthic fish such as sand gobies, a rich meiofaunal resource is available (Miller 1979b; Miller & Fouda 1986). A wide range of benthic feeding by small fish is found in coral communities and in cichlid species flocks. Apart from small prey, algal resources of benthic habitats, much utilized by mbuna cichlids (Reinthal 1990), are also exploited by a variety of other small teleosts, from hill-stream homalopterid loaches and sicydiine gobies to intertidal blennies (Heymer 1982). More specialized opportunities for small epibenthic fish (reviewed by Gerking 1994) include mucus picking, scale feeding (lepidophagy), fin biting, and the widely studied 'cleaning' of other fish, in turn promoting mimicry of harmless or cleaner species by fin biters and scale feeders. Some South American trichomycterid catfish even enter the opercular chambers of large catfish and characins to browse on gill filaments and blood (Machado & Sazima 1983). Other cryptobenthic species are in effect vertebrates ectoparasitic on invertebrates. Gobiodontine gobies include in their diet coral polyps of the colonies within which they live and urchin-dwelling clingfish and gobies eat tube feet (Miller 1979b).

As well as relating to a dedicated niche, small size might also be advantageous in situations of low productivity, with selection for flexibility in growth pattern. The small size of many mesopelagic and bathypelagic fish has been attributed to trophic limitation, with corresponding reduction in skeleton and other tissues (Marshall 1984).

Surviving

Individual survival is facilitated by small size in a variety of ways, perhaps most simply when a small fish is passed over by a predator optimally foraging for a larger morsel. In small fish, burst- and sprint-swimming, associated with escape, are powered chiefly by aerobic metabolism and recovery is faster than in larger fish, where there is greater emphasis on an anaerobic component (Goolish 1991).

Reduced dimensions permit refuge in places inaccessible to larger predators, from crevices to shallow water (Kneib 1987), even without further cover in the latter (Ruiz, Hines & Posey 1993). Soft substrates offer opportunities for small fish to bury themselves, construct a burrow, or use an existing one (Atkinson & Taylor 1991). Cover can protect against physical sources of mortality such as spates (Ross & Baker 1983) and, because it tends to be patchily distributed, may reduce efficiency of predation (Gilinsky 1984). In an unstable habitat, individual survival may be achieved in a dormant state. The galaxioid *Lepidogalaxias salamandroides* aestivates over the dry season encased in mucus in a small burrow (Pusey 1989).

Growing

Using Von Bertalanffy parameters, it is suggested as an invariant of life history that growth rate is inversely related to asymptotic size (Charnov 1993). A corollary of this is that small fish should reach maturity and mature size at an earlier age than their ancestral forms. Earlier maturation has a number of life-history consequences, including enhanced rate of reproduction (Stearns 1992). More immediately, in the context of small fish, an increased growth rate has been viewed as advantageous in guppy populations subject to predation aimed at smaller juveniles (Reznick 1983).

Reproducing

Within a clade, small animals tend to be more abundant than larger ones (Griffiths 1992), so that the possibility of large populations and the probability of finding a mate are retained even within smaller habitats. Not only does earlier age at maturity confer the advantages noted above, but reproductive effort also varies inversely with age at first maturity (Roff 1992). As a scaling-down benefit, reproduction involving demersal brood care can take advantage of spawning sites more inaccessible to predators, from merely spatial separation in the intertidal zone (Almada & Santos 1995), to the location of eggs within cavities such as the empty barnacle tests used by the angel blenny (Hastings 1986) or even the gill chamber of a living crab, used by the liparid *Careproctus* (Peden & Corbett 1973).

Functional costs of small size

Feeding

Although individual small fish require less energy and material in absolute terms, an allometric increase in relative metabolic rate is a corollary of reducing body size, clearly seen within small species themselves, such as *Gambusia* (Mitz & Newman 1989) and *Gasterosteus* (Wootton 1994). Part of this may be due to the higher relative growth rate accompanying smaller limiting size (Charnov 1993). Under these circumstances, requirements of food acquisition could increase work and risk costs, as well as reduce ability to withstand starvation. Other size-related constraints on feeding energetics recorded for small fish may accrue from limitation of territory (Grant & Kramer 1990) or foraging area (Dill 1990). A relatively higher metabolic rate combined with inadequate absolute capacity for energy storage probably influences the ability of small fish to survive or reproduce under seasonal deprivation, whether in winter (Conover 1992) or summer (Pusey 1990).

Surviving

Age-related structural changes and mortality in a protected environment have been recorded for a number of small fishes (Markofsky & Milstoc 1979; Woodhead 1979). An invariant of fish life histories is that, when relative growth rate increases with lower limiting size, lifespan is decreased (Charnov 1993), as Craig & Fletcher (1984) demonstrated for zebrafish, *Brachydanio rerio*. Small fish also appear to be more susceptible to environmental toxins (Patwardhan & Gaikwad 1991), a tendency that is probably linked to a relatively higher metabolic rate.

If the ratio of size at first maturity to asymptotic size is also invariant, then maturity must be reached at an earlier age when the growth rate increases in relation to a decreased limiting size. Age at maturity is inversely proportional to mortality rate (Charnov 1993) and, although promoting instantaneous reproductive value, this effect could reduce future fitness (Roff 1992). In the killifish *Nothobranchius*, earlier maturing males had a much higher death rate in captivity (Markofsky & Perlmutter 1972). Following maturity, reproductive effort is high in many small species and, for the small cyprinodont *Oryzias latipes*, Hirschfield (1980) has demonstrated a negative correlation between overall reproductive effort and survival. Another disadvantage of enhanced primary reproductive effort in small fish could be mechanical, associated with increased girth, especially in females, affecting flight or refuge (Roff 1992). The metabolic demands of enlarged ovaries tend to make gravid female sticklebacks approach the surface for respiration (Giles 1987) and thereby perhaps incur greater risk of bird predation.

In small cryptobenthic fish of temperate waters, such as the blennioids and gobies noted by Miller (1979b), age composition suggests predominantly deterministic sources of adult mortality, with no indication that, in this ecotope, small size causes departure from the classic pattern of late maturation and relatively long life described for much larger teleosts (Stearns 1992). Here, stochastic influences on mortality appear to operate predominantly on the planktonic phase, believed to relate to annual variation in productivity (Roff 1992). Appropriate fluctuation in year-class success has been noted in the small but long-lived *Lesueurigobius friesii* (Gibson & Ezzi 1978; Nash 1982).

However, in open epibenthic and nektonic conditions, without physical cover as an integral part of the habitat, many studies indicate that adult survival of little fish is markedly unpredictable, as a result of the inevitable extrinsic risk from top-down predation on smaller organisms. As well as bigger fish, piscivorous birds, reptiles, and mammals contribute to this onslaught. Scaling down also brings fish within the reach of many invertebrate predators (Miller 1979b). Long lists of predators can be compiled for common small fish, such as silversides (Middaugh 1981) and sticklebacks (Reimchen 1994). In estuarine sand-gobies, Doornbos & Twisk (1987) recorded an annual mortality rate of over 99%, much of which was due to predation by a few species of fish and birds. The significance of predation pressure for small fish is indicated by the major changes in population size which follow alterations in predator regime, as by introduction or through fisheries (Miller 1979b; Barel *et al.* 1991). Even where cover seems available, finely divided habitats with impermeable boundaries may promote the evolution of small-sized specialized predators (Tokeshi 1994). Thus, a small cardinal fish *Apogon exostigma* pursues tiny gobies (*Eviota*) within coral heads, one of many examples of predation pressure in coral fish communities (Doherty & Williams 1988). Other cover such as vegetation may be penetrated for predation as well as being used by sit-and-wait predators (Savino & Stein 1989).

Instability of extreme habitats made exploitable by small size can be another source of stochastic mortality throughout adulthood; for instance, in savannah pools (Simpson 1979), desert waters (Constantz 1981), and hill streams small fish may be at risk from drought, wave action, and flash flooding, all events with an appreciable variance in return interval (Grossman, Moyle & Whitaker 1982; White & Pickett 1985). Although floodplains offer a seasonally predictable resource for the New Guinea rainbowfish *Glossolepis multisquamatus*, trapping in pools as the water recedes can cause random mortality (Coates 1990). In such habitats, the temporal pattern of habitat stability clearly fluctuates within potential lifespan.

In stable ecosystems, for small epibenthic and cryptobenthic fishes with more stringent habitat requirements, the random availability of such niches, either species-specific or guild-specific, has been regarded as a special source of mortality at the transition between planktonic and benthic life (discussed by Sale (1980) and Doherty & Williams (1988)). Mortalities of around 25% per week have been reported for several coral fish within the first week of recruitment, falling markedly thereafter (Doherty & Williams 1988). In the analogous Malawi mbuna communities, there is no planktonic stage and young fish of approximately 11 mm are deposited among rocks by the mouthbrooding female (Trendall 1988). Over the next few weeks, mortality among these recruits may exceed 90%, including a priority effect of exclusion by established young, which is also found on coral reefs (Shulman *et al.* 1983). Trendall (1988) regarded this as an example of stochastic influences on recruitment success which was less at risk from deterministic influences of community structure. However, within a species, such as anemone fishes, significance of habitat (host) availability may vary between localities and have less influence on population numbers than mortality in the planktonic stage (Fautin 1992).

Reproducing

Given the strong positive relationship of ovarian biomass (Miller 1984a; Wootton 1984) and fecundity (Charnov 1993) with body size, small species may be expected to suffer a resultant constraint on absolute number of offspring per clutch (Roff 1992). Thus, although absolute egg size may remain comparable across a wide size range in the family Cyprinidae, a monophyletic family with similar post-larval ecotopes, individual females of one of the smallest species, *Boraras micros*, have only 30–50 eggs at 12.5–13.0 mm (Kottelat & Vidthayanon 1993) while larger species will produce tens or hundreds of thousands per clutch (Banarescu & Coad 1991). Male size may also influence reproductive success, as in the correlation with hatching success in the monogamous coral goby *Paragobiodon echinocephalus* (Kuwahara, Yogo, & Nakashima 1993).

Functional responses

Surviving

A direct relationship between asymptotic length and lifespan appears in teleost groups such as the clupeids, gadids, and flatfish (Beverton & Holt 1959), all fairly uniform in basic ecotope within each family. However, the gobioid fishes, with wide ecological radiation (Miller 1984a), can show great disparity in longevity at the same small size, suggesting that selection in response to extrinsic sources of mortality may find a genomic capability to override any intrinsic curb on lifespan. Between the eastern Atlantic epibenthic sand goby, *Pomatoschistus minutus*, and the cryptobenthic *Lesueurigobius friesii*, there may be an order of magnitude difference in longevity (1.6 and 11 years), even though maximum standard length is virtually the same (7.3 and 7.4 cm, respectively) (Miller 1979b). As a response to the link between growth rate and mortality, with faster-growing small fish more likely to die earlier, there may be short-term selective value in the increased reproductive allocation seen when growth rate is enhanced in *Gambusia* at higher temperatures (Trendall 1983; Vondracek, Wurtsbaugh, & Cech 1988).

As part of the accepted trade-off between reproduction and longevity (Roff 1992; Stearns 1992), Charnov (1993) suggested for fishes a mean of approximately 0.5 for the product of average duration of adult life over age at maturity. However, stretching this invariant, annual killifish have an inherent capability to attain an adult lifespan considerably longer than half the time taken to reach maturity (4–6 weeks) (Simpson 1979). For 82 species of gobies (data cited by Miller 1984a), it is evident (Fig. 2) that, while most species, above and below the size boundary of 100 mm, fall within a limited range around 0.5, there are some small cryptobenthic gobiids which have a relatively much longer reproductive life. Apart from the most extreme case, that of the Californian blind goby, *Typhlo-*

gobius californiensis, for which age determination was not verified, these include *Lesueurigobius friesii* and *Lepidogobius lepidus*, both burrowers, maturing at 1–2 years and with potential reproductive lives of approximately 6–9 years (Grossman 1979; Nash 1982). Two *Gobius* species, one small (*Gobius paganellus*), approach such values; they are also cryptobenthic but distant phyletically from the previous species.

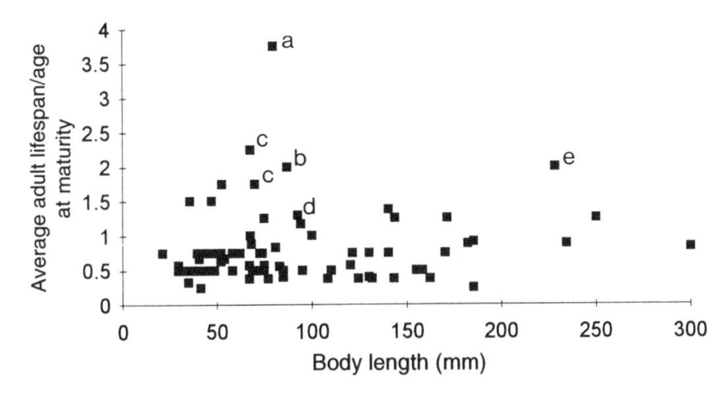

Fig. 2. Ratio of average adult lifespan to age at maturity against body length for 82 gobioid species. Data from Miller (1984a); average adult lifespan taken as 0.5 (maximum recorded lifespan minus age at maturity). (a) *Typhlogobius californiensis*, (b) *Lepidogobius lepidus*; (c) *Lesueurigobius friesii* (two populations), (d) *Gobius paganellus*, (e) *Gobius cobitis*.

Another case of unexpected viability in conjunction with reproduction is shown by the growth pattern of the common goby, *Pomatoschistus microps*. After the first winter of life, this small epibenthic species enjoys a long breeding season of multiple spawning, with growth checks in both sexes and, in females, a cumulative primary reproductive effort which may exceed twice body weight (Miller 1979a, b). However, contrary to what might be predicted from such intensive reproductive effort, the end of the breeding season is followed by vigorous growth (Miller 1979a; Fouda & Miller 1981), perhaps supporting the idea that a deleterious effect of reproduction is not inescapable (Reznick 1985). The subsequent disappearance of most of this generation in the second autumn of life (Miller 1975; Fouda & Miller 1981) must, in the absence of debilitated or dead individuals, be attributed to extrinsic sources of mortality, chiefly predation (Doornbos & Twisk 1987). From field observations, it is difficult to demonstrate the predicted trade-off between reproductive effort and intrinsic sources of mortality in small fish. In Waddensee sand gobies, the severe winter of 1962–1963 was followed by a delayed start to the next breeding season and a higher than average survival into a second breeding season by the year class involved (Fonds 1973). This could be interpreted as longevity promoted by lower reproductive effort, since the age group involved would presumably have spawned on fewer occasions in the shortened season, or as an effect on predators over the winter.

Energy and material expenditure by small fish on individual protection against predators is usually a combination of tactics, which may vary over lifespan with the nature of predation (Endler 1986) and within species according to local forces (Rodd & Reznick 1991). Work is done in simply escaping as well as in the more complex activities of commensalism with crustaceans (Karplus 1987) or schooling, a much-studied behaviour pattern with other functional advantages (Real & Caraco 1986). Structural frameworks range from cryptic coloration and mimicry (Randall & Kuiter 1989) to defensive mechanisms such as bony plates and spines, most famously deployed in the sticklebacks to suit specific predators, whether fish, bird, snake, or invertebrate (Reimchen 1994). The occurrence of injured sticklebacks and experimental work suggest that post-attack defences can be effective, but behavioural techniques do have the advantage that they can be switched on and off as required.

Reproducing

In small fishes where mortality is high and unpredictable across the entire lifespan, as in the case of that due to predation, the expected functional response is early maturity and high reproductive effort, with a short average adult lifespan resulting from external sources of mortality, against which intrinsic reproductive costs are less significant (Roff 1992; Stearns 1992).

Extensive field work and laboratory studies (summarized by Reznick, Bryga, & Endler 1990) have tested such predictions in the guppy, *Poecilia reticulata*. Trinidadian guppy populations in which adults were selectively eaten by a cichlid predator have been shown to have genetically determined earlier maturity and smaller size at maturity, more offspring in the first clutch, shorter intervals between clutches, and greater reproductive effort, than those populations in which a small cyprinodont preys on juveniles rather than adults. By transplantation, the direction of predation was changed from adults to juveniles and significant life-history evolution in the appropriate traits was observed within 11 years, over 30–60 generations. However, other environmental factors are believed to contribute towards about one-fifth of the life-history variation in the Trinidadian guppies (Strauss 1990). Where habitat instability may be the cause of unpredictable survival, early maturation is seen in desert fishes (Constantz 1981) and hill-stream gobies (Kinzie 1988).

Reproductive effort in small fish may achieve high values of gonadosomatic index as a result of repeat or batch spawning within a breeding season, a means of compensating for the size-dependent constraint on primary reproductive effort and instantaneous fecundity (Miller 1979b). Thus, breeding from late March to mid-July, the Mississippi sand smelt *Menidia audens* may spawn daily and, over a conservative estimate of 85 days of reproductive activity, could produce in total approximately 4.8–6.4 times its individual biomass (Hubbs 1976). Among small fish, the interval between successive batches may stretch from 1 day, as in the rice medaka *Oryzias latipes* (Hirschfield 1980), to 4 days, as in the lemon tetra, *Hyphessobrycon pulchripinnis* (Burt, Kramer, Nakatsuru, & Spry 1988), or to the several days or longer between many goby broods (Miller 1984a). This intensive strategy requires a food supply for repeated gonad maturation by the reproducer, with spawning female sticklebacks requiring up to four times the amount needed for maintenance (Wootton 1994). Although reproductive efficiency (conversion of food to gonad energy) may reach 40% or more in small fishes (Rogers 1988), in the common goby and other species the period of most intensive reproduction is associated with a growth check (Miller 1979a; Fouda & Miller 1981), while fish kept immature reach a larger size than ripening individuals by the start of the breeding season (Miller 1973). Although viviparity is associated with diminishing fecundity, overall reproductive effort may be comparable to that found among highly fecund oviparous teleosts (Miller 1979b); there is a trade-off between reproduction and growth in guppies (Reznick 1983). Both for the repeat spawners and also for their progeny, suitable conditions are a prerequisite, so that this pattern of breeding is most appropriate for low-latitude ecosystems characterized by moderate but lengthy periods of productivity (Miller 1979b; Burt *et al.* 1988).

Very early maturation among small teleosts is seen in annual killifish and gambusias (Miller 1979b; Simpson 1979) but selection for earliest maturation may be compromised by selection for reproduction at an appropriate season for larval survival. In the common goby, a suitable size for sexual maturation is reached in many individuals by the first winter of life (Miller 1975; Fouda & Miller 1981), but reproduction does not begin until the following March or April, presumably in relation to temperature and availability of food for spawners and progeny. Over winter, the waiting subadults are probably subject to predation before the opportunity to reproduce. In fact, an early start to successful reproduction in this species is facilitated by production of larger eggs in the first batch (Rogers 1989), which may compensate for less predictable larval nutrition at the start of plankton production. In temperate populations of *Gambusia*, a similar fuelling of spring breeding from stores is believed to be related to higher offspring fitness and higher fecundity in the first brood of the season (Reznick & Braun 1987).

Early maturation with semelparity is an answer to uncertain expectation of reproducer life if success of reproduction is assured. Among small fish, the 'annual' killifish are in fact typically daily iteropares (Simpson 1979) but true semelpares may be found among the aphyine gobies (Miller 1984a), a progenetic group perhaps stemming from the characteristically protracted post-larval stage of many gobies (Kinzie 1988). The related anadromous Japanese ice goby, *Leucopsarion petersi*, also semelparous, may find a predictable nursery for its offspring by parasitizing the enrichment of fresh waters by salmonids, being unlikely itself to provide a substantial biomass of dead spawners sufficient to fertilize streams and provide larval food.

Where larval mortality appears higher than adult and only semi-predictable for seasonal reasons, later maturation and limited instantaneous reproductive effort can be interpreted as adaptive for this schedule in promoting iteroparity and lifetime fecundity by minimizing any intrinsic effect of reproduction on adult survival (as discussed by Stearns 1992). In small temperate fish, this pattern is seen in the rock goby *Gobius paganellus* and other cryptobenthic species under temperate seasonality near the northern limit of geographical range (Miller 1979b). Female gonadosomatic index does not exceed 20% (Miller 1979b; Nash 1982) and there are probably no more than two spawnings per year. In the cryptobenthic freshwater bullhead, *Cottus gobio*, populations of moorland streams, with varying annual recruitment, spawn once a year and have a longer lifespan than fish in more productive lowland chalk streams, where repeat spawning occurs over a shorter adult life (Fox 1978).

A similar pattern of moderate reproductive effort and protracted iteroparity, as well as meeting the planktonic vicissitudes of a complex life history, would also be appropriate if random availability of patchy habitats made demersal recruitment a particularly critical phase for small species. In warmer ecosystems, there is the additional benefit of maturation unimpeded by winter. Anemonefishes may live for a decade and several years may be required before recovery of the population after artificial defaunation of anemones (Fautin 1992). In this form of iteroparity among coral fishes, periods of intensive breeding often coincide with mesoscale current eddies and increased probability of larval retention at oceanic islands (Lobel & Robinson 1986; Lobel 1989). There may also be a facility for delaying metamorphosis if a suitable habitat is not forthcoming (Victor 1986).

Bestowal and fecundity

The transduction of energy and allocation of material per offspring is an important aspect of functional ecology. In teleosts, for a given ovarian biomass, propagule size (bestowal), and number (fecundity) are inversely related (Elgar 1990). Among small fish, a range of fecundity and bestowal tactics exist within the constraint that the gonadosomatic index and fecundity reduce in absolute values with diminishing size. The basic teleost single-generation complex life cycle, with a planktonivorous post-larval stage and more or less radical metamorphosis (Youson 1988), eliminates the burden of rearing offspring and aids dispersal (Moran 1994) but the young are placed in a risky environment (Pepin 1991). It is associated with low bestowal and high fecundity, for which the large cystovarian ovary of teleosts is obviously adapted. High fecundity enhances the probability of some offspring (1) avoiding predation and (2) finding food in the large-scale patchiness of the rich planktonic ecosystem, then, at metamorphosis, (3) finding the adult habitat, (4) entering a vacant niche, and, ultimately, (5) finding a mate (Barlow 1981; Duarte & Alcaraz 1989; Winemiller & Rose 1993).

For small species with stochastic adult survival and population size kept below saturation, the last of these requirements for fitness may further exacerbate the fecundity disadvantage of small size. However, compensatory enhancement of fecundity in small fish can be achieved by reduced bestowal (thereby producing more eggs per unit biomass) and the multiplying effect of repeat spawning within a season (see above). Down to limits probably set by ontogenetic processes, smaller eggs also permit more rapid maturation of the oocyte (Pauly & Pullin 1988; Duarte & Alcaraz 1989) and thus greater frequency of batch spawning, which further contributes to cumulative fecundity (Miller 1979b; Burt *et al.* 1988). Ware (1975) suggested that optimum sizes for pelagic fish eggs should range from 0.5 to

3.5 mm, depending on incubation time, but sizes down to 0.26 mm have been noted for demersal eggs of small gobies (Miller 1984a). This need not be merely an automatic accompaniment of small adult size but can be interpreted as a positive adaptation for rapid development and fecundity. Given survival and low bestowal, impressive spawning frequency and multiples of instantaneous fecundity may be achieved. Spawning from 100 to 500 eggs (at 0.3 mm) per clutch, a female zebrafish is reported to have produced 5530 eggs in 5 months, with a relative fecundity of over several thousand eggs per gram of adult somatic weight (Miller 1979b).

Tiny eggs with fast development are also seen in small species, as in New Guinean rainbowfish (Coates 1990), spawning on flooded ground, where waters are temporary but may offer the advantages of plentiful food and fewer predators. Maximizing fecundity is also seen when adult and juvenile habitats are spatially well separated and the juvenile phase protracted. Tropical hill-stream sicydiine gobies produce large numbers of tiny larvae which are washed down to the estuary, where they grow for some months (Radtke, Kinzie & Folsom 1988) and then ascend the river to the adult habitat (Miller 1984a; Kinzie 1993).

In a number of small fishes, such as *Gasterosteus*, populations differ in mean egg size (Baker 1994) and, in other species, selection for larger offspring has progressed, despite reduced fecundity and the disadvantage of longer brood intervals. This process has entailed abandonment of the complex type of life history with the elimination of a distinctive pre-metamorphic post-larval stage. Among several benefits (Roff 1992), there are underlying metabolic advantages of increasing larval size (Wieser 1991). More bestowal and larger but fewer offspring are associated with a limited food resource for post-larvae (Winemiller & Rose 1993) or risk from displacement to an unsuitable environment, in particular in fresh waters. A striking example is shown by the Australian desert goby *Chlamydogobius eremius* and the closely related freshwater/estuarine *Mugilogobius* species (Miller 1987), both with typical gobioid breeding behaviour of nest and egg care by the male (Miller 1984a) and similar female gonadosomatic indices. In the Central Australian basin, the former species inhabits mound springs, of limited extent before the water disappears into the sands, and with little if any plankton, both features raising the strong likelihood of stranding and starvation for midwater offspring. As expected and also seen in other desert spring fishes (Constantz 1981), egg size is large, with yolk diameter 1.42–1.5 mm and hatchling length approximately 5 mm, and fecundity is low, between 150 and 250, 201 eggs being counted in a fish of 44.5 mm (Miller 1987; Horsthemke 1989). In *Mugilogobius abei*, with a planktotrophic marine phase, yolk diameters range from 1.42 to 1.5 mm and fecundity from 1407 to 6920 in females of 40–60 mm (sources in Miller 1987). In an entirely different environment, but one that probably also lacks a food resource for planktonic young, small bathyal and abyssal seasnails (liparids) produce eggs up to 5.3 mm in diameter, the latter recorded from female *Osteodiscus cascadiae* not more than 81 mm in length and with no more than seven ripe eggs (Stein 1980); length at hatching could be approximately 15 mm, almost one-fifth of the adult length. Part of the reproductive energetics of both gobiids and liparids is expenditure on nest guarding by the male. Exemplifying a common trade-off effect on egg size (Roff 1992), 'guarding' species of percid darters have larger, fewer eggs, from a lower primary reproductive effort, which receive parental care, in contrast to the many smaller eggs scattered over plants or stones by repeat-spawning species with higher gonadosomatic indices (Paine 1990) but no post-zygotic secondary expenditure (Miller 1979a).

Among small fish, the relatively largest offspring at commencement of independent life are produced by viviparity, most studied in three New World cyprinodontiform families, which exhibit contrasting techniques (Wourms, Grove, & Lombardi 1988; Meyer & Lydeard 1993) and a range of brood fecundity/bestowal tactics from ovoviviparity to advanced viviparity but no care of progeny once born (Miller 1979b).

Although the gonadosomatic index of their full-term ovary may be comparable to that of small oviparous fishes (Miller 1979b), viviparous fishes tend to be less fecund than their closest oviparous relatives. Apart from adaptation to external conditions, selection for larger offspring in viviparous species may involve a saving in maternal metabolism if larger embryos present relatively smaller interfaces and have relatively lower metabolism than would a particular biomass of brood divided into smaller units. Nevertheless, given the size and oocyte productivity of the teleost ovary, fecundity in viviparous small fish can still be two orders of magnitude greater than that of the most prolific small mammals.

In the ovarian type of teleost viviparity, the lumen of the cystovarian ovary acts like that of a uterus (Fig. 3a). In the goodeids and *Jenynsia*, the ovarian lining secretes nutritive material absorbed through embryonic trophotaeniae or dispensed via maternal trophonemata (Wourms *et al.* 1988). However, the alternative technique of follicular gestation (Fig. 3b) offers a unique opportunity for small fish to resolve the conflict between number and size of offspring. In poeciliids, as well as quite independently in a number of other small viviparous fishes, with direct access of spermatozoa to the surface of the follicles, fertilization of the egg can take place within the follicle and development can continue to full term before the follicle ruptures, each follicle thus acting as a separate brood chamber. Across the poeciliids, there is a range in the timing of energy and material transfer to the intrafollicular embryo, from mostly pre-zygotic ovoviviparity, variously facultative, to highly effective post-zygotic viviparity (Wourms *et al.* 1988; Reznick & Miles 1989). Follicular gestation of the last sort provides the basis for a reproductive technique which combines high bestowal with rapid production of young.

In oviparous small teleosts, to maximize clutch frequency rather than instantaneous clutch number, it is possible to ovulate eggs at frequent intervals, well within the time needed for vitellogenesis, without endangering other oocytes at earlier stages, because all are isolated in their individual follicles. In viviparous species with ovarian gestation, even with spermatozoa already stored and progressive availability of fertilized eggs, it is mechanically impossible to express just

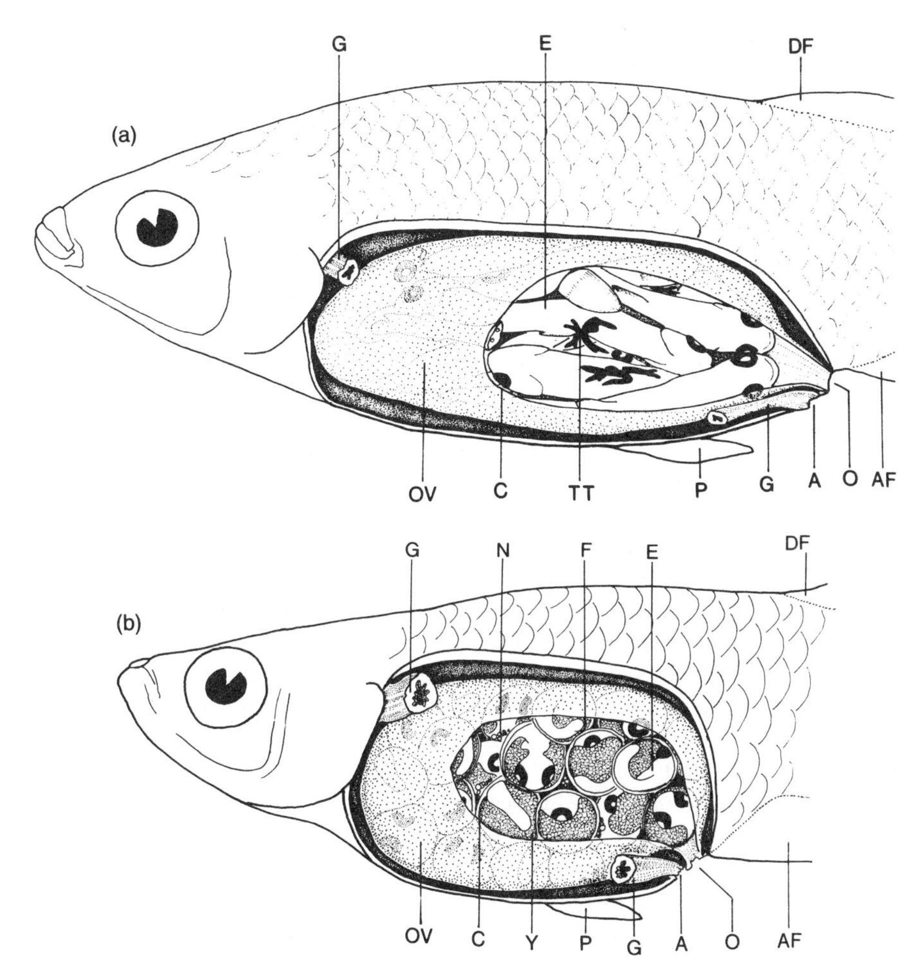

Fig. 3. Modes of gestation in viviparous cyprinodontiforms: (a) ovarian gestation in *Ilyodon whitei*, 68 mm standard length (Goodeidae); (b) follicular gestation in *Gambusia affinis*, 34 mm (Poeciliidae). A, anus; AF, anal fin; C, cut edge of ovary; DF, dorsal fin; E, embryo; F, follicle wall; G, gut (cut end); N, immature oocytes; O, genital opening; OV, ovary; P, right pelvic fin; TT, trophotaeniae; Y, yolk. After Miller (1984b).

full-term embryos from the ovary without discharging any others that might be present. Follicular gestation, however, permits a return to the oviparous level of spawning frequency by the technique of superfoetation (superembryonation), essentially a form of clutch overlap discussed by Burley (1980). Since each embryo develops in a separate intact follicle, release of a full-term offspring need not prejudice further development of earlier stages in other follicles, an arrangement which also permits switching of energy allocation between embryos according to developmental stage (Constantz 1980). The characteristic feature of superfoetation is thus the presence of embryos representing several stages at any one time in the ovary. Superfoetation appears to have evolved independently at least four times within the Poeciliidae, where it is facultative in some species (Reznik & Miles 1989) and is also found in the entirely unrelated viviparous clinids (Veith 1979; Prochazka & Griffiths 1992).

Superfoetation permits release of young at intervals less than individual gestation time. These offspring need not be of relatively large size if small size is not disadvantageous, as in the case of the comparatively fecund *Clinus superciliosus* (Veith 1979). However, if larger offspring size is at a premium, superfoetation can also permit the frequent release of the relatively largest young among small fish. The most striking example is the least killifish, *Heterandria formosa*, which produces newborn young of approximately one-third maternal length, yet at minimal cost to the mother and without the delay of a long gestation time between successive births (Miller 1979b). Although there is a high degree of viviparity in this genus, to gestate the same number of large young simultaneously in one clutch would be spatially impossible for the mother and inflict impossible nutritive costs (Travis, Farr, Henrich, & Cheong 1987). The constraints of the cost-expansion, variant-energy, and limit hypotheses of Burley (1980) for clutch size as an alternative to clutch overlap would be insurmountable.

The frequency of offspring production by superfoetatious species is advantageous if adult survival is unpredictable (Downhower & Brown 1975; Miller 1979b) and *Heterandria* does experience high adult mortality in relation to that of juveniles (Travis *et al.* 1987). It has been suggested that the minimal body girth of superfoetatious females would be advantageous for streamlining in fast-flowing habitats, where large young would also be better able to avoid displacement (Thibault & Schultz 1978). In addition, in the superfoetatious poeciliids, such as *Poeciliopsis* and *Heterandria*, where cannibalism (Polis 1981) appears widespread, large offspring could be of value in being more difficult to swallow and faster at escaping. In small sand-dwelling clinids with large offspring (Prochazka & Griffiths 1992), superfoetation would seem to answer the conflict between selection for slim mothers that would be better able to bury themselves and for young large enough to overcome the mechanical difficulties in burying themselves that very small fish might encounter (Atkinson & Taylor 1991).

Functional interactions with mankind

The functional ecologies of small fish and mankind interact in a number of ways, with important consequences for the former. Most directly, some small species are predated in artisanal fisheries, as for the tiny goby *Mistichthys* and juvenile sicydiines, and at a larger scale in industrial fishing for sandeels or as bycatches, as well as by capture for aesthetic purposes and research; mortality is high among coral fish exploited for the aquarium trade (Wood 1985). Indirect ecosystem alteration can also influence small fishes by altering predation pressure and direction (Bruton 1990), by removal of a food source, such as insects and plant material by deforestation (Angermeier & Karr 1983), or by damage to habitats such as coral reefs (Rubec 1988). Ultimately, there may be complete ecosystem destruction by loss of water! Watertable changes appear to have eliminated the type streams for two species of Greek freshwater gobies (Miller 1990) and desert pupfish are threatened in a similar way (Pister 1990).

Conversely, small fish have benefited from human activities, as measured by extension of area and population. Creation of new habitats, as by dredging, in new dams and via major ship canals, has favoured spread of small species. Use of *Gambusia* as a larviphage has resulted in the spread of these fish throughout the tropical and warm temperate world (Haas & Pal 1984; Morton, Beumer, & Pollock 1988). Accidentally, small size has facilitated transfer of fish in ballast water of large ships (Carlton 1985), an ongoing process (Miller, Wright, & Wongrat 1989). Releases from aquaria, while opening new opportunities for mollies and swordtails, may not be advantageous to native ecosystems (Courtenay & Meffe 1989).

Small size is also a prerequisite for a special kind of commensalism with mankind. This involves the maintenance of such fishes in captivity for ornamental purposes, typically in an indoor container of water for which Philip Henry Gosse introduced the familiar term of 'aquarium' (Freeman & Wertheimer 1980). The immediate advantage to an individual fish, depending on the skill of its keeper, is security and, to the latter, aesthetic satisfaction and even psychotherapy (McBride, this volume, pp. 293–308). The aquarium hobby, with maintenance of marine and freshwater tropical and coldwater species, now uses a variety of sophisticated techniques (Adey & Loveland 1991) and supports a world-wide trade of over $7 billion per year (Andrews 1990). It is an important source of currency for some developing countries. In operation at any one time, there may be at least 30 million aquaria world-wide.

Captive breeding of endangered species is a serious aspect of aquarium technology, as applied, for example, to Lake Victoria cichlids (Reid 1990) and may conserve natural diversity. However, as an inevitable consequence of domestication, populations of ornamental fishes, such as guppies and angel-fish, *Pterophyllum scalare*, have been subject to a high degree of directional selection for features of coloration and finnage markedly different from the natural condition. It is also possible to speculate about what functional traits might be emphasized among aquarium stocks.

As an overall environment, the global aquarium ecosystem is exceptionally patchy, with innumerable small ephemeral habitats stocked and linked by human agency, formerly directly from the wild but more usually nowadays from a number of more persistent larger populations kept in being by commercial breeders. At best these habitats are only semi-predictable in terms of reproductive success, depending on the whim or expertise of the aquarist. Over time, the evolutionary history of any one species in captivity could thus be visualized as a complex multiple raceme, comparable on a much smaller time scale to the model envisaged for the history of freshwater colonization by marine sticklebacks (Bell & Foster 1994). Apart from generalized feeding, functional selection in this ecosystem may be thought to favour longevity, maximizing the chance of meeting a mate or being presented with conditions for breeding. However, high growth rate, early maturation and high reproductive effort, all features reducing lifespan, would appeal to the human commensal and would therefore increase chances of mate encounter because suitable individuals would be more likely to be spread through the hobby. Also, in a system geared for the feeding of juvenile and adult fish, and with the impossibility of independent dispersal by small planktonic post-larvae, it might be expected that large offspring size would favour both partners.

Conclusion

The evolution of small size among teleosts has permitted the phyletic exploration of extreme niches, from the narrowest of physical spaces to the dangerous front line of open water. Across this broad range of ecotopes there is a corresponding diversity in whole-organism functional ecology of small fishes, as varied as any among vertebrates. Whatever intrinsic effects on function result from small body size, selection appears to have directed the allocation of energy and material expenditure into life histories to suit differing patterns of mortality generated by ecological interaction within an ecotope (Partridge & Harvey 1988). When this is further complicated by human activities, in the biota of a future world entirely disposed by mankind, the basic attribute of being small and the capacity for functional adaptation might perpetuate at least some small species long after the disappearance of cod or bluefin tuna.

References

Adey, W. H. & Loveland, K. (1991). *Dynamic aquaria: building living ecosystems*. Academic Press, San Diego.

Almada, V. C. & Santos, R. S. (1995). Parental care in the rocky intertidal: a case study of adaptation and exaption in Mediterranean and Atlantic blennies. *Rev. Fish Biol. Fish.* 5: 23–37.

Andrews, C. (1990). The ornamental fish trade and conservation. *J. Fish Biol.* 37 (Suppl. A): 53–59.

Angermeier, P. L. & Karr, J. R. (1983). Fish communities along environmental gradients in a system of tropical streams. *Envir. Biol. Fishes* 9: 117–135.

Atkinson, R. J. A. & Taylor, A. C. (1991). Burrows and burrowing behaviour of fish. *Symp. zool. Soc. Lond.* No. 63: 133–155.

Baker, J. A. (1994). Life history variation in female threespine stickleback. In *The evolutionary biology of the threespine stickleback*: 144–187. (Eds Bell, M. A. & Foster, S. A.). Oxford University Press, Oxford.

Baltz, D. M., Rakocinski, C., & Fleeger, J. W. (1993). Microhabitat use by marsh-edge fishes in a Louisiana estuary. *Envir. Biol. Fishes* 36: 109–126.

Banarescu, P. & Coad, B. W. (1991). Cyprinids of Eurasia. In *Cyprinid fishes. Systematics, biology and exploitation*: 125–155. (Eds Winfield, I. J. & Nelson, J. S.). Chapman & Hall, London. (*Fish Fish. Ser.* No. 3.)

Barel, C. D. N., Ligvoet, W., Goldschmidt, T., Witte, F., & Goudswaard, P. C. (1991). The haplochromine cichlids in Lake Victoria: an assessment of biological and fisheries interests. In *Cichlid fishes. Behaviour, ecology and evolution*: 258–279. (Ed. Keenleyside, M. H. A.). Chapman & Hall, London.

Barlow, G. W. (1981). Patterns of parental investment, dispersal and size among coral-reef fishes. *Envir. Biol. Fishes* **6**: 65–85.

Bell, M. A. & Foster, S. A. (1994). Introduction to the evolutionary biology of the threespine stickleback. In *The evolutionary biology of the threespine stickleback*: 1–27. (Eds Bell, M. A. & Foster, S. A.). Oxford University Press, Oxford.

Berge, J. A. & Hesthagen, I. H. (1981). Effects of epibenthic macropredators on community structure in an eutrophicated shallow water area, with special reference to food consumption by the common goby *Pomatoschistus microps*. *Kiel. Meeresforsch*. No. 5: 462–470.

Beverton, R. J. & Holt, S. J. (1959). A review of the lifespans and mortality rates of fish in nature, and their relation to growth and other physiological characteristics. In *Ciba Foundation colloquia on ageing* **5**. *The life-span of animals*: 142–180. (Eds Wolstenholme, G. E. W. & O'Connor, M.). Churchill, London.

Bruton, M. N. (1990). The conservation of the fishes of Lake Victoria, Africa: an ecological perspective. *Envir. Biol. Fishes* **27**: 161–175.

Bruun, A. F. (1940). A study of a collection of the fish *Schindleria* from South Pacific waters. *Dana Rep*. No. 21: 1–12.

Buckup, P. A. (1993). Phylogenetic interrelationships and reductive evolution in Neotropical characidin fishes (Characiformes, Ostariophysi). *Cladistics* **9**: 305–341.

Burley, N. (1980). Clutch overlap and clutch size: alternative and complementary reproductive tactics. *Am. Nat.* **115**: 223–246.

Burt, A., Kramer, D. L., Nakatsuru, K., & Spry, C. (1988). The tempo of reproduction in *Hyphessobrycon pulchripinnis* (Characidae), with a discussion on the biology of 'multiple spawning, in fishes. *Envir. Biol. Fishes* **22**: 15–27.

Calow, P. (1987). Towards a definition of functional ecology. *Funct. Ecol.* **1**: 57–61.

Carle, K. J. & Hastings, P. A. (1982). Selection of meiofaunal prey by the darter goby, *Gobionellus boleosoma* (Gobiidae). *Estuaries* **5**: 316–318.

Carlton, J. T. (1985). Transoceanic and interoceanic dispersal of coastal marine organisms: the biology of ballast water. *Oceanogr. mar. Biol. a. Rev.* **23**: 313–371.

Charnov, E. L. (1993). *Life history invariants: some explorations of symmetry in evolutionary ecology*. Oxford University Press, Oxford.

Coates, D. (1990). Biology of the rainbowfish, *Glossolepis multisquamatus* (Melanotaeniidae) from the Sepik River floodplains, Papua New Guinea. *Envir. Biol. Fishes* **29**: 119–126.

Conover, D. O. (1992). Seasonality and the scheduling of life history at different latitudes. *J. Fish Biol.* **41** (Suppl. B): 161–178.

Constantz, G. D. (1980). Energetics of viviparity in the gila topminnow (Pisces: Poeciliidae). *Copeia* **1980**: 876–878.

Constantz, G. D. (1981). Life history patterns of desert fishes. In *Fishes in North American deserts*: 237–290. (Eds Naiman, R. J. & Soltz, D. L.). Wiley, New York.

Courtenay, W. R. & Meffe, G. K. (1989). Small fishes in strange places: a review of introduced poeciliids. In *Ecology and evolution of livebearing fishes (Poeciliidae)*: 319–331. (Eds Meffe, G. K. & Snelson, F. F.). Prentice Hall, Eaglewood Cliffs, New Jersey.

Craig, J. F. & Fletcher, J. M. (1984). Growth and mortality of zebra fish, *Brachydanio rerio* (Hamilton Buchanan), maintained at two temperatures and on two diets. *J. Fish Biol.* **25**: 43–55.

Damuth, J. (1987). Interspecific allometry of population density in mammals and other animals: the independence of body mass and population energy use. *Biol. J. Linn. Soc.* **31**: 193–246.

Dill, L. M. (1990). Distance-to-cover and the escape decisions of an African cichlid fish, *Melanochromis chipokae*. *Envir. Biol. Fishes* **27**: 147–152.

Doherty, P. J. & Williams, D. McB. (1988). The replenishment of coral reef fish populations. *Oceanogr. mar. Biol. a. Rev.* **26**: 487–551.

Doornbos, G. & Twisk, F. (1987). Density, growth and annual food consumption of gobiid fish in the saline Lake Grevelingen, The Netherlands. *Neth. J. Sea Res.* **21**: 45–74.

Downhower, J. F. & Brown, L. (1975). Superfoetation in fishes and the cost of reproduction. *Nature, Lond.* **256**: 345–346.

Duarte, C. M. & Alcaraz, M. (1989). To produce many small or few large eggs: size-independent reproductive tactic of fish. *Oecologia* **80**: 401–404.

Elgar, M. A. (1990). Evolutionary compromise between a few large and many small eggs: comparable evidence in teleost fish. *Oikos* **59**: 283–287.

Endler, J. A. (1986). Defense against predators. In *Predator–prey relationships*: 109–134. (Eds Feder, M. E. & Lauder, G. V.). University of Chicago Press, Chicago.

Fautin, D. G. (1992). Anemonefish recruitment: the roles of order and chance. *Symbiosis* **14**: 143–160.

Fonds, M. (1973). Sand gobies in the Dutch Wadden Sea (*Pomatoschistus*, Gobiidae, Pisces). *Neth. J. Sea Res.* **6**: 417–478.

Fouda, M. M. & Miller, P. J. (1981). Age and growth of the common goby *Pomatoschistus microps*, on the south coast of England. *Estuar. cstl Shelf Sci.* **12**: 121–129.

Fox, P. J. (1978). Preliminary observations on different reproduction strategies in the bullhead (*Cottus gobio* L.) in northern and southern England. *J. Fish Biol.* **12**: 5–11.

Freeman, R. B. & Wertheimer, D. (1980). *Philip Henry Gosse. A bibliography.* Dawson, Folkestone.

Gerking, S. D. (1994). *Feeding ecology of fish.* Academic Press, San Diego.

Gibson, R. N. & Ezzi, I. A. (1978). The biology of a Scottish population of Fries' goby, *Lesueurigobius friesii*. *J. Fish Biol.* **12**: 371–389.

Gibson, R. N. & Ezzi, I. A. (1992). The relative profitability of particulate- and filter-feeding in the herring, *Clupea harengus* L. *J. Fish Biol.* **40**: 577–590.

Giles, R. J. (1987). A comparison of the behavioural responses of parasitized and non-parasitized three-spined sticklebacks, *Gasterosteus aculeatus* L., to progressive hypoxia. *J. Fish Biol.* **30**: 631–638.

Gilinsky, E. (1984). The role of fish predation and spatial heterogeneity in determining benthic community structure. *Ecology* **65**: 455–468.

Goolish, E. M. (1991). Aerobic and anaerobic scaling in fish. *Biol. Rev.* **66**: 33–56.

Grant, J. W. A. & Kramer, D. L. (1990). Territory size as a predictor of the upper limit to population density of juvenile salmonids in streams. *Can. J. Fish. aquat. Sci.* **47**: 1724–1737.

Greenfield, D. W. & Johnson, R. K. (1990). Community structure of western Caribbean blennioid fishes. *Copeia* **1990**: 433–448.

Griffiths, D. (1992). Size, abundance, and energy use in communities. *J. Anim. Ecol.* **61**: 307–315.

Grossman, G. D. (1979). Demographic characteristics of an intertidal bay goby (*Lepidogobius lepidus*). *Envir. Biol. Fishes* **4**: 207–218.

Grossman, G. D., Moyle, P. B., & Whitaker, J. O. (1982). Stochasticity in structural and functional characteristics of an Indiana stream fish assemblage: a test of community theory. *Am. Nat.* **120**: 423–454.

Haas, R. & Pal, R. (1984). Mosquito larvivorous fishes. *Bull. ent. Soc. Am.* **30**: 17–25.
Hanken, J. & Wake. D. B. (1993). Miniaturization of body size: organismal consequences and evolutionary significance. *A. Rev. Ecol. Syst.* **24**: 501–519.
Hastings, P. A. (1986). Habitat selection, sex ratio and sexual selection in *Coralliozetus angelica* (Blennioidea: Chaenopsidae). In *Indo-Pacific fish biology*: 785–793. (Eds Uyeno, T., Arai, R., Taniuchi, T., & Matsuura, K.). Ichthyological Society of Japan, Tokyo.
Heymer, A. (1982). Le comportement pseudo-amphibie de *Coryphoblennius galerita* et *Blennius trigloides*. *Revue fr. Aquariol. Herpetol.* **9**: 91–96.
Hirschfield, M. F. (1980). An experimental analysis of reproductive effort and cost in the Japanese medaka, *Oryzias latipes*. *Ecology* **61**: 282–292.
Horsthemke, H. (1989). Die Australische Wustengrundel *Chlamydogobius eremius* (Zietz, 1896). *Aquarien-Terrarien* **42**: 288–293.
Hubbs, C. (1976). The diel reproductive pattern and fecundity of *Menidia audens*. *Copeia* **1976**: 386–388.
Johnson, G. D. & Brothers, E. B. (1993). *Schindleria*: a paedomorphic goby (Teleostei: Gobioidei). *Bull. mar. Sci.* **52**: 441–471.
Karplus, I. (1987). The association between gobiid fishes and burrowing alpheid shrimps. *Oceanogr. mar. Biol. a. Rev.* **25**: 507–562.
Kinzie, R. A., III (1988). Habitat utilization by Hawaiian stream fishes with reference to community structure in oceanic island streams. *Envir. Biol. Fishes* **22**: 179–192.
Kinzie, R. A., III (1993). Reproductive biology of an endemic, amphidromous goby *Lentipes concolor* in Hawaiian streams. *Envir. Biol. Fishes* **37**: 257–268.
Kneib, R. T. (1987). Predation risk and use of intertidal habitats by young fishes and shrimp. *Ecology* **68**: 379–386.
Kottelat, M. & Vidthayanon, C. (1993). *Boraras micros*, a new genus and species of minute freshwater fish from Thailand (Teleostei: Cyprinidae). *Ichthyol. Explor. Freshw.* **4**: 161–176.
Kuwahara, T., Yogo, Y., & Nakashima, Y. (1993). Size-assortative monogamy and paternal egg care in a coral goby *Paragobiodon echinocephalus*. *Ethology* **95**: 65–75.
Levin, S. A. (1992). The problem of pattern and scale in ecology. *Ecology* **73**: 1943–1967.
Lindsey, C. C. (1966). Body size of poikilotherm vertebrates at different latitudes. *Evolution* **20**: 456–465.
Lobel, P. S. (1989). Ocean current variability and the spawning season of Hawaiian reef fishes. *Envir. Biol. Bishes* **24**: 161–171.
Lobel, P. S. & Robinson, A. R. (1986). Transport and entrapment of fish larvae by ocean mesoscale eddies and currents in Hawaiian waters. *Deep-Sea Res. (A)* **33**: 483–500.
Machado, F. A. & Sazima, I. (1983). Comportamento alimentar do peixe hematofago *Branchioica bertonii* (Siluriformes, Trichomycteridae). *Ciênc. Cult., S. Paulo* **35**: 344–348.
Markofsky, J. & Milstoc, M. (1979). Histopathological observations of the kidney during aging of the male annual fish *Nothobranchius guentheri*. *Expl Geront.* **14**: 149–155.
Markofsky, J. & Perlmutter, A. (1972). Age at sexual maturity and its relationship to longevity in the male annual cyprinodont fish, *Nothobranchius guentheri*. *Expl Geront.* **7**: 131–135.
Marshall, N. B. (1984). Progenetic tendencies in deep-sea fishes. In *Fish reproduction: strategies and tactics*: 91–101. (Eds Potts, G. W. & Wootton, R. J.). Academic Press, London.
Meyer, A. & Lydeard, C. (1992). The evolution of copulatory organs, internal fertilization, placentae and viviparity in killifishes (Cyprinodontiformes) inferred from a DNA phylogeny of the tyrosine kinase *X-src*. *Proc. R. Soc. (B)* **254**: 153–162.

Middaugh, D. P. (1981). Reproductive ecology and spawning periodicity of the Atlantic silverside *Menidia menidia* (Pisces: Atherinidae). *Copeia* **1981**: 766–776.

Miller, P. J. (1973). Effects of photoperiod on growth in gobiid fish. *Proc. Challenger Soc.* **4**: 197–198.

Miller, P. J. (1975). Age-structure and life-span in the Common goby, *Pomatoschistus microps*. *J. Zool., Lond.* **177**: 425–448.

Miller, P. J. (1979a). A concept of fish phenology. *Symp. zool. Soc. Lond.* No. 44: 1–28.

Miller, P. J. (1979b). Adaptiveness and implications of small size in teleosts. *Symp. zool. Soc. Lond.* No. 44: 263–306.

Miller, P. J. (1984a). The tokology of gobioid fishes. In *Fish reproduction: strategies and tactics*: 119–153. (Eds Potts, G. W. & Wootton, R. J.). Academic Press, London.

Miller, P. J. (1984b). A brief review of livebearing. *Aquarist Pondkpr* **49**: 16–18.

Miller, P. J. (1987). Affinities, origin and adaptive features of the Australian Desert Goby, *Chlamydogobius eremius* (Zietz, 1896) (Teleostei: Gobiidae). *J. nat. Hist.* **21**: 687–705.

Miller, P. J. (1990). The endurance of endemism: the Mediterranean freshwater gobies and their prospects for survival. *J. Fish Biol.* **37** (Suppl. A): 145–146.

Miller, P. J. & Fouda, M. M. (1986). Notes on the biology of a Red Sea goby, *Silhouettea aegyptia* (Chabanaud, 1933) (Teleostei: Gobiidae). *Cybium* **10**: 395–409.

Miller, P. J., Wright, J., & Wongrat, P. (1989). An Indo-Pacific goby (Teleostei: Gobioidei) from West Africa, with systematic notes on *Butis* and related eleotridine genera. *J. nat. Hist.* **23**: 311–324.

Mitz, S. V. & Newman, M. C. (1989). Allometric relationship between oxygen consumption and body weight of mosquitofish, *Gambusia affinis*. *Envir. Biol. Fishes* **24**: 267–273.

Moran, N. A. (1994). Adaptation and constraint in the complex life cycles of animals. *A. Rev. Ecol. Syst.* **25**: 573–600.

Morton, R. M., Beumer, J. P., & Pollock, B. R. (1988). Fishes of a subtropical Australian saltmarsh and their predation upon mosquitoes. *Envir. Biol. Fishes* **21**: 185–194.

Nash, R. D. M. (1982). The biology of Fries' goby, *Lesueurigobius friesii* (Malm), in the Firth of Clyde, Scotland, and a comparison with other stocks. *J. Fish Biol.* **21**: 69–85.

Nelson, J. S. (1994). *Fishes of the world*. (3rd edn). John Wiley & Sons, New York.

Paine, M. D. (1990). Life history tactics of darters (Percidae: Etheostomatiini) and their relationshp with body size, reproductive behaviour, latitude and rarity. *J. Fish Biol.* **37**: 473–488.

Partridge, L. & Harvey, P. H. (1988). The ecological context of life history evolution. *Science* **241**: 1449–1455.

Patwardhan, S. A. & Gaikwad, S. A. (1991). Size dependent toxicity of sumithion EC 50 on *Gambusia affinis affinis* (Baird & Girard). *Pollut. Res.* **10**: 43–45.

Pauly, D. & Pullin, R. S. V. (1988). Hatching time in spherical, pelagic, marine fish eggs in response to temperature and egg size. *Envir. Biol. Fishes* **22**: 261–271.

Peden, A. E. & Corbett, C. A. (1973). Commensalism between a liparid fish, *Careproctus* sp., and the lithodid box crab, *Lopholithodes foraminatus*. *Can. J. Zool.* **51**: 555–556.

Pepin, P. (1991). Effect of temperature and size on development, mortality, and survival rates of the pelagic early life history stages of marine fish. *Can. J. Fish. aquat. Sci.* **48**: 503–518.

Pister, E. P. (1990). Desert fishes: an interdisciplinary approach to endangered species conservation in North America. *J. Fish Biol.* **37** (Suppl. A): 183–187.

Polis, G. A. (1981). The evolution and dynamics of intraspecific predation. *A. Rev. Ecol. Syst.* **12**: 225–251.

Prochazka, K. & Griffiths, C. L. (1992). Observations on the distribution patterns, behaviour, diets and reproductive cycles of sand-dwelling clinids (Perciformes: Clinidae) from South Africa. *Envir. Biol. Fishes* **35**: 371–379.

Pusey, B. J. (1989). Aestivation in the teleost fish *Lepidogalaxias salamandroides* (Mees). *Comp. Biochem. Physiol.* **92A**: 137–138.

Pusey, B. J. (1990). Seasonality, aestivation and the life history of the salamanderfish *Lepidogalaxias salamandroides* (Pisces: Lepidogalaxiidae). *Envir. Biol. Fishes*: **29**: 15–26.

Radtke, R. L., Kinzie, R. A., III & Folsom, S. D. (1988). Age at recruitment of Hawaiian freshwater gobies. *Envir. Biol. Fishes* **23**: 205–213.

Randall, J. E. & Kuiter, R. H. (1989). The juvenile Indo-Pacific grouper *Anyperodon leucogrammicus*, a mimic of the wrasse *Halichoeres purpurescens* and allied species, with a review of the recent literature on mimicry in fishes. *Revue fr. Aquariol. Herpetol.* **16**: 51–56.

Real, L. & Caraco, T. (1986). Risk and foraging in stochastic environments. *A. Rev. Ecol. Syst.* **17**: 371–390.

Reid, G. M. (1990). Captive breeding for the conservation of cichlid fishes. *J. Fish Biol.* **37** (Suppl. A): 157–166.

Reimchen, T. E. (1994). Predators and morphological evolution in threespine stickleback. In *The evolutionary biology of the threespine stickleback*: 240–276. (Eds Bell, M. A. & Foster, S. A.). Oxford University Press, Oxford.

Reinthal, P. N. (1990). The feeding habits of a group of herbivorous rock-dwelling cichlid fishes (Cichlidae: Perciformes) from Lake Malawi, Africa. *Envir. Biol. Fishes* **27**: 215–233.

Reznick, D. A. (1983). The structure of guppy life histories: the tradeoff between growth and reproduction. *Ecology* **64**: 862–873.

Reznick, D. A. (1985). Costs of reproduction: an evaluation of the empirical evidence. *Oikos* **44**: 257–267.

Reznick, D. A. & Braun, B. (1987). Fat cycling in the mosquitofish (*Gambusia affinis*): fat storage as a reproductive adaptation. *Oecologia* **73**: 401–413.

Reznick, D. A., Bryga, H. & Endler, J. A. (1990). Experimentally induced life-history evolution in a natural population. *Nature, Lond.* **346**: 357–359.

Reznick, D. N. & Miles, D. B. (1989). Review of life history patterns in poeciliid fishes. In *Ecology and evolution of livebearing fishes (Poeciliidae)*: 125–148. (Eds Meffe, G. K. & Snelson, F. F.). Prentice Hall, Eaglewood Cliffs, NJ.

Rodd, F. H. & Reznick, D. N. (1991). Life history evolution in guppies: III. The impact of prawn predation on guppy life histories. *Oikos* **62**: 13–19.

Roff, D. A. (1992). *The evolution of life histories*. Chapman & Hall, London.

Rogers, S. I. (1988). Reproductive effort and efficiency in the female common goby. *Pomatoschistus microps* (Kroyer) (Teleostei: Gobioidei). *J. Fish Biol.* **33**: 109–119.

Rogers, S. I. (1989). Seasonal variations in fecundity and egg size of the common goby, *Pomatoschistus microps*. *J. mar. biol. Ass. U.K.* **69**: 535–543.

Ross, R. T. & Baker, J. A. (1983). The response of fishes to periodic spring floods in a southeastern stream. *Am. Midl. Nat.* **109**: 1–14.

Rubec, P. J. (1988). The need for conservation and management of Philippine coral reefs. *Envir. Biol. Fishes* **23**: 141–154.

Ruiz, G. M., Hines, A. H., & Posey, M. H. (1993). Shallow water as a refuge habitat for fish and crustaceans in non-vegetated estuaries: an example from Chesapeake Bay. *Mar. Ecol. Progr. Ser.* **99**: 1–16.

Sale, P. F. (1980). The ecology of fishes on coral reefs. *Oceanogr. mar. Biol. a. Rev.* **18**: 367–421.

Sale, P. (1988). Perception, pattern, chance and the structure of reef fish communities. *Envir. Biol. Fishes* **21**: 3–15.

Savino, J. F. & Stein, R. A. (1989). Behavior of fish predators and their prey: habitat choice between open water and dense vegetation. *Envir. Biol. Fishes* **24**: 287–293.

Schlosser, I. J. & Toth, L. A. (1984). Niche relationships and population ecology of rainbow (*Etheostoma caeruleum*) and fantail (*E. flabellare*) darters in a temporally variable environment. *Oikos* **42**: 229–238.

Shulman, M. J., Ogden, J. C., Ebersole, J. P., McFarland, W. N., Miller, S. L., & Wolf, N. G. (1983). Priority effects in the recruitment of juvenile coral reef fishes. *Ecology* **64**: 1508–1513.

Simpson, B. R. C. (1979). The phenology of annual killifishes. *Symp. zool. Soc. Lond.* No. **44**: 243–261.

Southwood, T. R. E. (1981). Bionomic strategies and population parameters. In *Theoretical ecology: principles and applications* (2nd edn): 30–52. (Ed. May, R. M.). Blackwell, Oxford.

Stearns, S. C. (1992). *The evolution of life histories*. Oxford University Press, Oxford.

Stein, D. L. (1980). Aspects of reproduction of liparid fishes from the continental slope and abyssal plain off Oregon, with notes on growth. *Copeia* **1980**: 687–699.

Strauss, R. E. (1990). Predation and life-history variation in *Poecilia reticulata* (Cyprinodontiformes: Poeciliidae). *Envir. Biol. Fishes* **27**: 121–130.

Thibault, R. E. & Schultz, R. J. (1978). Reproductive adaptations among viviparous fishes (Cyprinodontiformes: Poeciliidae). *Evolution* **32**: 320–333.

Tokeshi, M. (1994). Community ecology and patchy freshwater habitats. *Symp. Br. ecol. Soc.* **34**: 63–91.

Townsend, C. R. & Winfield, I. J. (1985). The application of optimal foraging theory to feeding behaviour in fish. In *Fish energetics: new perspectives*: 67–98. (Eds Tytler, P. & Calow, P.). Croom Helm, London.

Travis, J. T., Farr, J. A., Henrich, S., & Cheong, R. T. (1987). Testing theories of clutch overlap with the reproductive ecology of *Heterandria formosa*. *Ecology* **68**: 611–623.

Trendall, J. T. (1983). Life history variation among experimental populations of the mosquitofish, *Gambusia affinis*. *Copeia* **1983**: 953–963.

Trendall, J. T. (1988). Recruitment of juvenile mbuna (Pisces: Cichlidae) to experimental rock shelters in Lake Malawi, Africa. *Envir. Biol. Fishes* **22**: 117–131.

Veith, W. J. (1979). Reproduction in the live-bearing teleost *Clinus superciliosus*. *S. Afr. J. Zool.* **14**: 208–211.

Victor, B. C. (1986). Delayed metamorphosis with reduced larval growth in a coral reef fish (*Thalassoma bifasciatum*). *Can. J. Fish aquat. Sci.* **43**: 1208–1213.

Vondracek, B., Wurtsbaugh, W. A., & Cech, J. J. (1988). Growth and reproduction of the mosquitofish, *Gambusia affinis*, in relation to temperature and ration level: consequences for life history. *Envir. Biol. Fishes* **21**: 45–57.

Warburton, K. (1989). Ecological and phylogenetic constraints on body size in Indo-Pacific fishes. *Envir. Biol. Fishes* **24**: 13–22.

Ware, D. M. (1975). Relation between egg size, growth, and natural mortality of larval fish. *J. Fish. Res. Bd Can.* **32**: 2503–2512.

Weitzmann, S. H. & Vari, R. P. (1988). Miniaturization in South American freshwater fishes; an overview and discussion. *Proc. biol. Soc. Wash.* **101**: 444–465.

White, P. S. & Pickett, S. T. A. (1985). Natural disturbance and patch dynamics: an introduction. In *The ecology of natural disturbance and patch dynamics*: 3–13. (Eds Pickett, S. T. A. & White, P. S.). Academic Press, San Diego.

Wieser, W. (1991). Limitations of energy acquisition and energy use in small poikilotherms: evolutionary implications. *Funct. Ecol.* 5: 234–240.

Winemiller, K. O. & Rose, K. A. (1993). Why do most fish produce so many tiny offspring? *Am. Nat.* **142**: 585–603.

Winterbottom, R. & Emery, A. R. (1981). A new genus and two new species of gobiid fishes (Perciformes) from the Chagos Archipelago, central Indian Ocean. *Envir. Biol. Fishes* **6**: 139–149.

Wood, E. (1985). *Exploitation of coral reef fishes for the aquarium trade.* Marine Conservation Society, Ross-on-Wye.

Woodhead, A. D. (1979). Senescence in fishes. *Symp. zool. Soc. Lond.* No. 44: 179–205.

Wootton, R. J. (1984). Introduction: strategies and tactics in fish reproduction. In *Fish reproduction: strategies and tactics*: 1–12. (Eds Potts, G. W. & Wootton, R. J.). Academic Press, London.

Wootton, R. J. (1994). Energy allocation in the threespine stickleback. In *The evolutionary biology of the threespine stickleback*: 114–143. (Eds Bell, M. A. & Foster, S. A.). Oxford University Press, Oxford.

Wourms, J. P., Grove, B. D. & Lombardi, J. (1988). The maternal–embryonic relationship in viviparous fishes. In *Fish physiology* **11**. *The physiology of developing fish. Part B. Viviparity and posthatching juveniles*: 1–134. (Eds Hoar, W. S. & Randall, D. J.). Academic Press, San Diego.

Youson, J. H. (1988). First metamorphosis. In *Fish physiology* **11**. *The physiology of developing fish. Part B. Viviparity and posthatching juveniles*: 135–196. (Eds Hoar, W. S. & Randall, D. J.). Academic Press, San Diego.

Symp. zool. Soc. Lond. (1996) No. 69: 201–224

Small size in amphibians—its ecological and evolutionary implications

B. T. CLARKE

Department of Zoology
The Natural History Museum
Cromwell Road
London SW7 5BD, UK

Synopsis

An overview of the ecological and evolutionary implications of reduced body size in amphibians is presented. Three significant formative influences in the evolution of the modern lissamphibian morphotypes—the limbless, worm-like caecilians, the tailed tetrapod urodeles, and the squat, tailless anurans—are outlined. These influences are the shift from an aquatic to a terrestrial lifestyle, the development of concealment and water conservation strategies leading to fossoriality, and a reduction in body size. The last of these ultimately leads to the evolution of extremely small adult body size and is examined in greater detail. Miniaturization may occur sporadically in distantly related forms or in more closely related species, within a monophyletic lineage. Two kinds of miniaturization are identified—primary and secondary. Primary miniaturization involves some form of radical morphological reorganization. Secondary miniaturization produces morphologically similar forms differing mainly in size. A systematic survey of small and miniature amphibians is presented and comment is provided on the morphological consequences of size reduction for each of the three amphibian orders. The ecological, behavioural, and evolutionary advantages of reduced body size to small and miniature amphibians are discussed. The need for further data, particularly accurate ecological and behavioural observational work, is noted.

Introduction

The amphibians are the oldest group of extant terrestrial vertebrates, having their origin back in the Devonian, over 360 million years ago (Duellman & Trueb 1986). Whether one considers the dipnoan lungfish or the panderichthyid osteolepiforms as ancestral to the Amphibia, it is evident that both the ancestral stock and the early amphibians were considerably larger than their modern descendants. There are only three main living groups of lineages of modern amphibians, each a separate order with a distinctive morphotype. The three comprise the limbless, worm-like Gymnophiona (caecilians), the primitive, tailed Urodela (newts and salamanders),

ZOOLOGICAL SYMPOSIUM No. 69
ISBN 0–19–857787–7

and the highly derived, tailless, squat bodied Anura (frogs and toads). There have been at least three major evolutionary trends which have operated on all three lineages throughout their history and continue to operate today; each has had a role in shaping the three distinctive morphotypes. The three trends are (1) a primary ecological shift from an aquatic to a terrestrial environment, which may be continued as a move toward a progressively more terrestrial habit, (2) the evolution of fossoriality, initially developed as a means of concealment, in leaf litter and the upper soil layers, which in some cases led to burrowing, and (3) a recurrent evolutionary tendency toward the production of forms with a reduced body size. This paper concerns the last of these—size reduction in amphibians and its ecological and evolutionary significance, especially the evolution of extreme small size or miniaturization. Contributions on the other two major evolutionary trends are in preparation and will appear elsewhere (B. T. Clarke in prep. a, b).

This paper is concerned with physical size in amphibians and uses the terms 'small' and 'miniature' to describe states of reduced body size. However, Hanken & Wake (1993), and D. B. Wake (in litt.) have also introduced the concept of 'biological size' in reference to organisms with a large cell/large genome size. There is a complex relationship between an organism's genome size, its cell size and its metabolic and developmental rates. Cell size and genome size are positively correlated and species with large cells/a large genome size frequently exhibit the adult morphology and developmental and growth patterns of physically smaller forms. These relationships produce some interesting effects; D. B. Wake (in litt.) regards *Necturus maculosus*, the mudpuppy, at over 150 mm snout–vent length (SVL), as miniature because it has very large cells and a large genome size (in pg DNA). The smallest known species of salamanders, genus *Thorius*, have a genome size of *c*. 25 pg DNA and hence individually large cells and are thus profoundly small in Wake's terms, being both physically and biologically miniaturized.

I regard biological size as a separate issue from physical size. Biological size is a developmental consequence of increased genome size—a factor independent of physical size. Possession of physically small size has definite ecological, behavioural, and evolutionary consequences which are not found in physically normal or large-sized species. A new term is needed to replace 'biological size', one which stresses the relationship between genome size and its developmental consequences. Biological smallness *sensu* D. B. Wake is a special case which is not relevant to a consideration of the effects of reduced physical size. The difference in viewpoint is clear: Wake is concerned specifically with development (organism–internal environment relationships), whereas I am concerned with the interaction between an organism's physical size and its habitat (organism–external environment relationships). In the context of this paper, 'size' refers to physical size unless otherwise stated.

Size concepts and categories—looking for boundaries

The largest and smallest species of any type of animal are the most readily identifiable of size categories; between these two extremes lies the total known variation in amphibian body size. The problem is, how do we partition this range of size variation into groups or classes to facilitate meaningful discussion and further investigation of size-related issues? What do we mean by 'large', 'normal' (or 'medium'), 'small', and 'miniature' with respect to amphibians, when most amphibians are of small or miniature size relative to most other vertebrates? Are these merely arbitrary designations or is there a correlation between size category, morphology, and other biological attributes such as ecology (especially factors like resource partitioning) and reproductive behaviour in amphibians? Body size may be expressed in two main ways: either as a measurement (or set of measurements), or in terms of the relationships between size and the animal's mode of life. Hanken & Wake (1993: 502), for example, use the idea of animals having a body size with a value along the 'same body-size continuum', yet go on to discuss the concept of a 'critical size' for the smallest or miniature size category. The difficulty for the biologist working on miniature forms is drawing a distinction between taxa that are merely smaller than 'normal' and ones which are in some sense truly miniature. In other words, for amphibians, we are looking for clearly identifiable size-correlated changes in their overall biology, specifically those which signal a transition from medium to small and, in particular, from small to miniature size ranges (Hanken 1982, 1983a, b, 1993). These changes in lifestyle attributes possess the quality of objective reality (i.e. the existence of such change is independent of the presence of an observer). It is in this sense that Hanken & Wake (1993) raise the issue of the concept of a 'critical size' for significantly small forms. They point out that reduction in body size has related effects on anatomy, ecology, life history, and behaviour, and comment on a correlation between reduced size and physiological and ecological functions, reproductive strategies, and other major changes in 'the way an organism deals with its ancestral adaptive zone'. Ideally, size categorization should identify a group, in terms of a range of size values in millimetres, whose members share an exclusive or near-exclusive set of size-related biological attributes.

Nonetheless, arbitrary size categories can be useful and their importance is increased when correlations emerge between specific linear measurements and qualitative changes. Linear size measurements (arbitrary and correlated) provide a starting point for estimating the evolutionary frequency of body-size reduction. Even using arbitrary values enables us to make statements like those given in Hanken & Wake (1993), that using a standard length of 30 mm or less as a criterion for miniature status for plethodontid salamanders suggests that miniature forms independently evolved at least ten times in the evolutionary history of the family.

Small or miniature? A problem of definition

Hanken & Wake (1993), albeit unintentionally, provide two slightly different definitions of miniaturization. In the introduction to their paper, they take it to mean the 'evolution of extremely small *adult* body size', while on the following page under the heading 'What is miniaturization?' they regard it as 'the evolution of extremely small body size *within a lineage*' (my italics). These are, in fact, two quite separate usages of the same term. In the first, it is used in a general sense; here 'miniaturization' describes a size relationship with larger, comparable forms. This need not indicate comparison with a genealogically close relative or relatives; the comparison may be with a form(s) which is merely structurally sufficiently similar to permit comparison. Miniaturization may be used in this sense to refer, for example, to all minute frogs as a group within the context of the Anura, regardless of their taxonomic relationships. The second sense is a restricted usage, specifically requiring close phylogenetic relationship of group members under consideration. In my estimation, the qualification that we are dealing with adult body size should either be understood or included in both definitions. The Hanken/Wake definitions draw attention to some important implications of the concept of miniaturization: (1) the concept is relational—to regard a species as miniature implies the existence of a similar larger form; (2) miniaturization is an evolutionary process which gives rise to miniature species, but in the general case—for example, all miniature frogs within the Anura—possession of extreme small size does not necessarily imply close phylogenetic relationship; (3) as a description, 'extremely small' presumably indicates a grouping of species with low size values rather than *the* single smallest species. Consequently, the distinction between miniature forms and ones which are merely small is left unspecified. Yet, at the same time, distinguishing miniaturization as an identifiable evolutionary process and miniature amphibians as a separate group implies a distinction between miniature and non-miniature forms. We need to ask a range of questions when addressing apparent size-related change. Where is the boundary line between miniature and small species? Can we provide objective evidence that the small/miniature boundary is a real one, citing biological attributes which are correlated with size, or is it only possible to give an approximate measurement or range of measurements to indicate small and miniature size-class membership? Are the observed qualitative character changes a function of reduced body size or are they a consequence of other ecological/environmental factors, independent of size considerations?

Miniaturization in amphibians—features of the evolutionary process

It is important to note that, when speaking of miniaturization as a process, it is possible to draw a distinction between two types—primary and secondary miniaturization. Primary miniaturization refers to the marked reduction in body size associated with major morphological and evolutionary change. It has been suggested that miniaturization (or more properly, size reduction) has played an important role in the evolution of the Lissamphibia (Milner 1988, 1993). This form of radical reduction in size, which may be accompanied by evolutionary novelty, characterizes primary miniaturization. Secondary miniaturization involves less radical change and is seen as a size reduction in one or more species within a lineage, which otherwise remain morphologically similar. It is important to specify taxonomic level when discussing miniaturization, particularly primary miniaturization where morphological novelty may be associated with a new ecological/evolutionary initiative for a particular group. Specifying the taxonomic level clearly distinguishes significant morphological change which takes place early in the history of a major group (like the Amphibia or the Urodela), responsible for shaping the lineage morphotype, from more minor changes to the morphotype seen in (relatively) lower level taxa like the Plethodontidae. This proviso is especially important at a time when phylogenies at family-level remain equivocal (see comments below in the section on size classes and systematic distribution of small and miniature amphibians).

A brief survey of small and miniature amphibians

Size classes and systematic distribution of small and miniature amphibians

Caecilians, urodeles, and anurans are sufficiently different in body form to require different criteria for small and miniature status. At 75 mm the smallest known caecilian, *Idiocranium russeli*, is too long to be regarded even as 'small' in comparison with urodeles and is three times my suggested maximum value for 'small' status in anurans (see below).

Unequivocal family-level phylogenies remain elusive for each of the three amphibian orders (Nussbaum & Wilkinson 1989; Hillis 1991; Nussbaum 1991; D. B. Wake 1991; M. H. Wake 1993; Larson & Dimmick 1993; Hedges, Nussbaum & Maxson 1993), but there is little doubt regarding the identities of the most primitive families in any of the three orders. This is particularly useful information in respect of the evolutionary significance of miniaturization, because we need to be able to identify the primitive families in order to distinguish forms with a primary reduction in body size from those which are secondarily miniaturized.

Caecilians

The caecilians present special problems in terms of resolving qualitative and quantitative criteria to derive any of the size-categories because of the lack of information on many aspects of caecilian biology and the inadequacy of using just a body length measurement as an indicator of size in primarily fossorial organisms with a cylindrical body and a short tail. Differences in the body width make categorizing small and miniature caecilians particularly difficult because long, narrow-bodied species may exhibit features associated with body-size reduction while short, wide-bodied species may not. These problems are especially acute in the case of the small/miniature dichotomy because of the effects of body width on cranial morphology, notably the size and degree of ossification of the various cranial elements and the level of contact between them (M. H. Wake 1986a)—features which may be used to provide an osteological perspective on size-class determination. I suggest that caecilians with a total length of 150 mm or less are miniature, while those >150 mm up to 250 mm be classed as small.

Only two to four species qualify as miniature—*I. russeli* and *Grandisonia brevis* and perhaps *Grandisonia diminutiva* (but this may be an invalid species *fide* Nussbaum & Wilkinson 1989) and possibly *Ichthyophis billitonensis* (only known from a single specimen at 135 mm, believed to be adult size by Taylor 1968). Many other species straddle the miniature/small categories; some caecilians seem to have the ability to carry on growing with age. Taylor records an astonishing size range for *Geotrypetes seraphini occidentalis* from 101 to 355 mm total length, where the 101 mm individual is described as apparently adult in dentition and in lacking gill slits. Taylor (1968) is not consistent in his descriptions of size versus length; *Gegenophis carnosus* at 162–172 mm (although 'greatest known length 170 mm' according to his diagnosis) is 'a small species' yet *Siphonops hardyi* and *Microcaecilia albiceps* are 'diminutive species' at 142–178 and 168–227 mm, respectively. In some cases, the ratio of body length to width is clearly a consideration (see the Appendix). Table 1 lists miniature and small species of caecilians. Other points of interest include the frequency and distribution of small species in families and within genera, and the high incidence of very small forms in some genera (see Table 1 and the Appendix). With a single exception, *Microcaecilia* species are <250 mm total length and include small but no miniature species as presently determined (the exception being *Microcaecilia supernumeraria* at 258 mm). *Uraeotyphlus* includes only one species, out of four, larger than 250 mm.

At family level, the evolutionary relationships of the caecilians are closer to a stable state of resolution than either the urodeles or the anurans (Hillis 1991). Nussbaum (1977, 1991) has shown that the rhinatrematids, the ichthyophiids, and the uraeotyphlids are the most primitive (plesiomorphous) of the caecilian families. Undoubtedly the most significant observation concerning caecilian size is the relatively small size of primitive rhinatrematid caecilians. Although based on few individuals, it is noticeable that, of the nine species of rhinatrematids, *Rhinatrema bivittatum*, and three *Epicrionops* spp. are all 220 mm or less in length (see Table 1). On this basis, I suggest that the caecilians may be the only group of extant amphibians to show signs of primary miniaturization associated with the genesis of the Lissamphibia—the rhinatrematids being the only living, small-sized primitive amphibians.

Urodeles

I follow the size classes of D. B. Wake & Lynch (1976) and regard their class 1, 35 mm or less, as miniature; their class 2, 40–55 mm, leaves an apparent discontinuity, so I regard small salamanders as being 36–55 mm. Phylogenies for urodeles at familial level remain equivocal (Hecht & Edwards 1977; Duellman & Trueb 1986; Larson & Wilson 1989; Hillis 1991; Larson 1991). None the less, as in the case of the caecilians and anurans, there is some general agreement regarding the identity of the primitive members of the order. Of the two primitive families, the Cryptobranchidae comprises three large species (Liu 1950; Dundee 1971) and the Hynobiidae includes species of 10 cm or more (Halliday & Verrell 1986). There is therefore no evidence, in the modern urodele families, of the primary miniaturization which Milner (1993) associates with the origin of the Lissamphibia.

Table 1. Miniature and small species of caecillians. Data from Taylor (1968) unless otherwise indicated.

Species	Total length (mm)
(a) Miniature	
Caeciliidae	
Idiocranium russeli	75–114[a]
Grandisonia brevis	112
Grandisonia diminutiva	64–95
Ichthyophiidae	
Ichthyophis billitonensis	135 ('approaching . . . if not completely adult')
(b) Small–miniature	
Rhinatrematidae	
Epicrionops b. bicolor	230 (one specimen)
Epicrionops b. subcaudalis	188–270
Epicrionops columbianus	161 (one specimen)
Epicrionops lativittatus	218 (one specimen)
Epicrionops parkeri	208 (one specimen)
Rhinatrema bivittatum	207; 188 (two specimens)
Caeciliidae	
Boulengerula changamwensis	147–234
Brasilotyphlus brasiliensis	191–260
Gegenophis carnosus	162-172 ('a small species')
Gegenophis fulleri	220 (3rd sp. *Gegenophis ramaswamii* to 340)
Geotrypetes angeli	202–234
Geotrypetes congoensis	159–234 ('a small species')
Geotrypetes pseudoangeli	157–234
Grandisonia sechellensis	135–175 (190)
Grandisonia larvata	97–220
Herpele multiplicata	239 (second sp. in genus: *Herpele squalostoma* is large)
Indotyphlus battersbyi	170–220
Microcaecilia albiceps	168–227 ('a diminutive species')
Microcaecilia rabei	150–185
Microcaecilia taylori	to 172[b]
Microcaecilia unicolor	170–235
Parvicaecilia pricei	143–191 ('a diminutive species')
Praslinia cooperi	168–230
Siphonops hardyi	142–178 ('a diminutive species')
Siphonops insulatus	145–200
Ichthyophiidae	
Caudacaecilia asplenia	191; 202; 207 (three specimens)
Uraeotyphlidae	
Uraeotyphlus malabaricus	145–234
Uraeotyphlus menoni	207–245 (yet described as a 'medium sized species')
Uraeotyphylus narayani	199–237
[*Uraeotyphlus oxyurus*	Large to 305 – the fourth species in the genus]
(c) Small or miniature to >250 mm	
Caeciliidae	
Boulengerula uluguruensis	180–266
Boulengerula boulengeri	123–275 (to 308)
Geotrypetes s. seraphini	195–309 (yet 'a moderately small species')
Geotrypetes s. occidentalis	101–355 (the 101mm individual apparent adult in dentition and absence of gill slits: a very wide size range)
Parvicaecilia nicefori	198–263 (four specimens: 198, 221, 242, 263; 'a slender elongate diminutive species'
Ichthyophiidae	
Caudacaecilia larutensis	169; 252 (two specimens)
Caudacaecilia weberi	209–258

[a] Data from M. H. Wake (1986a)
[b] Data from Nussbaum & Hoogmoed (1979)

Only one family of modern urodeles includes miniature species—the lungless salamanders, family Plethodontidae (see Table 2 for data on *Thorius* and a representative sample of other plethodontid salamanders). In most phylogenetic analyses, notably those based on morphological characters, the plethodontids are seen as an advanced family. However, using data from ribosomal RNA sequence variation, the urodele phylogeny is virtually inverted and the plethodontids emerge as a basal taxon (Larson & Wilson 1989; Larson 1991). If true, this would reopen the question of their status as a group exhibiting primary miniaturization. Combining the available data sets, Larson & Dimmick (1993) concluded that the plethodontids, along with the amphiumids and rhyacotritonids, were early offshoots of the clade including all other salamanders except the basal sirenids, cryptobranchids, and hynobiids. Thus, in both the morphological and combined data models the plethodontids, and hence the living salamanders, are secondarily miniaturized. Unless it can be shown that the plethodontids have evolved morphological novelties which take them on a new evolutionary trajectory, they remain as a secondarily miniaturized group. In the latter case, they could only be thought of as exhibiting primary miniaturization in the context of the Plethodontidae but not of the order Urodela. It is important that taxonomic level is clearly stated or understood when a group is described as primarily or secondarily miniature. The Plethodontidae, containing approximately two-thirds of all the known urodele species (D. B. Wake 1966), has a high incidence of miniaturization, with 20 species at 30 mm standard length or below (D. B. Wake 1992).

Anurans

For anurans the issue of small size is more straightforward, since the body form is compact and tailless. There is not the problem of a slender versus a stout body influencing size category designations as in caecilians, nor the difficulty of large cell/large genome size giving rise to a concept of 'biologically' as opposed to physically small species as in urodeles. I regard 25–30 mm snout–vent length (SVL) as a critical division in anurans, at which there begin to be found physiological and ecological modifications of the kind mentioned by Hanken & Wake (1993), in particular in the area of their reproductive biology. For anurans, I would adopt 20–25 mm as 'small' and 20 mm or less as 'miniature'. Interestingly, it seems that Lynch is of a similar opinion; in a paper describing two new species of minute leptodactylids from the Andes of Peru and Ecuador he defines 'minute' as meaning 'adults less than 20 mm SVL' (Lynch 1986). It seems that 20 mm SVL is a critical point in anuran size; many species have females which just get to 21–22 mm—which on a strict application of the 20 mm rule removes them from miniature status. Trueb & Alberch (1985) suggested that anurans less than 13 mm SVL be regarded as 'dwarf', 14–25 mm as 'small'—values which are similar to those given above, but ones which I consider too restrictive in the case of the extreme-small-size category. The anurans show a higher incidence of, and more extreme, miniaturization than any other terrestrial vertebrate group (see Table 3 for a representative sample of low-end size variation in the Anura). Both factors make them a particularly suitable group for the study of miniaturization.

Table 2. Urodeles
(a) Miniature species: Plethodontidae, genus *Thorius*

Species	Males			Females		
	Standard length (mm)	$\bar{x}$ (mm)	n	Standard length (mm)	$\bar{x}$ (mm)	n
Thorius aureus[a]	21.1–29.3	25.9	21	22.6–34.9	29.1	20
Thorius arboreus[a]	16.1–18.4	17.0	4	15.2–20.0	17.2	7
Thorius boreas[a]	25.1–29.9	27.4	21	24.7–34.9	29.7	20
Thorius smithi[a]	Not known			20.3; 20.4	20.4	2
Thorius insperatus[a]	Not known			19.5	19.5	1
Thorius macdougalli[a]	17.5–22.6	20.2	12	17.6–26.5	18.9	11
Thorius dubitus[b]	21 (type; maximum size 22)			Not given		
Thorius troglodytes[b]	26 (type + maximum size)			Not given		
Thorius maxillabrochus[c]	24.3–24.5	24.4	2	Not known		
Thorius minutissimus[d]	21.5–24.0	22.6	9	22.5–27.0	24.1	9
Thorius narisovalis[d]	23.0–27.0	25.4	10	26.5–31.0	27.8	10
Thorius pennulatus[d]	17.0–21.0	18.9	10	17.0–21.0	18.9	10
Thorius pulmonaris[e]	27.2–27.5	27.4	2	24.4	24.4	1
Thorius schmidti[d]	22.5–24.5	23.5	8	23.5–26.5	25.0	10

Data from: [a]Hanken & Wake (1994); [b]Taylor (1941); [c]Gehlbach (1959); [d]Hanken (1982); [e]Taylor (1939).

(b) Miniature salamanders other than *Thorius* (D. B. Wake & Lynch, 1976) size class 1–35 mm or less

Bolitoglossa chica, Bolitoglossa minutula ('smallest species of the genus'), *Bolitoglossa occidentalis, Bolitoglossa peruviana, Bolitoglossa riletti, Bolitoglossa rufescens.*

Chiropterotriton barbouri, Chiropterotriton nasalis, Chiropterotriton picadoi, Chiropterotriton richardi [all now *Nototriton* D. B. Wake & Elias, 1983]:
Chiropterotriton bromeliacia, Chiropterotriton cuchumatanus, Chiropterotriton megarhinus, Chiropterotriton rabbi, Chiropterotriton xolocalae [=*Dendrotriton* D. B. Wake & Elias, 1983] *Chiropterotriton chondrostega, Chiropterotriton dimidiatus.*

Parvimolge praecellens [=*Pseudoeurycea*], *Parvimolge townsendi.*

Nototriton tapanti (only known from the holotype; 23.5 mm SL), *Nototriton guanacaste* ('a moderately large species of *Nototriton*' at max. of 29.7 mm), *Nototriton abscondens* (similar; to 33 mm, $\bar{x}$ 10 males 27.5 mm, 10 females 28.0 mm), *Nototriton picadoi* (similar; 'largest known is male at 32 mm'), *Nototriton richardi* ('diminutive'; to about 24 mm SL, $\bar{x}$ 4 males 22.4 mm, 4 females 22.5 mm).

Data on *Bolitoglossa, Chiropterotriton* and *Parvimolge* from D. B. Wake & Lynch (1976); *Nototriton* from Good & Wake (1993).

Table 3. Anura: a representative sample of miniature anuran species 20 mm SVL or less (m, male; f, female)

Species	SVL (mm)	Reference
Brachycephalidae		
Psyllophryne didactyla	m 8.6; f 10.2	Izecksohn (1971)
Brachycephalus ephippium	'less than 16 mm'	Duellman & Trueb (1986)
Bufonidae		
Didynamipus sjostedti	m 15.5; f 17.0–19.3	Grandison (1981)
Dendrobatidae		
Dendrobates bombetes	m 16.7–18.5; f 17.2–19.8	Myers & Daly (1980)
Dendrobates fulguritus	m 13.5–15.0; f 14.0–16.5	Silverstone (1975)
Dendrobates minutus	m 12.0–15.0; f 12.0–15.5	Silverstone (1975)
Dendrobates opisthomelas	m 14.5–18.5, f 14.5–19.5	Silverstone (1975)[a]
Dendrobates viridis	m to 14.0; f to 15.2	Myers & Daly (1976)
Hylidae		
Limnoaedus ocularis	SVL '11–20 mm'	Franz & Chantell (1978)
	11–17 mm	Behler & King (1979)
Leptodactylidae		
Sminthillus limbatus	'less than 15 mm SVL'	Lynch (1971)
Adelophryne adiastola	m 13.0–13.7; f 13.9	Hoogmoed & Lescure (1984)
Adelophryne baturitensis	m 11.8–13.4; f 12.2–16.3	Hoogmoed, Borges & Cascon (1994)
Adelophryne gutturosa	m 12.6–14.5; f 12.4–13.0	Hoogmoed & Lescure (1994); Hoogmoed *et al.* (1994)
Adelophryne maranguapensis	m 11.9–12.6; f 17.4	Hoogmoed *et al.* (1994)
Adelophryne pachydactyla	m 11.1; f ?	Hoogmoed *et al.* (1994)
Eleutherodactylus sisyphodemus	m 12.0–13.8; f 15.6–17.9 (12.0–17.9)	
Phyllonastes heyeri	m 12.9–14.1; f 13.1–15.9	Lynch (1986)
Phyllonastes lochites	m ?; f 14.9	Lynch (1976)
Phyllonastes myrmecoides	m ?; f 12.0–13.6	Lynch (1976)
Phrynopus bagrecito	m 13.8–16.3; f 14.4–18.6	Lynch (1986)
Phyzelaphryne miriamae	m 14.6–15.1; f 19.4–20.0	Hoogmoed & Lescure (1984)
Microhylidae		
Synapturanus rabus	16.2–19.0; f holotype 17.2	Pyburn (1976)
Myobatrachidae		
Uperoleia minima	16 mm	*fide* Hanken (1993)
Ranidae		
Batrachylodes minutus	m 15.8–17.7; f 16.7–19.3	Brown & Parker (1970)
Cacosternum nanum parvum	m 14.3–15.7; f ?	Poynton (1963)
Phrynobatrachus alticola	m 16; f 20	Guibé & Lamotte (1963)
Phrynobatrachus ghanensis	m 13.2–13.8; f ?	Schiøtz (1964)
Phrynobatrachus cornutus	m 14–16; f 18–20	Perret (1988)
Phrynobatrachus taiensis	m 14; f ?	Perret (1988)
Phrynobatrachus villiersi	m 12–13; f 14–14.5	Perret (1988)

Many *Phrynobatrachus spp.* are in the 16–22 mm size range.

[a] Silverstone also lists *Dendrobates quinquevittatus* as a member of the *minutus* group; but it just exceeds the 20 mm SVL criterion (males 14.5–20.0 mm, females 15.0–21.5 mm). Myers (1987) transferred the above-mentioned *Dendrobates* species to *Minyobates*.

A stable family-level phylogeny for the anurans is still conspicuously lacking (see e.g. Hillis 1991; Ford & Cannatella 1993; Hillis, Ammerman, Dixon, & de Sa 1993). However, once again the identity of the relatively primitive family is less open to dispute. Most authorities would recognize the ribbed frog families, the Leiopelmatidae and Discoglossidae (where the Leiopelmatidae includes the Ascaphidae, and the Discoglossidae is a monophyletic family including the Bombinatoridae *fide* Clarke 1988). The primitive anurans, the discoglossoid frogs (Leiopelmatidae and Discoglossidae), are represented by species which are 30 mm SVL or larger (Clarke 1988) which places them in the medium or 'normal' size category. Consequently, there is no evidence, in the modern Anura, of the primary miniaturization which Milner (1993) associates with the origin of the Lissamphibia. Thus small and miniature anuran species, being found only in the relatively more advanced families, exhibit secondary miniaturization in the context of the Anura as a whole. Miniature forms occur sporadically throughout the Anura, being found in both 'advanced' and 'transitional' families (*sensu* Lynch 1973), including the Arthroleptidae, Brachycephalidae, Bufonidae, Dendrobatidae, Leptodactylidae, Microhylidae, Myobatrachidae, and Ranidae (Clarke 1989, with additions). Small bufonids are apparently rare, but this may be sampling error; indeed it is likely that we underestimate the number of small and miniature species in particular (Clarke 1989). It seems likely that there is an increased incidence of miniature anurans in and around grass tussocks, in leaf litter, and certain kinds of arboreal habitats (leaf surfaces, leaf axils, and holes in e.g. bamboo stems), but further data are needed to verify the significance of the relationship. Miniature anurans, like miniature caecilians and urodeles, are more prevalent in hotter, damper areas of the world (see section on water balance, p. 217).

There are many anuran species which have miniature-sized males but small-sized females ('mixed species'). Table 3 includes only species with miniature-sized males and females. This does not imply that mixed species are less important; on the contrary, they may well be particularly significant in understanding the mechanism(s) by which miniature species have arisen.

Morphotype, lifestyle, and trends in body size reduction

It has been suggested that the ancestors of the modern orders of amphibian diverged, each taking different ecological pathways; the caecilians became worm-like burrowers, the urodeles remained on the soil surface taking slow-moving invertebrate prey, while the early anurans became sit-and-hide strategists, pressing themselves into the soil (later on digging themselves in backwards), emerging rapidly to snap at fast-moving prey (Hillenius 1976). This theory, outlined by Hillenius, has much to commend it, particularly if more recent data on Triassic and Jurassic amphibians are considered (Rage & Roček 1986, 1989; Jenkins & Walsh 1993; also Clarke in prep. a, b). A reduction in body size would have enabled early lissamphibian lineages to enter new, physically smaller niches, probably under cover, e.g. in leaf litter on the forest floor or in spaces under logs and rocks; situations which may have stimulated the development of burrowing behaviours. The caecilians, for example, have evolved a novel mode of burrowing, termed vermiform locomotion by Gaymer (1971). I suggest that miniaturization and burrowing were, and remain, significant factors in the evolution of all three modern amphibian orders—anurans, urodeles, and caecilians. Thus while Hillenius (1976) was mainly concerned with the evolution of the anuran body form (and here some of his more detailed ideas are often highly questionable), his ecological divergence theory is more important because it provides an explanation for the evolution of all three modern lissamphibian morphotypes.

Caecilians

Caecilians have a highly derived vermiform body shape and are primarily adapted to burrowing in soft, damp earth. Evolutionary changes in their head morphology probably reflect a move to relatively drier habitats, away from the immediate vicinity of standing water. Members of the more primitive families have terminal or subterminal mouths, a sensory tentacle which is located closer to the eye than to the nostril (within or adjacent to the eye socket in the most primitive group, the Rhinatrematidae), an eye which is externally visible, not hidden beneath bone, and a higher number of skull bones than other caecilians. The eye is covered by skin and even, in some cases, by bone; in all caecilians the eye is reduced and is functionally progressively replaced by the tentacle, which is formed from 'recycled' structural components of the eye (Billo & Wake 1987; M. H. Wake 1992). In addition, the number of skull bones has been progressively reduced, either by loss or by fusion (Taylor 1969). In the more advanced (derived) families the mouth is strongly subterminal (Nussbaum 1991), taking on a shark-like appearance, the tentacle is more fully developed (small in typhlonectines), and evolutionarily has migrated, ventrally beneath the level of the eye, and anteriorly to a position closer to the nostril (this is also seen in ichthyophiids and uraeotyphlids; in caeciliidids the character is variable, but the composition and relationships of this family need further investigation).

The more primitive caecilian families live near streams and other bodies of water and lay large eggs which are guarded by the female. The eggs hatch into a gilled larval stage which becomes terrestrial on metamorphosis (although the change in form from larva to adult is notably less marked than in anurans). In more advanced caecilians, development changes to direct development with metamorphosis occurring entirely within the egg, the young hatching as small adult morphs, up to and including viviparity (M. H. Wake 1986b). These morphological and reproductive changes may reflect a move to life in more terrestrial habitats, still in damp soil but away from close proximity to streams and free-standing bodies of water.

These changes in head morphology and reproductive mode may be attributed solely to ecological factors associated with increased terrestriality (fossoriality) or to miniaturization following increased utilization of physically small ecological niches, or a combination of the two. Modifications may affect part of the body, as in the highly fossorial scolecomorphids which are characterized by reduction in skull size as well as number of cranial elements (Nussbaum 1985), or the whole body as in *Idiocranium russeli*, where females guard their eggs in small spaces under matted vegetation (Parker 1936; Sanderson 1937). In principle, instances of these kinds of body-size reduction are cases of secondary miniaturization, although they may (as in these two examples) involve an element of morphological novelty.

Urodeles

The urodeles have retained the basic body shape of the early tetrapods and the associated relatively slow-moving gait, where the centre of gravity is low and body–ground contact is maintained. Miniaturization is associated with habitats with reduced niche size; examples include leaf litter, under bark, in bromeliads and other crevices. Modifications to the palmar and plantar surface of the hand and feet are associated with arboreality and log dwelling (increased surface area) and fossoriality (decreased surface area) (D. B. Wake & Lynch 1976). The presence of a highly vascularized throat and buccal pump mechanism which functionally compensate for the loss of the lungs in the plethodontids, and their elongate, low-cross-sectional/high-surface-area body pattern probably facilitated the development of miniaturized forms. D. B. Wake (1966), D. B. Wake & Lynch (1976), and Hanken & Wake (1994) provide useful data on morphologies and habitat partitioning.

Anurans

The anuran body shape is extraordinarily adaptable. I suggest that the wide head, broad body, and absence of a tail is an adaptation to crevice dwelling (i.e. entering limited-space environments) in a wide variety of situations. Body posture is also important when backing into limited spaces—in anurans the anterior half of the body is usually raised clear of the underlying substrate (at approximately a 30–45° angle) and the entire body axis is inclined, with the skull and vertebral column making an angle with the pelvic girdle in the sacral region—an attitude which is made possible by the shortened vertebral column, the ilial-ischial-urostylar configuration, and the powerful associated musculatures. Coupled with hind limbs which are relatively larger than those seen in the urodeles, adapted for saltatory locomotion (even in the more spindly-limbed small and miniature anuran species), these factors produce amphibians which are especially well suited to exploiting small crevices. From such a broad category of ecological niches, anurans are free to lunge forward to feed and retreat into a safe humidity and temperature-stable microenvironment. Small wonder that the anuran morphotype has been so successful in terms of species numbers and diversity.

Small size is more important in crevice-dwelling anurans, while low weight is more important in arboreal forms. The development of toe discs and the presence of a moist, granular/areolate belly skin which functions as an additional adhesive surface, as well as a reduced body weight, all enable tree-living species of anurans to adapt to the rigours of arboreal life at medium to small body size. That is to say, life in the arboreal zone may be achieved by modifications to the anuran model at small size without the necessity for going to the extreme of miniaturization.

Miniature size in arboreal frogs is normally only required by an additional factor—the need to enter especially small spaces. The only miniature 'tree' frog in North America is *Limnaoedus ocularis*, the little grass frog (approximately 17 mm SVL), which as its common name suggests is found in grassy meadows or amongst sedges, exploiting cover at the base of grass and sedge clumps (Ashton & Ashton 1988)—another form of limited space/crevice-like habitat.

Ecological and behavioural implications

Small size confers a variety of ecological advantages, primarily increased habitat and food availability, leading to increased population densities. Small amphibians occupy a narrow band of microhabitats, a range of terrestrial niches similar to those of terrestrial insects and small mammals, which may even share them. Extreme reduction in body size may prove to be as important in the evolution of the Amphibia as the ecological transition of the first amphibians from an aquatic to a terrestrially based existence. In both cases there has been a break-through from a zone of high-level competition to one with a wide range of new, relatively unoccupied niches. Body-size reduction enables species to move into new microhabitats previously unavailable because of incompatibility between animal and niche size. The dual advantages of finding new microecosystems to exploit, where they approach the status of top-end predator, while retreating from their own predators in their earlier, more open environment attend an ecological shift from a large to a spatially-restricted niche.

Understanding the significance of a change in the organism–environment relationship between an ancestral amphibian species and its dwarfed descendant requires a radical shift in our perceptions of the issues of scale, habitat, and niche availability. The degree of change increases as body size decreases, that is to say the effect of scale upon niche availability is greater in a lineage which has undergone a phylogenetic transition from 'small' to 'miniature' size than one in which the change is from 'normal' to 'small' size. For normal-sized amphibians, their environment comprises two components: their main environment, i.e. their habitat—the world at large—and their immediate ecological niche or microhabitat with its own microclimate. As amphibians attain small and, more especially, miniature size they effectively move toward a one-component environment—the microhabitat assumes greater significance because of the 'closer fit' between body size and niche size. This 'closer fit' produces a very stable environment immediately surrounding the small/miniature amphibian's body—a significant advantage for an ectothermic terrestrial vertebrate with a naked, permeable skin. Thus, the microclimate provides a 'buffer' zone between the amphibian and general climatic conditions in the world at large which consequently exert a lesser effect on miniature amphibians than on their larger relatives. For the small and the miniature amphibian, the microclimate effectively becomes the total climate as its microhabitat becomes its whole habitat.

Niche availability

Reduced body size enables amphibians to exploit a wider range of retreat sites, for example, in leaf litter, at the base of sedges and grass tussocks (important sites), in small crevices and holes in and under fallen and rotting trees, tree roots, rocks and boulders, in holes in bamboo stems, or in other suitably sized spaces in natural cover like leaf axils in banana plants and bromeliads; some may hide in burrows of their own making or made by other animals (Clarke 1989). Pending an accurate assessment of the incidence of burrowing modifications to the hands, feet, and snouts of miniature species of amphibians, I suggest that as a group they are less likely to possess such modifications because they have less need for them, given the increased availability of suitable retreat sites. Indeed, Hoogmoed & Lescure (1984: 108) have pointed out that minute species of leptodactylid frogs 'do not seem to be built for digging and might lack the strength for it'. Reduced size and hence lower body weight (mass) also mean that miniature species are able to colonize more delicate arboreal habitats. Like *Limnaoedus ocularis, Dendrobates bombetes* and other members of the *Dendrobates minutus* group live amongst grass tussocks, which offer retreat sites at high density but of limited space (Myers & Daly 1980). *Dendrobates arboreus* and many of the bolitoglossine salamanders are able to live high up in the canopy in the cloud forests of Colombia and Ecuador. Many of the bolitoglossine salamanders have fully webbed feet enabling them to climb over shiny leaf surfaces, modifications which, coupled with reduced size, have contributed to the explosive speciation of these lungless (plethodontid) salamanders and their remarkable range extension into tropical Central and South America.

The potential for small size to lead to high species-density exploitation of favourable environments is an interesting feature of the ecology of small species and deserves further study. Hanken & Wake (1994) give a particularly interesting account of the ability of *Thorius* spp. to co-exist at close quarters with up to three species in one area. At a site in northern Oaxaca each species of *Thorius* was found in sympatry with one or more other species.

Nutrition

The advantages of being small are not confined to 'more homes in the habitat'. Small body size affects basal metabolic rate, nutritional requirements, and water balance, especially via evaporative loss, all of which have ecological implications. Reviews on feeding and feeding strategies are provided by Duellman & Trueb (1986) and Larsen (1992). Reduced body size is positively beneficial in nutritional terms, particularly for an ectothermic animal like an amphibian. Small amphibians probably spend less time on feeding activities, since their nutritional requirements are quickly and easily satisfied. Their food (mainly small insects and other arthropods) is abundant, which presents small-sized amphibians with two clear feeding-strategy options. The first is to take food items as large as physical limitations permit. In this case, the nutritional value of each item represents a high percentage of the total daily food requirement. The time spent feeding is reduced, leaving them more time to pursue other life activities. One problem with this strategy is that small species may be in competition with larger forms for the same food resource; another, that restraint of an insect prey which is large in relation to predator size may prove difficult or even hazardous to the predator. The second strategy is to utilize smaller food items than would be practicable (in energetic terms) for larger species. This means that small and miniature species could feed on very small insects and spiders—even as small as collembolans, thrips, and mites—for which there would be little competition from larger amphibians. There is some evidence to suggest that there is at least a loose correlation between predator size and prey size in amphibians (Christian 1982). Small-sized food resources often have the advantage of being exceptionally abundant. While the energetic effort required to secure a necessary daily intake may be realistic in cost–benefit terms for a small amphibian, it will not be so for a larger species (the energetic cost of obtaining the food being greater than the benefit gained from feeding on it). The ability to utilize an abundant food source, particularly in circumstances of little or no competition, could be a major factor which would directly contribute to the high population densities of small species found in some regions. In the mountainous forests of eastern North America, it has been estimated that the total mass of just the woodland salamanders exceeds that of both the birds and the mammals (Halliday & Verrell 1986). Such high population densities lead to an increase in intraspecific variation, the basis for future evolutionary change and an important factor in the long-term survival of a lineage. In the case of small and more especially miniature forms, food restriction resulting from increased amphibian population densities is even less likely to be a limiting factor than for larger species, because the food resource is less likely to become depleted. Clearly, the same concentration of insects in unit area can support large numbers of smaller amphibians but only a small population of larger forms.

Reproductive strategies

Size effects operate not only on the adult but at all stages of development. It should be noted that such effects are not always simple and straightforward, for example a smaller female body size does not necessarily mean fewer eggs of the same size nor the use of smaller bodies of water for the deposition of smaller eggs and the subsequent development of fewer, small tadpoles. In some cases it may be so, but in others extreme small size in itself may provide sufficient constraint to entail a marked departure from the many eggs/aquatic tadpole strategy (see Salthe & Mecham 1974; Duellman & Trueb 1986). The interaction of body size and ecological situation probably account for altered states in reproductive strategies, both adverse environmental conditions and reduced size acting together to produce a higher than normal incidence of novel reproductive modes. It is evident that adverse environmental factors can produce similar results in normal-sized amphibians (Duellman & Gray 1983; Del Pino 1989) and that reduced size is more likely to be an additional or secondary rather than a primary cause of, for example, direct development of small clutches of large eggs, deposited in a terrestrial site or subject to some form of parental care. More data on size versus environmental effects are needed.

Water balance

The surface area to volume ratio increases as body size (snout–vent length) decreases. This physical consequence of body size reduction is further enhanced by a slim, elongate shape as in the smaller species of caecilians and plethodontid salamanders. Increasing the surface area of their permeable skin brings both problems and benefits. Smaller species of amphibians are potentially more prone to die from desiccation than larger ones. In practice, the problem may be overcome by living in areas with a relatively high humidity, like cloud forests or in situations where evaporative loss is at a minimum (hence also the preference for small living spaces with medium-range humidity values). Alternatively, or in some cases additionally, some forms have developed appropriate methods or techniques for water conservation or efficient means of rehydration, e.g. the pelvic patch especially evident in toads of the genus *Bufo* (McClanahan & Baldwin 1969; Duellman & Trueb 1986)—a vital feature in a smaller, xeric-adapted species like *Bufo quercicus*.

Evolutionary implications

Extreme size reduction may produce two forms of miniaturized amphibian: primary and secondary. Secondarily miniaturized forms are usually scaled-down versions of a larger ancestral form. Primarily miniaturized forms are of greater interest, since this form of miniaturization results in significant morphological (structural) changes to the basic morphotype or 'bauplan' (Rieppel 1984). Fundamental modifications of the latter kind entail the production of evolutionary novelties, shifting the miniaturized species and (potentially at least) its descendants into a new evolutionary pathway. Basic changes of this type, resulting in taxa entering new adaptive zones, have been implicated in the origin of higher taxa (e.g. Wright 1982); which is merely another way of saying that miniaturization has produced an alteration in the future direction of the evolutionary pathway taken by a lineage.

A lineage undergoing either form of miniaturization may be thought of as passing through an adaptive 'bottleneck'; such changes are likely to occur under conditions of high selective pressure. Surviving lineages pass into a new adaptive zone providing new evolutionary opportunities. As the intensity of the selective pressure within the environment decreases, diversification and speciation are re-established. These processes result in the production of a new morphotype(s) followed by relatively minor modifications on the new type(s)—explosive speciation of the kind seen in African anuran genera *Hyperolius* and *Phrynobatrachus* and in the Central and South American salamander genus *Bolitoglossa*. If this 'miniature-founder' model is correct then the ancestral or founding species of the new lineage would be expected to be of small size; one of the smallest, if not the smallest species in the group comprising the lineage. The difficulty in identifying the founder/ancestral amphibian species (if still extant) is the extreme lability of size change in amphibians. If Cope's Rule were inviolate (Cope 1885; Stanley 1973; Hanken 1993), then the ancestral species would be the smallest, with successive speciation events producing (step-wise) larger species. Nature is seldom so obliging. Where the original (or new and different) selective pressures that produced the miniature form in the first place continue unabated, one might equally expect the production of forms even smaller than the founding ancestor.

Size differences in males and females and in neighbouring populations

For the majority of amphibian species there is a marked disparity in size between the sexes; males are smaller than females (Duellman & Trueb 1986). The reason is quite evident: the production of a large number of relatively very large gametes requires female amphibians to build up sufficient energy reserves to meet such a reproductive burden. Available evidence suggests that female amphibians do not grow faster than males, they merely take longer to reach sexual maturity. Males produce large numbers of minute gametes, which is far less of a reproductive burden, so there is no need for male amphibians to attain as large a size as the female. Indeed the exception to this 'rule' is where large males have evolved a particularly aggressive courtship behaviour. This may involve the defence of a territory but is more appropriate in species with particularly strenuous, physical male–male combat encounters, e.g. in *Adelotus brevis*, gladiator frogs—the *Hyla faber* group, *Pyxicephalus adspersus*, and some of the ambystomatid and desmognathine salamanders. Thus, under 'normal' circumstances a male amphibian will attain sexual maturity earlier and at a smaller body size than a female of the same species.

In small and miniature species, the size difference between the sexes seems to be absent or is noticeably less apparent. This may, of course, be a size effect; a 10% size disparity would give small species with females 20 mm, males 18 mm SVL, whereas larger species with females of 60 mm SVL would have males of 54 mm. The important point is, would such a difference in size be of sufficient magnitude to have ecological consequences? Specifically, does it lead to habitat or resource partitioning between the sexes? I suggest that even where a size difference still persists in smaller species, it may no longer be of any selective significance. At 60 and 54 mm females and males may take different size classes of insect food; at 20 and 18 mm this effect is likely to be negligible or absent. Taken in isolation, the loss in ability to draw on food resources of different size would be replaced by competition for the same food resource and, in this respect, miniaturization would become evolutionarily disadvantageous, leading to a reduction in inclusive fitness within the lineage. In practice, the advantage gained from being able to utilize small abundant food items would outweigh the disadvantage of losing the food resource partitioning capability possessed by larger amphibians. Factors like the relative advantages of different food resource utilization strategies may help account for size transformations of populations within evolutionary lineages. If such changes have a geographical basis and result in reproductive isolation from the parent population, then the result will be population differentiation leading to speciation.

Conclusions

Progress on understanding the significance of miniaturization and the underlying mechanisms responsible for body size reduction in amphibians (as in other groups) is seriously hampered by a lack of data. The main problem is lack of the right kind of data. We need basic measurements on good-sized samples of a wider range of species. Such measurements should differentiate between males and females, the sexual maturity of specimens measured must be established and it is desirable that a mean value be included. Data on *Thorius* given by Hanken (1982) and Hanken & Wake (1994) show the standard required. Further ecological and behavioural data are vital: we need a much better idea of how, and under what conditions, most species of amphibians live. This latter is probably the greatest desideratum in amphibian biology and the greatest bar to future progress on miniaturization. My aim in writing this contribution is to provide a basis for further discussion, and to stimulate a wider interest in amphibians as subjects for studying miniaturization and its evolutionary mechanisms.

Acknowledgements

I would like to take this opportunity to thank Marvalee Wake, David Wake, and Jim Hanken for sending me copies of their recent papers and for answering requests for their opinions on various aspects of amphibian miniaturization. I am grateful to Mark Wilkinson for nominating me to present a paper on small amphibians at the Miniature Vertebrates Symposium and to Peter Miller for inviting me to take part. The manuscript benefited from constructive criticism provided by Peter Miller, Jim Hanken, John Poynton, David Wake, and Mark Wilkinson (on caecilian sections). However, all opinions, errors and omissions remain my responsibility. My family, Wendy, Amy, and Amber, provided a happy environment in which to work, which more than anything else made this contribution possible.

References

Ashton, R. E., Jr. & Ashton, P. S. (1988). *Handbook of the reptiles and amphibians of Florida. Part Three. The amphibians.* Windward Publishing Inc., Miami.

Behler, J. L. & King, F. W. (1979). *The Audubon Society field guide to North American reptiles and amphibians*. Alfred A. Knopf, New York.

Billo, R. & Wake, M. H. (1987). Tentacle development in *Dermophis mexicanus* (Amphibia, Gymnophiona) with an hypothesis of tentacle origin. *J. Morph.* **192**: 101–111.

Brown, W. C. & Parker, F. (1970). New frogs of the genus *Batrachylodes* (Ranidae) from the Solomon Islands. *Breviora* No. 346: 1–31.

Christian, K. A. (1982). Changes in the food niche during postmetamorphic ontogeny of the frog *Pseudacris triseriata*. *Copeia* **1982**: 73–80.

Clarke, B. T. (1988). *Evolutionary relationships of the discoglossoid frogs—osteological evidence*. PhD thesis: CNAA.

Clarke, B. T. (1989). Real vs apparent distributions of dwarf amphibians: *Bufo lindneri* Mertens 1955—a case in point. *Amphibia-Reptilia* **10**: 297–306.

Clarke, B. T. (In preparation a.). *Amphibian origins and evolution.*

Clarke, B. T. (In preparation b.) *Burrowing and the evolution of the anuran body form.*

Cope, E. D. (1885). On the evolution of vertebrates. *Am. Nat.* **19**: 140–148, 234–247, 341–353.

Del Pino, E. M. (1989). Marsupial frogs. *Scient. Am.* **260**: 76–84.

Duellman, W. E. & Gray, P. (1983). Developmental biology and systematics of the egg-brooding hylid frogs, genera *Flectonotus* and *Fritiziana*. *Herpetologica* **39**: 333–359.

Duellman, W. E. & Trueb, L. (1986). *Biology of amphibians*. McGraw-Hill Book Company, New York.

Dundee, H. A. (1971). *Cryptobranchus* and *C. alleganiensis*. *Cat. Am. Amphib. Rept.*: 101.1–101.4.

Ford, L. S. & Cannatella, D. C. (1993). The major clades of frogs. *Herpet. Monogr.* **7**: 94–117.

Franz, R. & Chantell, C. J. (1978). *Limnaeodus, L. ocularis*. *Cat. Am. Amphib. Rept.*: 209.1–209.2.

Gaymer, R. (1971). New method of locomotion in limbless terrestrial vertebrates. *Nature, Lond.* **234**: 150–151.

Gehlbach, F. R. (1959). New plethodontid salamanders of the genus *Thorius* from Puebla, Mexico. *Copeia* **1959**: 203–206.

Good, D. A. & Wake, D. B. (1993). Systematic studies of the Costa Rican moss salamanders, genus *Nototriton*, with descriptions of three new species. *Herpet. Monogr.* **7**: 131–159.

Grandison, A. G. C. (1981). Morphology and phylogenetic position of the West African *Didynamipus sjoestedti* Andersson, 1903 (Anura Bufonidae). *Monit. zool. ital.* (N.S.) (Suppl.) **15**: 187–215.

Guibé, J. & Lamotte, M. (1963). La réserve naturelle intégrale du Mont Nimba. XXVIII. Batraciens du genre *Phrynobatrachus*. *Mém. Inst. fr. Afr. noire* No. 66: 601–627.

Halliday, T. R. & Verrell, P. (1986). Salamanders and newts. In *The encyclopaedia of reptiles and amphibians*: 18–35. (Eds Halliday, T. R. & Adler, K.). George Allen & Unwin, London.

Hanken, J. (1982). Appendicular skeletal morphology in minute salamanders, genus *Thorius* (Amphibia, Plethodontidae): growth regulation, adult size determination, and natural variation. *J. Morph.* **174**: 57–77.

Hanken, J. (1983a). Miniaturzation and its effects on cranial morphology in plethodontid salamanders, genus *Thorius* (Amphibia, Plethodontidae). II. The fate of the brain and sense organs and their role in skull morphogenesis and evolution. *J. Morph.* **177**: 255–268.

Hanken, J. (1983b). Genetic variation in a dwarfed lineage, the Mexican salamander genus *Thorius* (Amphibia: Plethodontidae): taxonomic, ecologic, and evolutionary implications. *Copeia* **1983**: 1051–1073.

Hanken, J. (1993). Adaptation of bone growth to miniaturization of body size. In *Bone* 7. *Bone growth*: 79–104. (Ed. Hall, B. K.) CRC Press, Boca Raton.

Hanken, J. & Wake, D. B. (1993). Miniaturization of body size: organismal consequences and evolutionary significance. *A. Rev. Ecol. Syst.* **24**: 501–519.

Hanken, J. & Wake, D. B. (1994). Five new species of minute salamanders, genus *Thorius* (Caudata: Plethodontidae), from northern Oaxaca, Mexico. *Copeia* **1994**: 573–590.

Hecht, M. K. & Edwards, D. L. (1977). The methodology of phylogenetic inference above the species level. In *Major patterns of vertebrate evolution*: 3–51. (Eds Hecht, M. K., Goody, P. C., & Hecht, B. M.). Plenum Press, New York & London. (*NATO adv. Stud. Inst. Ser. Ser. A (Life Sci.)* **14**.)

Hedges, S. B., Nussbaum, R. A., & Maxson, L. R. (1993). Caecilian phylogeny and biogeography inferred from mitochondrial DNA sequences of the 12S rRNA and 16S rRNA genes (Amphibia: Gymnophiona). *Herpet. Monogr.* **7**: 64–76.

Hillenius, D. (1976). On the origin of the anuran body-form (Amphibia: Anura). *Beaufortia* **25**: 63–77.

Hillis, D. M. (1991). The phylogeny of amphibians: current knowledge and the role of cytogenetics. In *Amphibian cytogenetics and evolution*: 7–31. (Eds Green, D. M. & Sessions, S. K.). Academic Press, San Diego.

Hillis, D. M., Ammerman, L. K., Dixon, M. T., & de Sa, R. O. (1993). Ribosomal DNA and the phylogeny of frogs. *Herpet. Monogr.* **7**: 118–131.

Hoogmoed, M. S., Borges, D. M., & Cascon, P. (1994). Three new species of the genus *Adelophryne* (Amphibia: Anura: Leptodactylidae) from northeastern Brazil, with remarks on the other species of the genus. *Zool. Meded., Leiden* **68**: 271–300.

Hoogmoed, M. S. & Lescure, J. (1984). A new genus and two new species of minute leptodactylid frogs from northern South America, with comments upon *Phyzelaphryne* (Amphibia: Anura: Leptodactylidae). *Zool. Meded., Leiden* **58**: 85–115.

Izecksohn, E. (1971). Nôvo gênero e nova espécie de Brachycephalidae do Estado do Rio de Janeiro, Brasil. *Bolm Mus. nac. Rio de J. (N.S. Zool.)* No. 280: 1–12.

Jenkins, F. A., Jr & Walsh, D. M. (1993). An early Jurassic caecilian with limbs. *Nature, Lond.* **365**: 246–250.

Larsen, L. O. (1992). Feeding and digestion. In *Environmental physiology of the amphibians*: 378–394. (Eds Feder, M. E. & Burggren, W. W.). University of Chicago Press, Chicago & London.

Larson, A. (1991). A molecular perspective on the evolutionary relationships of the salamander families. *Evol. Biol.* **25**: 211–277.

Larson, A. & Dimmick, W. W. (1993). Phylogenetic relationships of the salamander families: an analysis of congruence among morphological and molecular characters. *Herpet. Monogr.* **7**: 77–93.

Larson, A. & Wilson, A. C. (1989). Patterns of ribosomal RNA evolution in salamanders. *Molec. Biol. Evol.* **6**: 131–154.

Liu, C. -C. (1950). Amphibians of Western China. *Fieldiana (Zool.)*. **2**: 1–400.

Lynch, J. D. (1971). Evolutionary relationships, osteology, and zoogeography of leptodactyloid frogs. *Misc. Publs Mus. nat. Hist. Univ. Kans.* No. 53: 1–238.

Lynch, J. D. (1973). The transition from archaic to advanced frogs. In *Evolutionary biology of anurans: contemporary research on major problems*: 133–182. (Ed. Vial, J. L.). University of Missouri Press, Columbia.

Lynch, J. D. (1976). Two new species of frogs of the genus *Euparkerella* (Amphibia: Leptodactylidae) from Ecuador and Peru. *Herpetologica* **32**: 48–53.
Lynch, J. D. (1986). New species of minute leptodactylid frogs from the Andes of Ecuador and Peru. *J. Herpet.* **20**: 423–431.
McClanahan, L. L. & Baldwin, R. (1969). Rate of water uptake through the integument of the desert toad, *Bufo punctatus*. *Comp. Biochem. Physiol.* **28**: 381–390.
Milner, A. R. (1988). The relationships and origin of living amphibians. In *The phylogeny and classification of the tetrapods.* **1.** *Amphibians, reptiles, birds*: 59–102. (Ed. Benton, M. J.). Clarendon Press, Oxford. (*Syst. Ass. spec.* Vol. No. 35A.)
Milner, A. R. (1993). The Paleozoic relatives of lissamphibians. *Herpet. Monogr.* **7**: 8–27.
Myers, C. W. (1987). New generic names for some Neotropical poison frogs (Dendrobatidae). *Papéis Dep. Zool. S. Paulo* **36**: 301–306.
Myers, C. W. & Daly, J. W. (1976). Preliminary evaluation of skin toxins and vocalizations in taxonomic and evolutionary studies of poison-dart frogs (Dendrobatidae). *Bull. Am. Mus. nat. Hist.* **157**: 173–262.
Myers, C. W. & Daly, J. W. (1980). Taxonomy and ecology of *Dendrobates bombetes*, a new Andean poison frog with new skin toxins. *Am. Mus. Novit.* No. 2692: 1–23.
Nussbaum, R. A. (1977). Rhinatrematidae: a new family of caecilians (Amphibia: Gymnophiona). *Occ. Pap. Mus. Zool. Univ. Mich.* No. 682: 1–30.
Nussbaum, R. A. (1985). Systematics of caecilians (Amphibia: Gymnophiona) of the family Scolecomorphidae. *Occ. Pap. Mus. Zool. Univ. Mich.* No. 713: 1–49.
Nussbaum, R. A. (1991). Cytotaxonomy of caecilians. In *Amphibian cytogenetics and evolution*: 33–66. (Eds Green, D. M. & Sessions, S. K.). Academic Press, San Diego.
Nussbaum, R. A. & Hoogmoed, M. S. (1979). Surinam caecilians, with notes on *Rhinatrema bivittatum* and the description of a new species of *Microcaecilia* (Amphibia, Gymnophiona). *Zool. Meded., Leiden.* **54**: 217–235.
Nussbaum, R. A. & Wilkinson, M. (1989). On the classification and phylogeny of caecilians (Amphibia: Gymnophiona), a critical review. *Herpet. Monogr.* **3**: 1–42.
Parker, H. W. (1936). The amphibians of the Mamfe Division, Cameroons. 1. Zoogeography and systematics. *Proc. zool. Soc. Lond.* **1936**: 135–163.
Perret, J. -L. (1988). Les espèces de *Phrynobatrachus* (Anura, Ranidae) à éperon palpébral. *Archs Sci., Genève* **41**: 275–294.
Poynton, J. C. (1963). Descriptions of southern African amphibians. *Ann. Natal Mus.* **15**: 319–332.
Pyburn, W. F. (1976). A new fossorial frog from the Colombian rain forest (Anura: Microhylidae). *Herpetologica* **32**: 367–370.
Rage, J. C. & Roček, Z. (1986). *Triadobatrachus* revisited. In *Studies in herpetology*: 255–258. (Ed. Roček, Z.). Charles University, Prague.
Rage, J. C. & Roček, Z. (1989). Redescription of *Triadobatrachus massinoti* (Piveteau, 1936) an anuran amphibian from the early Triassic. *Palaeontographia Abt. A Palaeozool.-Stratigr.* **206**: 1–16.
Rieppel, O. (19984). Miniaturization of the lizard skull: its functional and evolutionary implications. *Symp. zool. Soc. Lond.* No. 52: 503–520.
Salthe, S. N. & Mecham, J. S. (1974). Reproductive and courtship patterns. In *Physiology of the Amphibia* **2**: 309–521. (Ed. Lofts, B.). Academic Press, London & New York.
Sanderson, I. T. (1937). *Animal treasure*. Macmillan and Co., London.
Schiøtz, A. (1964). A preliminary list of amphibians collected in Ghana. *Vidensk. Medd. dansk. naturh. Foren.* **127**: 1–17.
Silverstone, P. A. (1975). A revision of the poison-arrow frogs of the genus *Dendrobates*

Wagler. *Sci. Bull. nat. Hist. Mus. Los Ang. Cty* No. 21: 1–55.

Stanley, S. M. (1973). An explanation for Cope's rule. *Evolution* **27**: 1–26.

Taylor, E. H. (1939). New salamanders from Mexico with a discussion of certain known forms. *Univ. Kans. Sci. Bull.* **26**: 407–439.

Taylor, E. H. (1941). Herpetological miscellany, No. II. *Univ. Kans. Sci. Bull.* **27**: 105–139.

Taylor, E. H. (1968). *The caecilians of the world: a taxonomic review.* University of Kansas Press, Lawrence.

Taylor, E. H. (1969). Skulls of Gymnophiona and their significance in the taxonomy of the group. *Univ. Kans. Sci. Bull.* **48**: 585–687.

Trueb, L. & Alberch, P. (1985). Miniaturization and the anuran skull: a case study of heterochromy. In *Functional morphology of vertebrates*: 113–121. (Eds Duncker, H. R. & Fleischer, G.). Gustav Fisher Verlag, Stuttgart.

Wake, D. B. (1966). Comparative osteology and evolution of the lungless salamanders, family Plethodontidae. *Mem. Sth. Calif. Acad. Sci.* **4**: 1–111.

Wake, D. B. (1991). Homoplasy: the result of natural selection, or evidence of design limitations? *Am. Nat.* **138**: 543–567.

Wake, D. B. (1992). An integrated approach to evolutionary studies of salamanders. *Contr. Herpet.* **9**: 163–177.

Wake, D. B. & Elias, P. (1983). New genera and a new species of Central American salamanders, with a review of the tropical genera (Amphibia, Caudata, Plethodontidae). *Contr. Sci.* No. 345: 1–19.

Wake, D. B. & Lynch, J. F. (1976). The distribution, ecology, and evolutionary history of plethodontid salamanders in Tropical America. *Sci. Bull. nat. Hist. Mus. Los Ang. Cty* No. 25: 1–65.

Wake, M. H. (1986a). The morphology of *Idiocranium russeli* (Amphibia: Gymnophiona), with comments on miniaturization through heterochrony. *J. Morph.* **189**: 1–16.

Wake, M. H. (1986b). Caecilians. In *The encyclopaedia of reptiles and amphibians*: 16–17. (Eds Halliday, T. R. & Adler, K.). George Allen & Unwin, London & Sydney.

Wake, M. H. (1992). 'Regressive' evolution of special sensory organs in caecilians (Amphibia: Gymnophiona): opportunity for morphological innovation. *Zool. Jb. (Anat.)* **122**: 325–329.

Wake, M. H. (1993). Non-traditional characters in the assessment of caecilian phylogenetic relationships. *Herpet. Monogr.* **7**: 42–55.

Wright, S. (1982). Character change, speciation, and the higher taxa. *Evolution* **36**: 427–443.

Appendix: notes

1. Size categorization in caecilians is a combination of slender build as well as length. Taylor (1968: 342) classes *Brasilotyphlus braziliensis* as 'a diminutive caecilian' yet gives its length as 260 mm. Presumably part of the reason for assigning it 'diminutive' status is the comment 'a very slender species'.

2. *Caudacaecilia* is an interesting genus in respect of the size categorization problem. Taylor (1968) recognizes five species; of these five *Caudacaecilia nigroflava* is large, reaching 425 mm. *Caudacaecilia paucidentula* Taylor describes as 'presumed to be a medium-small species' having 'a greatest known length of 295 mm'. *Caudacaecilia weberi* he gives as a 'small species', up to 258 mm, and we may presume, although he did not specifically comment on them, that Taylor would have regarded the remaining two, *Caudacaecilia asplenia* and *Caudecaecilia larutensis*, as also being small species. Thus, (i) if monophyletic, *Caudacaecilia* provides an example of a group with normal and dwarfed (sister) species and (ii) notwithstanding the designation of *Brasilotyphlus braziliensis* at 260 mm as diminutive (see above), they provide further evidence for regarding a total length in the region of 250–260 mm to be indicative of 'small caecilian' status.

3. *Microcaecilia* is an example of a genus comprising almost entirely small-sized caecilians; four of the five species are in the lower end of the small-size category. Taylor described *M. albiceps* as a 'diminutive species' with a maximum known length of 227 mm; *Microcaecilia rabei* 185 mm; *Microcaecilia unicolor* 235 mm. *Microcaecilia taylori* reaches 172 mm (Nussbaum & Hoogmoed 1979) and *M. supernumeraria* 258 mm (M. Wilkinson pers. comm.).

4. *Epicrionops columbianus* is known only from the type which, according to Taylor, is either a recently transformed individual or representative of a species which is much smaller than other *Epicrionops*. Therefore I regard it as a small, rather than a borderline miniature species. *Epicrionops marmoratus* at 278–299 mm is a medium-sized species, as is *Epicrionops nigrus* at 272 mm, *Epicrionops peruvianus* at 286 mm, *Epicrionops peruvianus petersi* at 217–287 mm and *Epicrionops peruvianus noblei* at 328 mm. As in the case of *Caudacaecilia*, *Epicrionops*, if monophyletic, provides another example of a genus of caecilians comprising normal-sized and dwarf sister species and therefore examples of dwarfing occurring within a group of species.

Symp. zool. Soc. Lond. (1996) No. 69: 225–237

Ecology of small reptile-grade sauropsids

R. A. AVERY

School of Biological Sciences
Bristol University
Bristol BS8 1UG, UK

Synopsis

Extremely small adult sizes in reptile-grade sauropsids are found only in lizards. Analysis of size distributions of 1200 species from seven geographical areas for which reasonably complete data can be assembled shows that lizards with adult snout–vent lengths (SVL) of less than 20 mm (body mass in these animals will be less than 0.5 g) constitute less than 0.1% of the total and are found only within tropical latitudes in the West Indies. Lizards with adult SVL less than 30 mm (body mass of these animals will be less than 1 g) are also uncommon, accounting for less than 5% of the total. They are not found in the two analysed geographical areas which lie entirely outside the tropics (north and east Argentina, and Europe). On the other hand, the modal size classes for lizards in tropical areas are 30.1–40.0 or/and 40.1–50.0 mm SVL. There is a progressive shift in modal size classes in lizard faunas with increasing distance from the Equator. These patterns are discussed. An attempt is made to assess the contribution of ecological studies of relative size, especially in island lizard faunas, to understanding the ecology and adaptive significance of very small size.

Introduction

The detailed phylogeny of the amniote radiation which occurred during the Permian has been a matter of great debate. Most authorities would agree, however, that the modern Chelonia, Squamata, Crocodylia, Sphenodontia, and birds are derived from a common sauropsid ancestry. The first four of these can be regarded as reptile-grade sauropsids.

The reptilian mode of organization allowed the evolution of the largest terrestrial animal ever to have lived (*Brontosaurus*). Many modern Squamata and Crocodylia are quite massive. The largest living snakes may be more than 9.5 m in length (*Python reticulatus, Liasis amethystinus*, and *Eunectes murinus*). The largest crocodiles may be more than 6 m in length (*Crocodylus porosus; Phobosuchus* from the Cretaceous may have reached 13 m). Individuals of the largest living lizard species (*Varanus komodoensis*) may reach 3 m, with a body mass of 160 kg. Most of the very large living reptiles have received a great deal of

ZOOLOGICAL SYMPOSIUM No. 69
ISBN 0–19–857787–7

both popular and scientific attention (as have the fossil dinosaurs and their relatives). Small size, however, has received almost no study.

In an absolute sense, by far the smallest reptiles are found among the lizards, since there are a number of species with total body lengths of less than 40 mm and total mass of less than 1 g. The smallest snake species may reach total body lengths of 200 mm (*Typhlops*). The smallest Testudines grow to more than 100 mm (*Homopus signatus* from terrestrial environments, *Clemmys muhlenbergii* and *Kinosternon signatus* from freshwater). The smallest crocodilian species (*Palaeosuchus palpebrosus* and *Osteolaemus osborni*) grow to 1.2 m. There are no very small amphisbaenians and the two living sphenodontian species grow to 60 cm. Thus, because lizards can be substantially smaller than any other reptiles, they form the basis for the greater part of the discussion in the remainder of this paper.

Body size distribution in lizards

Living lizard species range in adult body size from a total length of approximately 25 mm (some geckos in the genus *Sphaerodactylus*) to 3000 mm. Frequency distributions of sizes are not normal, but logarithmically skewed. This is demonstrated in data presented by Pough (1980), based on collection of lengths from the literature for 1780 species. The modal size range was 1–5 g (body mass was calculated from snout–vent length (SVL) using the relationship $M = 3.1 \times 10^{-2}\ SVL^{2.98}$). Twenty-eight percent of species fell within this category. A further 25.6% were within the range 5–10 g; 8.1% were less than 1 g. This general kind of relationship holds also for all reptile and amphibian orders except Testudinata (Pough 1980) and also for insects, birds, and mammals (May 1978; Blackburn & Gaston 1994).

In an effort to determine the extent to which relationships of this kind apply to lizards in local faunas, I plotted frequency distributions of size for species from six geographical areas. The ideal would have been to have plotted such distributions for all the major biomes in each zoogeographical region which contains lizards. This was not possible because the lizard faunas of some areas are insufficiently known, and the relevant data for others are so widely scattered in the literature that compiling them would have been a Herculean task. I chose six faunas for which there have been recent monographic descriptions of the lizards. These are the West Indies (Schwartz & Henderson 1991, with 303 species); Australia (Cogger 1992), further divided into rainforest with 59 species and non-rainforest habitats with 421 species; Africa south of the Zambesi and Cunene rivers (Branch 1988, with 230 species); north and east Argentina (Cei 1993, with 65 species); North America, excluding Mexico (Stebbins 1985 and Conant & Collins 1991, with 80 species) and Europe (Böhme 1981, 1984, 1986, with 50 species). The numbers of lizard species in each of these faunas are shown in Fig. 1 in the order of their distance from the Equator.

Several compromises and arbitrary decisions needed to be made in compiling these data. Size is given as SVL; in a few cases this was estimated from total length. Body mass would have been preferable, but appropriate data are not given in most

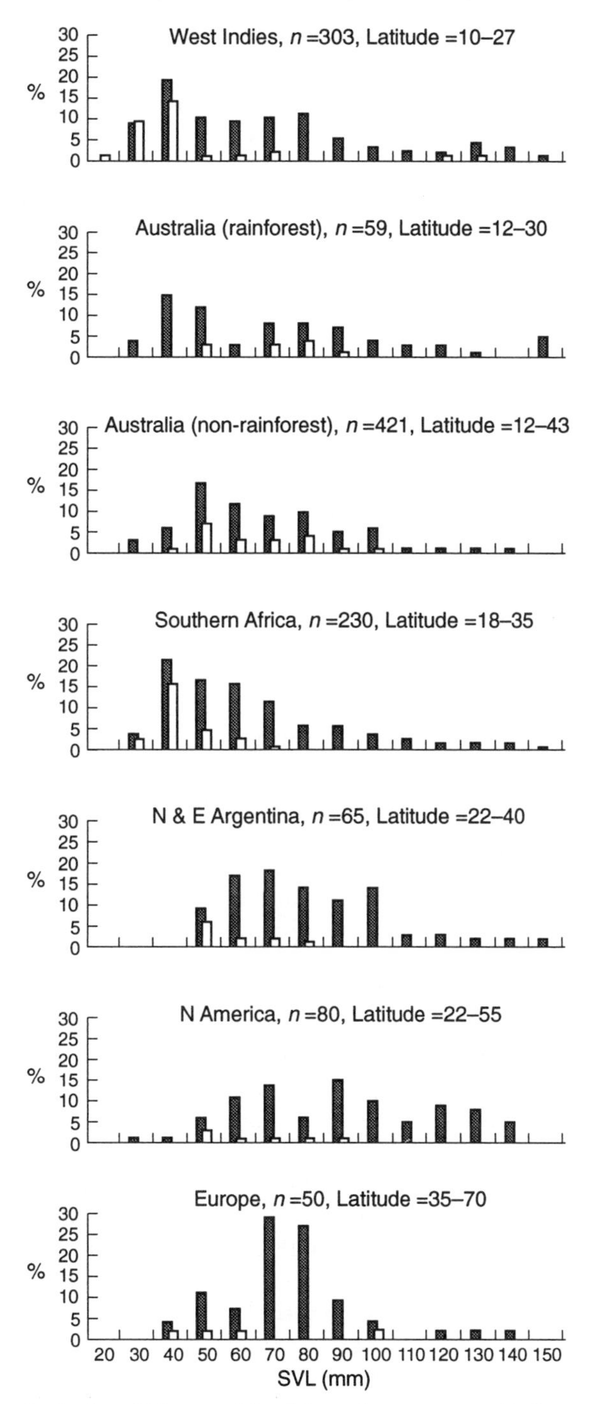

Fig. 1. Body size distributions (SVL, mm) of lizard faunas in seven geographical areas. Black histograms show all lizards, white histograms show Gekkonidae.

of the works listed above. Various authors give some combination of size ranges in the categories 'adult', 'typical', 'typical mean size of adults', or maximum size. None gives all of these, and I have estimated 'typical maximum size' (i.e. the size of typical large individuals, but not of the largest ever recorded), for each species. The errors involved must vary from species to species; they are perhaps greatest in little-known species for which only a few individuals (or even only one, as in *Sphaerodactylus elasmorhynchus* and *Sphaerodactylus williamsi*) are known. In the case of the Australian fauna, there are a number of species which occur in both rainforest and other habitats; I have assigned each of these to the habitat in which it appears to be commonest.

The results of this analysis are given in Fig. 1, which shows the frequency distributions of SVL, in 10 mm categories, for all lizards (black histograms) and for geckos only (family Gekkonidae; open histograms). Percentages do not add to 100 because lizards larger than SVL = 150 mm are not shown. Several patterns are apparent in these data.

The first point to be made is that the frequency distributions are all skewed to the right. They are not log-normal, since the skewness is retained—but of course reduced—on plotting the data on semi-logarithmic coordinates. This is similar to the pattern recorded for mammals, birds, and some insects (Blackburn & Gaston 1994).

Very small lizards, i.e. those with SVL less than 20 mm, are found only in the West Indies sample and they form a minute fraction of the total. They are all ($n = 3$) geckos. It is not clear, however, whether the fact that such small lizards are found in only one sample reflects the possibility that they are confined to the tropical West Indies for biological reasons, or simply that the sample sizes in other areas are small. Analysis based on the binomial distribution suggests a stochastic explanation of the latter kind: very small lizards form only 1% of the total species in the West Indies and the probability that they will not occur at all in the other samples is only greater than 0.05 for non-rainforest habitats in Australia and southern Africa.

Lizards with SVL from 20–29.9 mm are found in five of the seven samples, but overall comprise only 4.7% of species. Seventy percent of these are geckos. The next size class (30.0–39.9 mm), however, is modal in three of the seven samples. All of them are areas which extend well into the tropics. There is a significant ($r_s = 0.88$, $P < 0.01$) correlation between modal size and mean latitude (Fig. 2). This finding was unexpected. Very large reptiles are confined to hot tropical environments because their low ratios of surface area to body mass would not allow sufficiently rapid heat exchanges with the environment elsewhere (Stevenson 1985a; O'Connor & Spotila 1992). The size of the largest species in any lizard fauna outside the tropics varies inversely with latitude. This is illustrated for European lizards in Fig. 3, which shows the latitude at which species of increasing size (as defined above) appear along three transects over the shortest land routes from northern Scandinavia to the south of the Iberian, Italian, and Balkan peninsulas. It should be noted that the discontinuous lines drawn in Fig. 3 are to emphasize trends; they do not represent linear regressions, since the addition of

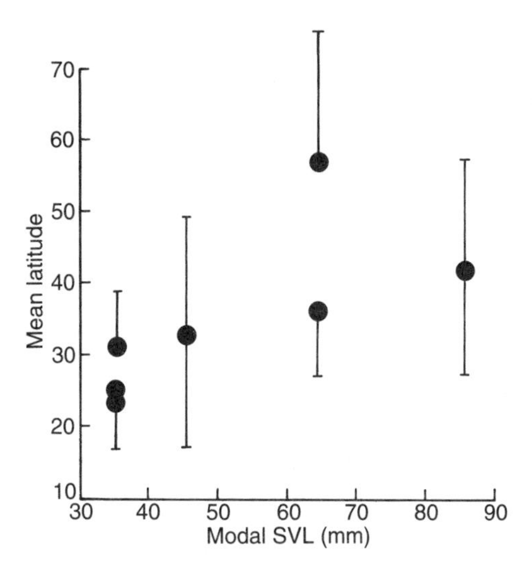

Fig. 2. Modal size class (SVL, mm) and mean latitude for lizard species in seven faunas (details in Fig. 1). The range of latitude is also indicated in some cases, but in others has been omitted for clarity.

north African lizards to these data suggests that the trends are not linear on semi-logarithmic coordinates. The relationship is highly significant ($r_s = 0.96$ for both SVL and body mass, d.f. = 9, $P < 0.001$). Plotting equivalent data for the smallest lizard species (also shown in Fig. 3) does not show an equivalent trend; although the correlations between SVL or body mass and latitude appear positive, they are not significant ($r_s = 0.55$ in both cases, d.f. = 6, n.s.).

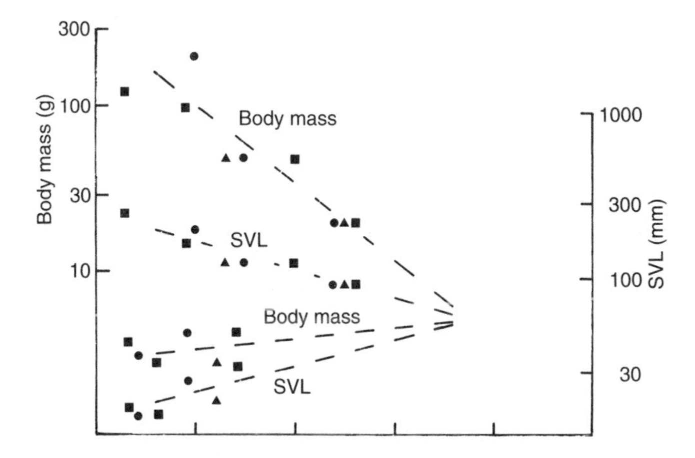

Fig. 3. Progressive appearance of lizard species of increasing size (body mass in grams and SVL in millimetres plotted logarithmically) along three transects radiating from northern Scandinavia to the southern tip of the Iberian (circles), Italian (triangles), and Balkan (squares) peninsulas. The figure also shows that the size of the smallest lizard species does not alter (the apparent trends are not significant, see text). Data from a variety of sources.

Why should the trend in *modal* size be in the opposite direction to that for the largest size? There can be no definitive answer to this question. One possibility is that small size is particularly adaptive in very densely forested tropical habitats, where direct sunshine is scarce or non-existent beneath the canopy. Many lizards in these habitats either abandon the precise behavioural thermoregulation which is so characteristic a feature of the biology of diurnal lizards elsewhere (Avery 1982), or they spend considerable periods tracking small patches of sunshine (Hertz, Fleishman, & Armsby 1994). Small size will be particularly advantageous in the second of these categories. There are, of course, disadvantages in small size (see below). Since lizards in non-forest areas do not need to be small for thermoregulatory reasons, small size may consequently not have evolved there.

The ecology of small size: do small lizard species have anything in common?

As has already been noted, the smallest known lizards are nocturnal geckos belonging to the genus *Sphaerodactylus* in the West Indies. Details of the four smallest species are shown in Table 1. Analysis of the ecology of these species is made difficult by the fact that two of them are known from only one specimen. Both may now be extinct. Even with the limited information available, however, it is apparent that these species are adapted for different habitats: in this respect at least, they have nothing in common. The smallest geckos in southern Africa are slightly larger (SVL = 20–30 mm; details are also shown in Table 1); they have few ecological features in common with *Sphaerodactylus* or with one another.

The ecology of small size: evolutionary dynamics of *Anolis* and *Cnemidophorus*

The genus *Anolis sensu lato* (there have been recent attempts to split this large genus: see Guyer & Savage 1987; Savage & Guyer 1989), which has radiated extensively among the numerous islands in the Caribbean and on the mainland of Central America, has received extensive study. Its evolutionary dynamics are probably better known than those of any other reptile group. The niches of *Anolis* may be described in terms of three major axes: size, perch (most species are arboreal ambush predators), and climate. There are a number of fairly distinct ecological types which are adapted to different combinations of values for these variables; they have been termed ecomorphs (Williams 1983). The smallest lizards in most assemblages are usually those which perch on small twigs, or in which the arboreal habit has been abandoned for life in bushes or amongst grasses or other short vegetation. A detailed analysis of the lizard faunas of islands within one group, the Lesser Antilles, leads to conclusions (Roughgarden, Heckel, & Fuentes 1983) that (1) 16 islands contain one native species, none being small; (2) eight islands contain two species, none with two species of equal size; and (3) six of these eight islands have one species which is small.

Table 1. Some ecological characteristics of four species of very small lizards from the West Indies and from southern Africa.

Location	Species	Characteristics
West Indies		
Haiti	*Sphaerodactylus elasmorhynchus*	Only one specimen known (female, SVL = 17 mm) Under a rotting tree Nocturnal
Haiti	*Sphaerodactylus ornoglaux*	Leaf litter, canopy, and in *Agave* plants Nocturnal
British Virgin Islands	*Sphaerodactylus parthenopion*	Possibly the smallest known lizard species Scrubby woodland Secretive, nocturnal
Haiti	*Sphaerodactylus williamsi*	Only one specimen known (female, SVL = 22 mm) Coastal oasis surrounded by *Agave*-cactus desert Nocturnal
Southern Africa	*Lygodactylus angolensis*	Dry deciduous woodland Diurnal
	Lygodactylus lawrencei	Arid rocky savannah Diurnal
	Narudasia festiva	Arid mountains Crepuscular
	Phyllodactylus lineatus	Very diverse habitats (the taxonomy is confused; there may be more than one species within this taxon) Nocturnal

Data derived from a variety of sources.

Phylogenetic analysis suggests that both species displacement and size assortment, which must have occurred subsequent to sympatry, have operated in producing the size distributions observed (Losos 1990a, b). Significant evidence for size assortment has also been found in detailed analyses of guilds of *Cnemidophorus* (which are ground-living, active, hunting predators) in the south-eastern states of the USA (Case 1983).

Studies of this kind help to identify and explain some of the selective forces which determine body size in lizards. They do not, however, directly address questions about very small size (indeed, the Lesser Antilles study leads to the conclusion that convergence to an intermediate size has been a major factor in the evolution of these island faunas).

Small size in lizards: a deductive approach

Much of the work described in the previous three sections has been largely inductive (although the studies of *Anolis* and *Cnemidophorus* in particular have also involved hypothesis testing). An alternative approach to understanding the evolution and significance of small size could be deductive. What factors might be significant in the evolution of small size? Under what circumstances might small size be advantageous or disadvantageous?

Advantages of small size

Body temperature and metabolism

The body temperature of an ectotherm influences its behaviour, physiology, and development (Stevenson 1985b). Since body temperatures fluctuate in relation to environmental conditions, the trajectories of body temperature over time are of crucial importance, as demonstrated in studies on the relationship between the time for which high body temperatures are experienced and growth (Avery 1984). Calculations based on a heat balance model have shown that animals with body masses between 0.1 and 1 kg have a potentially greater range of body temperatures than smaller or larger animals at any given air temperature (Stevenson 1985a). Very small animals can partly compensate by seeking small microenvironments at higher air temperatures (Stevenson 1985a) which, because of their small size, would be unavailable (or less effectively available) to larger individuals. Small lizards thus have two potential advantages. They can heat more rapidly because of the simple relationships of body surface area to mass (Bartholomew 1982; Huey 1982), although the magnitude of this advantage will depend on the ability to limit cooling rates by seeking appropriate microhabitats. The latter aspect of the thermal biology of lizards has been very little studied; there has been an undue emphasis on the biology of basking and warming up (but see Huey, Peterson, Arnold, & Porter 1989). Small lizards can more easily take advantage of small discontinuities in the distribution of air or surface temperatures (this aspect of thermal biology has received no systematic study in lizards, perhaps because the dynamics of the heat exchanges involved are complex; see Bakken 1989).

Growth

Small species might be at an advantage because their relative growth rates are more rapid (reviews: Andrews 1982; Avery 1994) and the difference between hatchling size and adult size relatively small, involving a relatively low investment of materials and energy. I know of no studies which directly address this hypothesis. Since small size can increase the period for which an individual is able to maintain high body temperatures (in those species which thermoregulate), it will result in an increase in growth rates (Avery 1984; Porter 1989; Sinervo & Adolph 1989).

Niche

Small species might be at an advantage because they can feed on smaller, more abundant prey. Although there is a huge literature on the relations between body size and prey size in lizards, no studies have directly addressed the question whether the smallest species in any guild obtain an advantage of this sort. One of the problems in answering a very general question of this kind is that the individual species in any guild tend also to have differing selectivities in dimensions other than size (see the extensive data in Pianka (1986), for example).

Competition

Small species might be at an advantage because competition with the species in the modal size classes (Fig. 1) is reduced. This would be a very difficult hypothesis to test. The fact that all of the very small species listed in Table 1 are from relatively large assemblages does not necessarily support it, because the correlation might simply be due to the greater probability of finding small species in large assemblages.

Crevices

Small size must increase the ability to escape predators by hiding, since a wider range of crevices and other refugia will be available. This may also increase prey availability. There are no studies which directly address this question, but recent experiments of Schlesinger & Shine (1994) demonstrate that crevice selection can be an active process. It must therefore have arisen by selection. The ability to penetrate into crevices may also convey thermoregulatory advantages (e.g. Bustard 1967, 1968; Dial 1978; Grant & Dunham 1988; Huey *et al.* 1989).

Disadvantages of small size

Predation

Small lizards must have a greater range of potential predators. I know of no direct evidence for this statement, but the fact that juvenile lizards of many species are usually more secretive than adults (e.g. Castilla & Bauwens 1991) and almost always have higher mortality rates (e.g. Avery 1975) provides circumstantial evidence for it. The conclusions from attempts to assess predation rates from tail-break frequencies must usually be so hedged with conditions that they must be regarded with very great caution (Arnold 1988).

Vision

Small size in plethodontid salamanders results in a reduction in retinal cell size and a denser packing of neurons in the visual areas of the brain (Roth, Rotluff, Grunwald, Henken, & Linke 1990). The same relative changes undoubtedly occur in lizards, and this may well impose constraints on the evolution of small size. In the juveniles of some fishes, the potential reduction in the size of retinal cells is compensated by a decrease in their density. This reduces visual acuity (Hairston, Li, & Easter 1982). Similar factors may place limits on the minimum sizes of lizards, since the smallest *Sphaerodactylus* (Table 1) are no smaller than the smallest Plethodontidae (Hanken & Wake 1994).

Reproduction

Clutch size in reptiles decreases with decreasing body mass. Since the shelled cleidoic eggs appear to have a minimum size threshold, relative clutch mass must limit the extent to which miniaturization is possible once mean clutch size reaches unity. This limit appears to have been met in small geckos (data in Shine (1992), Shine & Schwarzkopf (1992) and Griffith (1994) and references therein are relevant to this argument although they do not directly address this question).

Stamina

Comparative studies show that stamina and endurance capacity decrease with decreasing body size across a wide range of lizard taxa (Garland 1994; Garland & Losos 1994). Small lizards have poor capacities for sustained fast locomotion (e.g. Hailey, Gaitanaki & Loumbourdis 1987).

Phylogenetic considerations

There are two major classes of phylogenetic questions relating to small size in reptiles. The first is to ask why small size has evolved only in lizards. Since size reduction in Testudines and Crocodylia may be limited by their bony armour, the question is essentially why the smallest snakes (they are an extremely successful group; see Pough 1983) are larger than the smallest lizards. The answer may lie in the fact that there is a strong correlation between size and number of vertebrae (Lindell 1994). There must presumably be a minimum vertebral number which permits effective serpentine locomotion, although the question has not been directly investigated—but none of the legless tetrapods which have evolved since the Carboniferous have been very small. A second constraint relates to the problem of supplying energy to an elongate body via a relatively small mouth. The problem has been overcome in snakes by modifications of the skull which permit relatively enormous meals (Pough 1983). The smallest snakes (Typhlopidae, Leptotyphlopidae) are burrowing animals in which this mode of organization has perforce been abandoned, because the skull needs to be massive.

A second class of questions relates to phylogenetic conservatism. Most of the smallest lizard species are geckos (Fig. 1; Table 1). Studies of the evolution of sand burrowing in lizards show that independently developed traits have a general level of concordance that varies with the level of interrelatedness of the taxa (Arnold 1994, 1995). If similar principles apply to the evolution of small size, then there is no reason to assume that the actual size of the smallest species in different groups will be the same, either within Gekkonidae or within lizards as a whole. The fact that it is within geckos, and specifically within the genus *Sphaerodactylus*, that the smallest lizards are to be found, might be fortuitous. Whether this is indeed the case would require very careful phyletic studies.

Concluding comment

This chapter is essentially a catalogue of ignorance. The smallest lizards are little studied; even their systematics are poorly known (Table 1). Generalizations, even of an elementary kind, will only be possible when a great deal more knowledge has accumulated about lizard physiology, ecology, and evolution. Most of the generalizations in this chapter are, perforce, little more than speculation.

References

Andrews, R. M. (1982). Patterns of growth in reptiles. In *Biology of the Reptilia* 13. *Physiology D. Physiological ecology*: 273–330. (Eds Gans, C. & Pough, F. H.). Academic Press, London.

Arnold, E. N. (1988). Caudal autotomy as a defense. In *Biology of the Reptilia* **16**. *Ecology B. Defense and life history*: 235–273. (Eds Gans, C. & Huey, R. B.). Liss, New York.

Arnold, E. N. (1994). Do ecological analogues assemble their common features in the same order? An investigation of regularities in evolution, using sand-dwelling lizards as examples. *Phil. Trans. R. Soc. (B)* **344**: 277–290.

Arnold, E. N. (1995). Identifying the effects of history on adaptation: origins of different sand diving techniques in lizards. *J. Zool., Lond.* **235**: 351–388.

Avery, R. A. (1975). Age structure and longevity of common lizard (*Lacerta vivipara*) populations. *J. Zool., Lond.* **176**: 555–558.

Avery, R. A. (1982). Field studies of body temperatures and thermoregulation. In *Biology of the Reptilia* 12. *Physiology C. Physiological ecology*: 93–166. (Eds Gans, C. & Pough, F. H.). Academic Press, London.

Avery, R. A. (1984). Physiological aspects of lizard growth: the role of thermoregulation. *Symp. zool. Soc. Lond.* No. 52: 407–424.

Avery, R. A. (1994). Growth in reptiles. *Gerontology* **40**: 193–199.

Bakken, G. S. (1989). Arboreal perch properties and the operative temperature experienced by small animals. *Ecology* **70**: 922–930.

Bartholomew, G. A. (1982). Physiological control of body temperature. In *Biology of the Reptilia* 12. *Physiology C. Physiological ecology*: 167–211. (Eds Gans, C. & Pough, F. H.). Academic Press, London.

Blackburn, T. M. & Gaston, K. J. (1994). Animal body size distributions: patterns, mechanisms and implications. *Trends Ecol. Evol.* **9**: 471–474.

Böhme, W. (Ed.) (1981). *Handbuch der Reptilien und Amphibien Europas. Band 1. Echsen (Sauria) I.* Akademische Verlagsgesellschaft, Wiesbaden.

Böhme, W. (Ed.) (1984). *Handbuch der Reptilien und Amphibien Europas. Band 2/I. Echsen (Sauria) II.* AULA-Verlag, Wiesbaden.

Böhme, W. (Ed.) (1986). *Handbuch der Reptilien und Amphibien Europas. Band 2/II. Echsen (Sauria) III.* AULA-Verlag, Wiesbaden.

Branch, B. (1988). *Field guide to the snakes and other reptiles of southern Africa.* New Holland, London.

Bustard, H. R. (1967). Activity cycle and thermoregulation in the Australian gecko *Gehyra variegata*. *Copeia* **1967**: 753–758.

Bustard, H. R. (1968). The ecology of the Australian gecko *Heteronotia binoei* in northern New South Wales. *J. Zool., Lond.* **156**: 483–497.

Case, T. J. (1983). Sympatry and size similarity in *Cnemidophorus*. In *Lizard ecology: studies of a model organism*: 297–325. (Eds Huey, R. B., Pianka, E. R., & Schoener, T. W.). Harvard University Press, Cambridge, Ma. & London.

Castilla, A. M. & Bauwens, D. (1991). Thermal biology, microhabitat selection, and conservation of the insular lizard *Podarcis hispanica atrata*. *Oecologia* **85**: 366–374.

Cei, J. M. (1993). Reptiles del noroeste, nordeste y este de la Argentina. Herpetofauna de las selvas subropicales, puna y pampas. *Mus. reg. Sci. nat. Monogr. (Turin)* **14**: 1–949.

Cogger, H. (1992). *Reptiles and amphibians of Australia.* (5th edn). Reed, Chatsworth, NSW.

Conant, R. & Collins, J. T. (1991). *Reptiles and amphibians. Eastern/central North America.* Houghton Mifflin, Boston. (*Peterson Field Guide Ser.* No. 12.)

Dial, B. E. (1978). The thermal ecology of two sympatric, nocturnal *Coleonyx* (Lacertilia: Gekkonidae). *Herpetologica* **34**: 194–201.

Garland, T. (1994). Phylogenetic analysis of lizard endurance capacity in relation to body size and body temperature. In *Lizard ecology: historical and experimental perspectives*: 237–259. (Eds Vitt, L. J. & Pianka, E. R.). Princeton University Press, Princeton, NJ.

Garland, T. & Losos, J. B. (1994). Ecological morphology of locomotor performance in squamate reptiles. In *Ecological morphology: integrative organismal biology*: 240–302. (Eds Wainwright, P. C. & Reilly, S. M.). Chicago University Press, Chicago, Illinois.

Grant, G. W. & Dunham, A. E. (1988). Thermally imposed time constraints on the activity of the desert lizard *Sceloporus merriami. Ecology* **69**: 167–176.

Griffith, H. (1994). Body elongation and decreased reproductive output within a restricted clade of lizards (Reptilia: Scincidae). *J. Zool., Lond.* **233**: 541–550.

Guyer, C. & Savage, J. M. (1987). Cladistic relationships among anoles (Sauria: Iguanidae). *Syst. Zool.* **35**: 509–531.

Hailey, A., Gaitanaki, C., & Loumbourdis, N. S. (1987). Metabolic recovery from exhaustive activity by a small lizard. *Comp. Biochem. Physiol. (A)* **88**: 683–689.

Hairston, N. G., Li, K. T., & Easter, S. S. (1982). Fish vision and the detection of planktonic prey. *Science* **218**: 1240–1242.

Hanken, J. & Wake, D. B. (1994). Five new species of minute salamanders, genus *Thorius* (Caudata: Plethodontidae), from northern Oaxaca, Mexico. *Copeia* **1994**: 573–590.

Hertz, P. E., Fleishman, L. J., & Armsby, C. (1994). The influence of light intensity and temperature on microhabitat selection in two *Anolis* lizards. *Funct. Ecol.* **8**: 720–729.

Huey, R. B. (1982). Temperature, physiology, and the ecology of reptiles. In *Biology of the Reptilia* **12**. *Physiology C. Physiological ecology*: 25–91. (Eds Gans, C. & Pough, F. H.). Academic Press, London.

Huey, R. B., Peterson, C. R., Arnold, S. J., & Porter, W. D. (1989). Hot rocks and not-so-hot rocks: retreat-site selection by garter snakes and its thermal consequences. *Ecology* **70**: 931–944.

Lindell, L. E. (1994). The evolution of vertebral number and body size in snakes. *Funct. Ecol.* **8**: 708–719.

Losos, J. B. (1990a). Ecomorphology, performance capability, and scaling of West Indian *Anolis* lizards: an evolutionary analysis. *Ecol. Monogr.* **60**: 369–388.

Losos, J. B. (1990b). The evolution of form and function: morphology and locomotor performance in West Indian *Anolis* lizards. *Evolution* **44**: 1189–1203.

May, R. M. (1978). The dynamics and diversity of insect faunas. *Symp. R. ent. Soc. Lond.* No. 9: 188–204.

O'Connor, M. P. & Spotila, J. R. (1992). Consider a spherical lizard: animals, models and approximation. *Am. Zool.* **32**: 179–193.

Pianka, E. R. (1986). *Ecology and natural history of desert lizards: analyses of the ecological niche and community structure.* Princeton University Press, Princeton, NJ.

Porter, W. P. (1989). New animal models and experiments for calculating growth potential at different elevations. *Physiol. Zool.* **62**: 286–313.

Pough, F. H. (1980). The advantages of ectothermy for tetrapods. *Am. Nat.* **115**: 92–122.

Pough, F. H. (1983). Feeding mechanisms, body size, and the ecology and evolution of snakes. *Am. Zool.* **23**: 339–342.

Roth, G., Rotluff, B., Grunwald, W., Henken, J., & Linke, R. (1990). Miniaturization in plethodontid salamanders (Caudata: Pleurodontidae) and its consequences for the brain and visual system. *Biol. J. Linn. Soc.* **40**: 165–190.

Roughgarden, J., Heckel, D., & Fuentes, E. R. (1983). Coevolutionary theory and the biogeography and community structure of *Anolis*. In *Lizard ecology: studies of a model organism*: 371–410. (Eds Huey, R. B., Pianka, E. R., & Schoener, T. W.). Harvard University Press, Cambridge, Ma. & London.

Savage, J. M. & Guyer, C. (1989). Infrageneric classification and species composition of the anole genera, *Anolis, Ctenonotus, Dactyloa, Norops* and *Semiurus* (Sauria: Iguanidae). *Amphibia-Reptilia* **10**: 105–116.

Schlesinger, C. A. & Shine, R. (1994). Selection of diurnal retreat sites by the nocturnal gekkonid lizard *Oedura lesueurii*. *Herpetologica* **50**: 156–163.

Schwartz, A. & Henderson, R. W. (1991). *Amphibians and reptiles of the West Indies: descriptions, distributions, and natural history*. University of Florida Press, Gainesville.

Shine, R. (1992). Relative clutch mass and body shape in lizards and snakes: is reproductive investment constrained or optimized? *Evolution* **46**: 828–833.

Shine, R. & Schwarzkopf, L. (1992). The evolution of reproductive effort in lizards and snakes. *Evolution* **46**: 62–75.

Sinervo, B. & Adolph, S. C. (1989). Thermal sensitivity of growth rate in hatchling *Sceloporus* lizards: environmental, behavioral, and genetic aspects. *Oecologia* **78**: 411–419.

Stebbins, R. C. (1985). *A field guide to western reptiles and amphibians*. (2nd edn). Houghton Mifflin, Boston.

Stevenson, R. D. (1985a). Body size and limits to the daily range of body temperature in terrestrial ectotherms. *Am. Nat.* **125**: 102–117.

Stevenson, R. D. (1985b). The relative importance of behavioral and physiological adjustments controlling body temperatures in terrestrial ectotherms. *Am. Nat.* **126**: 362–386.

Williams, E. E. (1983). Ecomorphs, faunas, island size, and diverse end points in island radiations of *Anolis*. In *Lizard ecology: studies of a model organism*: 326–370. (Eds Huey, R. B., Pianka, E. R., & Schoener, T. W.). Harvard University Press, Cambridge, Ma. & London.

[illegible] relationships between [illegible] and [illegible]

[illegible]

[illegible] (1970). [illegible]

[illegible] University Press, Cambridge, MA [illegible]

[illegible]

[illegible] and [illegible]

[illegible]

[illegible] age and limits to the [illegible]

[illegible] behavioral and [illegible]

[illegible] (1971). Learning, [illegible] and [illegible]

[illegible] University Press, Cambridge, MA [illegible]

Symp. zool. Soc. Lond. (1996) No. 69: 239–258

Body size and the ecology of hummingbirds

PETER A. COTTON

Edward Grey Institute
Department of Zoology
University of Oxford
South Parks Road
Oxford OX1 3PS, UK

Synopsis

Hummingbirds (family Trochilidae) are among the smallest endothermic organisms and represent a physiological extreme of avian evolution. A foraging hummingbird has a metabolic rate, heart rate, and wingbeat frequency unparalleled among birds and matched by few other endotherms. Hummingbirds meet their energy requirements by feeding almost exclusively on floral nectar, and show many morphological and behavioural adaptations for this mode of life. Indeed their body size itself appears in part to be determined by co-evolution with flowering plants and competition with other nectar feeders. In spite of their high metabolism and narrow resource base, hummingbirds are an extremely successful group. The Trochilidae is the largest family of non-passerine birds and hummingbirds are found throughout the Americas in all habitats from sea-level to the snow-line.

Introduction

The body size of an organism has profound influences on virtually all aspects of its biology. The study of allometry in birds has involved describing relationships between body size and metabolic rate (Aschoff & Pohl 1970), the length of appendages (Greenewalt 1975), egg size and incubation time (Rahn & Ar 1974; Rahn, Paganelli & Ar 1975), longevity (Lindstedt & Calder 1976), and territory size (Schoener 1968). Scaling is widely used because empirical allometric relationships permit us to explore how different mechanical, physiological, and ecological constraints change in importance with body size. Hummingbirds are particularly interesting from this point of view as they are some of the smallest endothermic organisms.

The smallest hummingbird, the bee hummingbird (*Mellisuga helenae*), weighs little more than 2 g, approximately half the weight of the smallest passerine birds and one-tenth of the weight of the giant hummingbird, *Patagona gigas* (Dunning 1993). As its name suggests, the bee hummingbird is more similar in size to many large insects than to most bird species.

ZOOLOGICAL SYMPOSIUM No. 69
ISBN 0–19–857787–7

From data on the weights for 251 of the 319 species of hummingbird (Dunning 1993), the mean weight of hummingbird species is 5.1 g and the distribution of body weights is highly skewed; 70% of species weigh less than 6 g and very few species weigh more than 9 g (Fig. 1). The giant hummingbird, weighing on average 20.2 g, appears as an outlier, and only two other species (in *Ensifera* and *Topaza*) are heavier than 12 g. At the other extreme, only 7% of hummingbird species are lighter than 2.5 g.

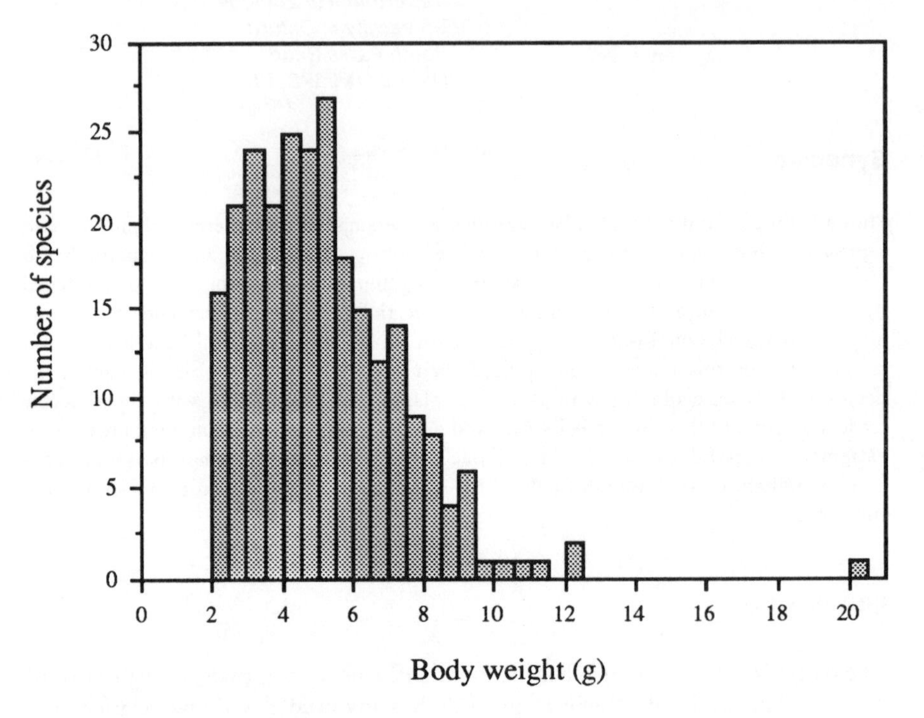

Fig. 1. Frequency distribution of mean body weights (g) of 251 hummingbird species.

Hummingbirds are a large, exclusively American family of birds, adapted for nectar feeding. They are found in most habitats from Alaska to Tierra del Fuego but reach their greatest diversity and abundance within 5° of the equator (Grant & Grant 1968; Skutch 1974). Their small size and somewhat unusual diet of nectar strongly affects all aspects of their lives, shaping physiology, locomotion, behaviour, and community structure. As pollinators, hummingbirds have co-evolved with many different plants, occasionally demonstrating quite remarkable relationships, but their dependence on nectar often forces them to compete with insects and, for such small birds, to undertake spectacular migrations.

The only previous review of the correlates and consequences of body size in nectar-feeding birds was published almost 20 years ago (Brown, Calder, & Kodric-Brown 1978); here I consider the relationship between body size and the ecology of hummingbirds in the light of more recent findings.

Physiological correlates of small size

Metabolic rate

Basal metabolic rate (BMR) is strongly related to body mass and in general birds have a higher BMR than mammals. Aschoff & Pohl (1970) showed that passerine birds have higher BMR than non-passerines, using approximately 55% more energy in both active and resting phases (passerines—active: $M = 140.9W^{0.704}$, resting: $M = 114.8W^{0.726}$; non-passerines—active: $M = 91.0W^{0.729}$, resting: $M = 73.5W^{0.734}$).

The allometric exponent for resting metabolic rate across families is 0.67, which accords with the surface area interpretation of metabolic scaling based on heat loss (Bennett & Harvey 1987). The allometric exponent for active metabolic rate (0.65) differs from that for resting metabolic rate, demonstrating that small-bodied birds need proportionately more energy per unit time of activity than large-bodied birds do. Bennett & Harvey (1987) concluded that taxonomic affiliation was the primary source of deviation from this allometric line. The food habits of an organism may also explain the residuals of BMR and body weight (Elgar & Harvey 1987). Passerines which feed on nectar (sunbirds, honeyeaters, and honeycreepers) all have very high BMR for their size. However, allometric equations across families do not appear to describe accurately the metabolism–weight relationship for hummingbirds. Krüger, Prinzinger & Schuchman (1982) measured the metabolic rates of 18 hummingbird species from different habitats and ranging from 2.7 to 17.5 g in weight. They found that the resting metabolic rate for hummingbirds was considerably higher than theoretically expected for non-passerine birds, even when torpor was taken into account. Their calculated equation ($M = 212W^{0.72}$ kJ h^{-1}) shows that hummingbirds have a higher BMR than all other bird species investigated. The value is similar to mammals of the same weight, such as shrews (Hanski 1984).

Because hummingbirds are extremely sensitive to energy stress, yet often face conditions when energy intake is reduced or expenditures must be increased, they should have flexible energy budgets. When faced with low food availability or high foraging costs, hummingbirds have been found to reduce their perching metabolic rate (PMR) in the short term (Tiebout 1991). This daytime reduction in metabolic rate was not as great as during nighttime torpor, but the body temperature of these birds was low (30–33°C) compared with normal (38–41°C), and PMR dropped by approximately 25%. This is the only report of reduced PMR solely in response to a foraging constraint, although non-torpid daytime reduction of body temperatures has been reported in response to low ambient temperatures and limited food availability (Beuchat, Chaplin & Morton, 1979; Schuchmann & Schmidt-Marloh 1979; Schmidt-Marloh & Schuchmann 1980).

Only three studies have measured the energy expenditure of free-living hummingbirds. Powers & Nagy (1988) used doubly labelled water to measure the field metabolic rate (FMR) of Anna's hummingbirds (*Calypte anna*). The estimated FMR of a hummingbird maintaining constant body mass (mean 4.48 g) was 32 kJ day^{-1} (297 kJ kg^{-1} h^{-1}), approximately 5.2 times the calculated BMR. During the day, metabolic rates were about 6.8 times BMR; at night this dropped to 2.1 BMR. To supply this energy, the Anna's hummingbird needs to eat about 180 meals per day, totalling approximately three times its own body weight. In comparison, a shrew (*Sorex* sp.) of similar size uses 238 kJ kg^{-1} h^{-1} (Hanski 1984). Stiles (1971) measured the time budgets of Anna's hummingbirds and calculated the FMR based on estimates of the energy expenditure for each activity. His figures for FMR are approximately 20% lower than those found in the doubly labelled water study of Powers & Nagy (1988). Doubly labelled water was also used to measure the FMR of crowned woodnymphs (*Thalurania colombica*, mean weight 4.9 g) and bronze-tailed plumeleteers (*Chalybura urochrysia*, mean weight 7.23 g) in Costa Rica (Weathers & Stiles 1989). Field metabolic rate averaged 37.9 kJ day^{-1} in woodnymphs and 57.9 kJ day^{-1} in a single plumeleteer. Converted to mass-specific values, these are 322.3 kJ kg^{-1} h^{-1} and 333.7 kJ kg^{-1} h^{-1}, respectively. In both studies the water flux was exceptionally high, as much as 366% of the bird's body weight. Most of this water comes from feeding on relatively dilute nectar; hummingbirds do not appear to drink water very often. Similar values were obtained for blue-throated hummingbirds (*Lampornis clemenciae*, mean mass = 8.77 g) and black-chinned hummingbirds (*Archilochus alexandri*, mean mass = 3.67 g) visiting a feeder (Powers & Conley 1994). Field metabolic rates averaged 81.7 kJ day^{-1} (388.2 kJ kg^{-1} h^{-1}) in blue-throated hummingbirds and 29.1 kJ day^{-1} (330.4 kJ kg^{-1} h^{-1}) in black-chinned hummingbirds. The allometric relationship for body weight and FMR (kJ day^{-1}) for all birds is $10.9M^{0.64}$, and for non-passerines $4.80M^{0.749}$ (Nagy 1987). Neither of these relationships adequately predicts the observed values of hummingbird FMR (Fig. 2). However, this apparent trend is based on a few studies with small sample sizes and more work is needed to validate the relationship.

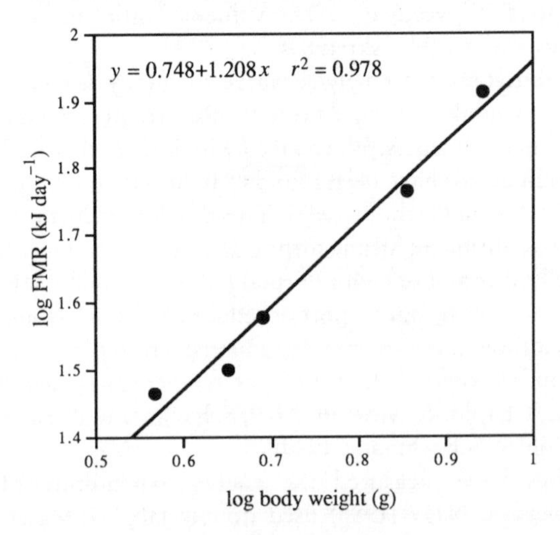

Fig. 2. Allometric relationship between log body weight (g) and log field metabolic rate (FMR, kJ day^{-1}).

Torpor

Torpor is a state resembling numbness with great reduction in body temperature and metabolism. Onset, maintenance, and arousal from torpor are active and regulated processes (Krüger *et al.* 1982; Bucher & Chappell 1992). Torpor has been described in some mammals, including bats, prosimians, and various rodents (Raths & Kulzer 1976). In birds torpor is known for some hummingbirds, nightjars, swifts, and mousebirds (Raths & Kulzer 1976; Krüger *et al.* 1982). Torpor in hummingbirds has been fairly well documented (e.g. French & Hodges 1959; Hainsworth & Wolf 1970; Wolf & Hainsworth 1972; Carpenter 1974; Hainsworth, Collins, & Wolf 1977). In these studies nighttime torpor was observed when the ambient temperature was experimentally lowered or when the bird could not gain its daily energy requirements. Carpenter (1976) found that torpor was frequent in free-living and captive Andean hillstars (*Oreotrochilus estella*) at all ambient temperatures from 0 to 30°C. More recently, Schuchmann, Krüger & Prinzinger (1983) demonstrated that hummingbirds regularly go into torpor at night, allowing body temperature to drop to ambient temperature. During torpor, metabolic rate is lowered by as much as 95% of mean active metabolism, independent of ambient temperature. Torpor even occurs in tropical species at high ambient temperatures (25°C) and with access to food. It appears therefore that torpor is a means to reduce energy expenditure and does not only occur in response to an emergency. Hiebert (1994) showed that the use of torpor by rufous hummingbirds (*Selasphorus rufus*) was highest in the autumn, but that they could enter torpor at any time. The ability to reduce nighttime energy consumption may be the only answer to the limited capacity of hummingbirds for storing reserves (Krüger *et al.* 1982). The plumage of hummingbirds does not seem particularly adapted to conserve heat; adults virtually lack down feathers, the body feathers being scale-like (Skutch 1974). The need to lose heat produced by the flight muscles may explain this (Schuchmann 1979).

High-montane environments may appear inhospitable to small birds with high energy demands, but hummingbirds are important members of many Andean ecosystems. Wolf & Gill (1986) reviewed the physiological constraints of low air density and low ambient temperature and concluded that high-altitude hummingbirds did not show specific physiological adaptations to the environment, such as larger body size. Rather, they used adaptations, such as torpor, characteristic of all hummingbirds, and were behaviourally adapted to track fluctuations in nectar supply. When conditions became unsuitable they escaped to other habitats.

Flight

Although hovering is the most expensive form of flight, most hummingbirds hover whilst foraging (Skutch 1974). The wings of hummingbirds are unique in their ability to rotate through 180°, allowing them great manoeuvrability and generating lift on both the upstroke and the downstroke; this enables hummingbirds to hover and fly backwards or upside-down (Norberg 1990). Hummingbirds are further adapted for flight by the enlargement of their flight muscles, which may comprise 21–35% of their body weight, twice that of many passerines (Rayner 1988; Norberg 1990). The pectoral muscles of hummingbirds are not especially large, but the supracoracoideus muscles are very large indeed. In most birds these muscles are not used to generate lift and are comparatively small, but in hummingbirds the supracoracoideus muscles make up 12% of total body weight (Norberg 1990). Hummingbirds scale in a unique manner with respect to flight costs (Calder 1984). Their wing area is scaled approximately linearly, so that wing loading is essentially size independent. The metabolic rate per gram of a hovering hummingbird is therefore essentially the same for all species studied (900 kJ kg^{-1} h^{-1}) (Hainsworth & Wolf 1972a; Epting 1980). Recently, the flight energetics of hovering hummingbirds were examined by simultaneous collection of metabolic and kinematic data in two species (Wells 1993b). Mass-specific oxygen consumption was found to be essentially the same for broad-tailed hummingbirds (*Selasphorus platycercus*, mean weight 3.97 g) and rufous hummingbirds (mean weight 4.28 g) and approximately 20% higher than reported in other studies. Recordings of oxygen consumption and wing kinematics for hummingbirds hovering at artificial flowers have shown that floral morphology can require hummingbirds to adjust their wingbeat frequencies and amplitude, making it more expensive metabolically to forage at wide flowers than at narrow flowers (Wells 1993a). When loaded by the addition of small weights, hovering hummingbirds increased their wingbeat amplitude to support the extra load while maintaining a nearly constant wingbeat frequency. A 10% increase in load required a 5.7% increase in mass-specific oxygen consumption (Wells 1993a). In forward flight, Gill (1985) found that long-tailed hermits (*Phaethornis superciliosus*) normally fly faster than the velocity predicted to minimize their cost of transport. The mean speed (11.5 m s^{-1}) was approximately twice the predicted speed, calculated according to Greenewalt (1975). Gill suggested that this could maximize visit rate to patches of flowers or reduce the time away from a lek. The speed of movement between flowers is much lower (0.2–1.2 m s^{-1}: Wolf, Stiles, & Hainsworth 1976; Montgomerie 1979). From the linear relation of flight speed and distance found by Gill (1985), it appears that the speed of movement between flowers is constrained by acceleration and deceleration. Flight in hummingbirds is discussed further by Rayner (this volume, pp. 83–109).

At high latitudes and altitudes many of the hummingbird species are migratory, with one species (rufous hummingbird) travelling as far as 7000 km each year (Calder 1976). The energetic cost of migration is very high (Kendleigh, Dol'nik & Gavrilov 1977), that per kilometre being $M_L = 0.1054W^{0.47}$ and in some cases must be met entirely by reserves carried by the bird. The rubythroat (*Archilochus colubris*) of eastern North America migrates to Central America for the winter, reaching as far as Panama. For some populations, this includes a 1000 km crossing of the Gulf of Mexico which must be fuelled with stored fat.

Reproduction and life history

The smallest bird egg known, that of the vervain hummingbird (*Mellisuga minima*), weighs only 0.37 g, but this amounts to more than 15% of the female's body weight. Although hummingbird eggs are relatively large, the total reproductive effort in terms of the clutch is smaller than for many other birds because hummingbirds almost always lay only two eggs (Skutch 1974). The relationship between both egg mass (g) and energy content (kJ) and female body weight is pronounced, but energy content ($E = 5.98M^{1.05}$) is a better predictor of the number of days required for incubation ($I = 6.35E^{0.283}$: Vleck & Vleck 1987). This relation predicts that an egg weighing 0.5 g would need 8.5 days for incubation. However, the shortest incubation period recorded for a hummingbird is 11 days, implying a limit to development rate (Calder 1984).

In almost all hummingbirds, the females provide the parental care. This is extended over a considerable period, since chicks are often dependent even after fledging. In many species, such as the scaly-breasted hummingbird (*Phaeochroa cuvierii*), the period of dependence may last as long as 65 days from hatching, making the total investment by the female hummingbird proportionately large by avian standards.

Nest failure in hummingbirds is high, usually as a result of predation. Estimates of the success of females at producing one fledgling range from 17% of clutches in hairy hermits (*Glaucis hirsuta*) breeding in Trinidad (D. W. Snow & Snow 1973), to 89% for Andean hillstars (Carpenter 1976). Most studies have reported values less than 50%. Following nest failure, female hummingbirds will often attempt to renest, but even in tropical regions the breeding season is seldom year-round (D. W. Snow & Snow 1973; Carpenter 1976; Stiles & Wolf 1979). The high rate of nest failures and slow rates of chick growth must mean that annual reproductive success is often very low. Nevertheless, hummingbirds are potentially long lived. Larger birds generally live longer ($\text{Lifespan}_{\text{years}} = 16.6M^{0.18}$: Lindstedt & Calder 1976), but ringing records of broad-tailed hummingbirds in North America show that they may reach 8.5 years, considerably longer than the six years predicted by the allometric equation for a 3.5 g bird (see Calder 1984). Average life expectancy is, however, only 1.3 years (Calder 1984), indicating that many individuals survive to breed for only one season.

Ecological consequences of small body size

Foraging

The diet of all hummingbirds consists of floral nectar and arthropods. Nectar from hummingbird-pollinated flowers is typically a solution of sucrose, with small concentrations of glucose and fructose (Hainsworth & Wolf 1976; Baker & Baker 1983a, b). Nectar viscosity increases with concentration, making extraction more difficult for a hummingbird, although, within the usual range of concentrations exploited (20–30%), hummingbirds extract nectar rapidly and efficiently (Hainsworth 1978; Diamond, Karasov, Phan, & Carpenter 1986).

Nectars used by hummingbirds contain very low levels of amino acids and lipids, so that the birds must obtain the majority of their proteins and fats from arthropods (Baker & Baker 1973). Almost all hummingbird species studied are known to eat arthropods (Remsen, Stiles, & Scott 1986), but the proportion of arthropods to nectar, and the methods of capture, vary considerably. The few time-budget studies available for foraging hummingbirds indicate that arthropod capture usually occupies only 2–12% of total foraging time (Gass & Montgomerie 1981). However, in periods when flowers are scarce, certain species feed almost exclusively on arthropods (Des Granges 1978) and these may be especially important for nesting hummingbirds and nestlings (Montgomerie & Redsell 1980). Schuchmann & Jacob (1981) investigated the time budgets of nine species of hummingbird, with a range of weights of 2.7–21.3 g, foraging under laboratory conditions. They demonstrated that larger species spent more time sitting, while the amount of time spent in flight and the frequency of visits to the feeder decreased with body weight. Duration of foraging bout was not correlated with body size, but the amount of nectar consumed on each bout increased with the weight of the bird. This is probably because crop volume increases with weight of bird ($V_{cp} = 10.23M^{0.68}$: Hainsworth & Wolf 1972b).

It is puzzling that, despite their high energy requirement, free-living hummingbirds spend only approximately 20% of the day feeding and 75% perched and apparently doing nothing (Stiles 1971; Hixon, Carpenter & Paton 1983). Diamond *et al.* (1986) found that hummingbirds have the highest rates of active glucose transport and the lowest passive glucose permeability of any vertebrate, allowing them almost 100% uptake efficiency. In spite of this rapid and efficient digestion, Diamond *et al.* suggest that crop-emptying time may limit feeding-bout frequency and could largely account for the time spent perched. Thus, although hummingbirds appear to minimize time spent foraging, in fact they may be maximizing energy intake, as quickly as the digestive system permits. Pyke (1981) observed that a hovering hummingbird can move more quickly between flowers than one which perches to feed and, consequently, can maximize net energy gain despite higher flight costs. The distance the bird travels to reach a foraging place may influence the decision whether to perch or to hover, but, as distance between flowers decreases, hovering becomes a more effective behaviour to maximize rate of net energy gain (Hainsworth 1986; Wolf & Hainsworth 1986). The size of the bird and its wing shape should also dictate whether a bird should perch or hover (Norberg 1990) as will floral architecture (Miller 1985).

The behaviour of foraging hummingbirds has been extensively studied from the perspective of optimal foraging theory (Hainsworth & Wolf 1976; DeBenedictis, Gill, Hainsworth, Pyke, & Wolf 1978; Kingsolver & Daniel 1983; Montgomerie, Eadie, & Harder 1984; Houston & Krakauer 1993). These studies have focused on the question of what function of energy intake is being maximized by foraging hummingbirds. Hainsworth & Wolf (1976) reported that hummingbirds did not maximize their rate of net energy intake when foraging under certain conditions, and concluded that the birds were not foraging optimally. Montgomerie *et al.* (1984) suggested that hummingbirds may be maximizing net energy per unit volume of nectar consumed, and concluded this to be an optimal choice on a time scale greater than a single foraging bout, because fewer foraging bouts would be needed to meet daily energy requirements. However, the authors did not take into account the costs incurred outside the feeding period or the increasing body mass during a meal. When mass-dependent costs are considered, maximizing net rate of energy gain can predict the choice of meal size and nectar concentrations shown by hummingbirds in the laboratory (Houston & Krakauer 1993). Although Houston & Krakauer's model shows that foraging hummingbirds may be maximizing energy, it does not consider the mechanics of nectar uptake or the optimization at different time scales such as during tongue loading or the licking cycle (see Kingsolver & Daniel 1983; Gass & Roberts 1992). In the field, factors such as territory ownership, competition and sexual behaviour all affect foraging activity in ways which cannot be explained through maximization of net rate of energetic gain (Houston & Krakauer 1993).

Most hummingbirds appear to track their food supplies over a period of time, adjusting their foraging and aggressive behaviour to changes on a time scale of one day or less (Kodric-Brown & Brown 1978; Gass 1979; Montgomerie 1979; Hixon *et al.* 1983).

For birds in general, Schoener (1968) found that territory size ($\propto M^{1.09}$) and home range ($\propto M^{1.16}$) increase in greater proportion with body mass, but for omnivorous and herbivorous birds, this was found to be less than linear ($\propto M^{0.35}$ and $\propto M^{0.70}$, respectively). From observations of four species of territorial hummingbirds defending patches of *Palicourea crocea* flowers (Cotton 1993), I obtained an exponent of 1.71 (Fig. 3). The predicted territory sizes for these species were also estimated from a cost–benefit model proposed by Pyke (1979) which resulted in an exponent of 1.56. Although based on a small sample size, this illustrates that feeding territory size in hummingbirds is strongly mass dependent.

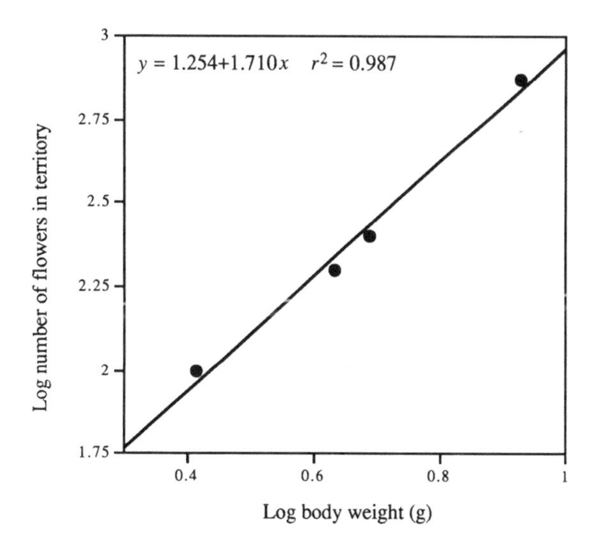

Fig. 3. The log number of flowers defended in feeding territories by four species of hummingbird of different body weights (g).

Territorial defence of flowers by hummingbirds appears to be an adaptive response to limited resource availability (Gass 1978; Kodric-Brown & Brown 1978). Interference competition in nectar-feeding birds involves aggressive interactions, the effectiveness of this action being determined by the dominance of relationships of individuals (Feinsinger 1976; Gass 1978; Gill & Wolf 1979; Ewald & Rohwer 1980; Ewald & Orians 1983).

Charnov (1976) assumed that foragers have a precise knowledge of the environment, an assumption implicit in the models of optimal territory size and time budget. Mitchell (1989) discusses the informational constraints on optimally foraging hummingbirds, but nectar-feeding birds are known to be able to use both prior information and patch-sampling information to track changes in resource availability, even if such changes are rapid (Gass 1979; Wolf & Hainsworth 1983; Tamm 1987; Valone 1992). Nectarivores defend feeding territories for short periods such as a day or a week and often abandon them as soon as nectar levels become unprofitable (Gill & Wolf 1975; Carpenter & MacMillen 1976; Trombulak 1990).

Few theoretical treatments have considered interspecific territorial interactions. Pimm (1978) predicted that where two species were involved in territoriality, the smaller species would take over any defence at lower profitability. In field experiments, using artificial feeding stations, he found that the smaller species defended unpredictable feeders, but was displaced from predictable feeders. More recently, ideal free distribution models have been used to predict the outcome of changes in species density, dominance, and habitat quality on territorial interactions (Pimm, Rosenzweig & Mitchell 1985; Rosenzweig 1986).

Plant–pollinator co-evolution

Hummingbirds depend on nectar for most of their energy needs and show a high degree of morphological, physiological, and behavioural adaptation to visiting flowers (Grant & Grant 1968; Stiles 1980; D. W. Snow 1981; Diamond *et al.* 1986). To a plant, bird pollination is energetically expensive, compared with insect pollination, but the advantages of a reliable long-range, long-lived pollinator must often outweigh the costs (Faegri & van der Pijl 1971; Carpenter 1988).

The structure of hummingbird flowers, especially the length and shape of the corolla tube, is often adapted to permit exploitation of pollinating hummingbirds (Grant & Grant 1968; Faegri & van de Pijl 1971). Stiles (1980) distinguishes three primary functions of the flower besides pollination itself: advertisement, reward, and filtering. In hummingbird-pollinated flowers, advertisement is primarily by flower colour; bird-flowers (ornithophilous) typically employ red, orange, and yellow in advertisement (Grant & Grant 1968; Raven 1972; Stiles 1976). This simultaneously minimizes their conspicuousness to bees, reducing nectar loss (Raven 1972) and contrasts strongly with the background of green foliage, increasing detectability by birds (Stiles 1976). The floral reward, nectar, must be rich enough to encourage pollinators to revisit, but poor enough to encourage movement between individual flowers to facilitate out-crossing. Nectar volumes of bird-pollinated flowers are greater than those of insect-pollinated flowers (Cruden, Herrmann, & Peterson 1983), but hummingbirds nevertheless must visit a large number of flowers to satisfy their energy requirements. A repeated foraging circuit is termed a trapline (B. K. Snow & Snow 1972; Stiles 1975; Gill 1988.

The flowers visited by traplining hummingbirds must, in some way, compensate for their low density by the rewards they offer. Thus, flowers occurring at low density typically have higher nectar energy content than high-density flowers defended by other hummingbirds. High energy rewards need to be protected from unwanted visitors, so flowers visited by high-reward trapliners often have long, tubular corollas, and are thickened at the base. Low-density, low-energy flowers do not need to invest in such elaborate floral structures to defend their resources (Stiles 1980).

Flowers must be protected from nectar thieves. Many hummingbird-pollinated flowers have a free-hanging corolla with no lip or perch (Faegri & van der Pijl 1971), preventing many insects from foraging but forcing hummingbirds to hover while foraging (Miller 1985). Additional features of bird-pollinated flowers are the absence of scent or nectar guides (Faegri & van der Pijl 1971; Proctor & Yeo 1973). A large number of plant families in the Americas have flowers adapted for pollination by hummingbirds, but most hummingbirds are extremely opportunistic and will forage at insect-pollinated flowers, often robbing the nectar without cross-pollinating the flower (Grant & Grant 1968; Colwell 1973; Skutch 1974). The most striking aspect of hummingbird–flower co-evolution is the close fit between the shape and length of the bill of some hummingbirds and the flowers on which they feed. Short-billed hummingbirds are often considered to be relatively unspecialized, feeding on a wide range of insect-pollinated and hummingbird-adapted flowers, often holding a feeding territory around a flowering tree or bush. At the other extreme, long-billed hummingbirds, such as the hermits (Phaethornithinae), are often described as specialized pollinators, traplining a few long-tubed, nectar-rich flowers which are widely scattered throughout the forest and which consequently make territorial defence uneconomical (B. K. Snow & Snow 1972; Wolf *et al.* 1976; Feinsinger 1978, 1983b; Feinsinger & Colwell 1978; Stiles & Wolf 1979; D. W. Snow 1981).

Occasionally the relationships between plants and birds are exceptionally intimate. The sword-billed hummingbird (*Ensifera ensifera*) has a culmen 105 mm long, allowing it exclusive access to the long, tubular flowers of *Passiflora mixta* (D. W. Snow & Snow 1980). Other hummingbirds have strongly curved bills, such as the white-tipped sicklebill (*Eutoxeres aquila*) which feeds from the curved flowers of *Heliconia pogonophora* and *Heliconia longa* (Stiles 1975, 1978, 1979). However, a closer examination of many mutualistic pollination and seed-dispersal systems has shown that tight co-evolution between birds and plants is uncommon (Howe & Vande Kerckhove 1979; Feinsinger 1983a; Herrera 1984; Moermond & Denslow 1985; Wheelwright 1985, 1988a, b).

Community organization and body size

Plants are strongly affected by competition for nectar and many have evolved mechanisms to prevent nectar loss to non-pollinators. As a result, communities of plants and hummingbirds are often organized on the basis of specificity relative to pollinator body size (Brown *et al.* 1978). Where several species of nectar-feeding birds co-exist in the same habitat, they often differ in body size, minimizing interspecific competition. Within hummingbird communities the division of the nectar resource usually involves bill–corolla size interactions, efficiency differences in exploiting resources, and dominance interactions (Wolf *et al.* 1976; Feinsinger & Colwell 1978). These give rise to a number of strategies or roles available to hummingbirds, such as territorial defence, territory parasitism, or traplining (Feinsinger & Colwell 1978).

Most hermit hummingbirds are high-reward trapliners and show a strong relationship between their culmen morphology and the corollas of the flowers visited (Stiles 1975). Some small, short-billed hummingbirds may be considered low-reward trapliners (Feinsinger & Colwell 1978), visiting dispersed flowers or insect-pollinated flowers but seldom defending territories. Clumped flowers, even with low or moderate amounts of nectar per flower, often attract medium-sized (4–6 g), short-billed hummingbirds which set up feeding territories and defend them against other hummingbirds (e.g. Stiles & Wolf 1970; Wolf *et al.* 1976; Gass 1978, 1979; Carpenter 1979; Hixon *et al.* 1983). Territorial defence by hummingbirds is usually aggressive, involving sometimes lengthy chases and high energetic costs (Wolf 1978). Territorial hummingbirds usually have high wing-disc loadings and so must expend relatively more energy to hover, but are highly manoeuvrable (Feinsinger, Colwell, Terborgh, & Chaplin 1979). Rich territories often attract many intruders, some of which are specialist territory parasites. Feinsinger & Colwell (1978) identify two classes of territory parasite: 'marauders' are large hummingbirds which forage with impunity in the territories of smaller hummingbirds, while 'filchers' are small hummingbirds which sneak into territories to steal nectar. One final strategy identified by Feinsinger & Colwell (1978) is that of the 'generalist' or 'facultative trapliner'. Most communities have available an opportunistic, loosely defined role that includes sporadic visits to dispersed flowers and filching from territories. This role is taken usually by small, short-billed hummingbirds. Thus, communities of hummingbirds are often considered to be organized on the basis of interference competition at defensible resources and co-evolved relationships between bill morphology and flower morphology (Feinsinger 1976; Wolf *et al.* 1976; Feinsinger & Colwell 1978; D. W. Snow & Snow 1980; Kodric-Brown, Brown, Byers & Gori 1984; Brown & Bowers 1985; Stiles 1985). Most studies (e.g. B. K. Snow & Snow 1972; Feinsinger & Colwell 1978) have found that, within any one community, only one species in a particular size class uses a particular foraging strategy, although hummingbirds are able to shift between roles depending on competition and the availability of resources. Even hummingbirds which usually act as high-reward trapliners, such as the magnificent hummingbird (*Eugens fulgens*), may become territorial when flowers are in unusually dense clumps (Wolf *et al.* 1976).

In many communities, hummingbird populations appear to be limited by the availability of nectar (Carpenter 1978; Gass & Montgomerie 1981; Montgomerie & Gass 1981). Most hummingbird populations appear to track their food supplies over a period of time, suggesting that the supply/demand economies of a co-evolved mutualism favour efficient resource use. Hummingbirds within these habitats appear to be maximizing net energy intake and minimizing time spent flying (Gass & Montgomerie 1981).

Daily activity patterns

Hummingbirds remain active and forage throughout the day, even in high latitudes where days are long (Calder 1975, 1976). In addition to the foraging decisions made on the time scale of single feeding bouts, hummingbirds must contend with diurnal changes in resource availability and competitive pressure. Most hummingbird-pollinated flowers are relatively rich in nectar when they open at dawn, but replace nectar at very low rates and senesce by the afternoon. However, the high metabolic requirements and flight costs of hummingbirds mean that storage of energy resources can be economical only over a short period of time (Montgomerie 1979). Thus, hummingbirds must track the availability of resources by moving within and between habitats (Montgomerie & Gass 1981). In the laboratory, hummingbirds have been shown to be capable of tracking both spatial and temporal changes in resource availability (Tamm 1987). Similarly, field experiments have shown that hummingbirds are very sensitive to differences in patch quality and renewal rates (Hixon *et al.* 1983; Montgomerie 1984; Montgomerie *et al.* 1984; Valone 1992).

Discussion

The life of hummingbirds, as of all other organisms, is clearly affected very strongly by body size, but what processes determine the upper and lower limits of body size? The relation between body size and BMR may be a compromise between the two objectives of providing locomotor power and minimizing heat loss (Kendeigh *et al.* 1977). The BMR of hummingbirds is above the Kleiber predictions (Kleiber 1932) for non-passerines, and may be higher even than those of small passerines (Krüger *et al.* 1982). Increasing basal metabolic rate may therefore restrict the evolution of still smaller body size.

Body energy reserves for overnight survival may also impose a lower size limit for hummingbirds (Brown *et al.* 1978), but the universal occurrence of torpor, even among tropical species, argues against this being a limiting factor, and there is no trend for smaller hummingbird species to be found in warmer climates, either at lower elevations or lower latitudes. The rufous hummingbird weighs 2.8 g but breeds regularly in Alaska (Calder 1976). In fact, torpor may allow hummingbirds, and other small endotherms, to sustain function at such small body sizes. Hummingbirds have exceptionally high field metabolic rates and an inability to maintain an FMR of between 9.2 and 10.46 kJ g^{-1} day^{-1} over an extended period may limit the evolution of small body size (Kendeigh *et al.* 1977). Digestive bottle-necks are known to restrict the rate of foraging in hummingbirds (Diamond *et al.* 1986) and other small vertebrates (Krebs & Harvey 1986), and may influence minimum body size as well as daily activity patterns.

A reduction in body size would also require hummingbirds to have a greater wingbeat frequency (Pennycuick 1972; Norberg 1990; Rayner, this volume, pp. 83–109). While hovering, the wings of a hummingbird beat 18–50 times per second (Norberg 1990), but neurogenic muscle (one impulse–one beat) is not capable of faster contractions, physiologically limiting the minimum weight of a vertebrate which hovers (Pennycuick 1989). Because of the structure of many hummingbird-pollinated flowers, hummingbirds are almost invariably forced to hover whilst feeding (Miller 1985). Cardiac physiology may impose a similar limitation on small body size (Jones & Johansen 1972). The resting heart beat of hummingbirds is approximately 8–9 s^{-1}; when active, the heart beat increases to over 20 s^{-1}, a similar rate to that found in some small bats. At such high rates there are fewer than 50 ms per beat, and in this time the heart must contract, relax, and then fill with blood.

The diet of hummingbirds often brings them into competition with insects (e.g. Willmer & Corbet 1981; Gill, Mack & Ray 1982; Laverty & Plowright 1985). Brown *et al.* (1978) suggest that competition for nectar resources may be as important as physiological constraints in limiting the evolution of body size in hummingbirds. Thus, it appears that a combination of physiological and ecological factors may prevent the evolution of even smaller size in hummingbirds. At the other extreme, the maximum body size of hummingbirds (at approximately 20 g still that of a miniature vertebrate) is probably determined by the nectar secretion of plants. Plants require pollinators to visit several flowers or individual plants to maximize pollen dispersal, and therefore offer only rewards which facilitate this. This is usually less than 100 ml of relatively dilute nectar, providing less than 100 J per flower (Stiles 1980; Cruden *et al.* 1983). Smaller birds have lower absolute costs for metabolism and activity and are able to specialize on the comparatively small rewards offered by flowers, while larger species must often rely on other sources of energy. Hummingbirds are the most specialized nectarivores and are the smallest birds; sunbirds, honeyeaters, and Hawaiian honeycreepers are all increasingly more dependent on insects for food and typically are larger than hummingbirds (Brown *et al.* 1978; Stiles 1980). The largest avian pollinators, the American orioles (Icteridae) and parrots (Psittacidae), are generally not highly dependent on nectar and show few morphological or behavioural traits for nectivory (Cruden & Toledo 1977; Forshaw 1989). Maximum body size in hummingbirds therefore appears to be dictated by the diminishing advantage of large pollinators to the plants (Brown *et al.* 1978).

Acknowledgements

I thank Peter Miller for his invitation to present a paper at the symposium. Chris Perrins and Peter Miller made helpful comments on the manuscript.

References

Aschoff, J. & Pohl, H. (1970). Der Ruheumsatz von Vögeln als Funktion der Tageszeit und der Körpergrösse. *J. Orn., Berl.* **111**: 38–47.

Baker, H. G. & Baker, I. (1973). Amino-acids in nectar and their evolutionary significance. *Nature, Lond.* **241**: 543–545.

Baker, H. G. & Baker, I. (1983a). A brief historical review of the chemistry of floral nectar. In *The biology of nectaries*: 126–152. (Eds Bentley, B. & Elias, T.). Columbia University Press, New York.

Baker, H. G. & Baker, I. (1983b). Floral nectar sugar constituents in relation to pollinator type. In *Handbook of experimental pollen biology*: 117–141. (Eds Jones, C. E. & Little, R. J.). Van Nostrand Reinhold, New York.

Bennett, P. M. & Harvey, P. H. (1987). Active and resting metabolism in birds: allometry, phylogeny and ecology. *J. Zool., Lond.* **213**: 327–363.

Beuchat, C. A., Chaplin, S. B., & Morton, M. L. (1979). Ambient temperature and the daily energetics of two species of hummingbirds, *Calypte anna* and *Selasphorus rufus*. *Physiol. Zool.* **52**: 280–295.

Brown, J. H. & Bowers, M. A. (1985). Community organization in hummingbirds: relationships between morphology and ecology. *Auk* **102**: 251–269.

Brown, J. H., Calder, W. A., & Kodric-Brown, A. (1978). Correlates and consequences of body size in nectar-feeding birds. *Am. Zool.* **18**: 687–700.

Bucher, T. L. & Chappell, M. A. (1992). Ventilatory and metabolic dynamics during entry into and arousal from torpor in *Selasphorus* hummingbirds. *Physiol. Zool.* **65**: 978–993.

Calder, W. A. (1975). Daylength and the hummingbird's use of time. *Auk* **92**: 81–97.

Calder, W. A. (1976). Energetics of small body size and high latitude: the rufous hummingbird in coastal Alaska. *Int. J. Biometeor.* **20**: 23–25.

Calder, W. A. (1984). *Size, function and life history*. Harvard University Press, Cambridge, Mass.

Carpenter, F. L. (1974). Torpor in an Andean hummingbird: its ecological significance. *Science* **183**: 545–547.

Carpenter, F. L. (1976). Ecology and evolution of an Andean hummingbird (*Oreotrochilus estella*). *Univ. Calif. Publs Zool.* **106**: 1–74.

Carpenter, P. L. (1978). A spectrum of nectar-eater communities. *Am. Zool.* **18**: 809–819.

Carpenter, F. L. (1979). Competition between hummingbirds and insects for nectar. *Am. Zool.* **19**: 1105–1114.

Carpenter, F. L. (1988). Pollen-transfer efficiency compensates for pollinator crashes in a specialized bird-pollinated plant. In *Acta XIX congressus internationalis ornithologici* **1**: 537–548. (Ed. Ouellet, H.). University of Ottawa Press, Ottawa.

Carpenter, F. L. & MacMillen, R. E. (1976). Threshold model of feeding territoriality and test with a Hawaiian honeycreeper. *Science* **194**: 639–642.

Charnov, E. L. (1976). Optimal foraging: the marginal value theorem. *Theor. Pop. Biol.* **9**: 129–136.

Colwell, R. K. (1973). Competition and coexistence in a simple tropical community. *Am. Nat.* **107**: 737–760.

Cotton, P. A. (1993). *The hummingbird–plant community of a lowland Amazonian rainforest.* DPhil thesis: University of Oxford.

Cruden, R. W., Hermann, S. M., & Peterson, S. (1983). Patterns of nectar production and plant–pollinator coevolution. In *The biology of nectaries*: 80–125. (Eds Bentley, B. & Elias, T.). Columbia University Press, New York.

Cruden, R. W. & Toledo, V. M. (1977). Oriole pollination of *Erythrina breviflora* (Leguminosae): evidence for a polytypic view of ornithology. *Plant Syst. Evol.* **126**: 293–403.

DeBenedictis, P. A., Gill, F. B., Hainsworth, F. R., Pyke, G. H., & Wolf, L. L. (1978). Optimal meal size in hummingbirds. *Am. Nat.* **112**: 301–316.

Des Granges, J.-L. (1978). Organisation of a tropical nectar feeding bird guild in a variable environment. *Living Bird* **17**: 199–236.

Diamond, J. M., Karasov, W. H., Phan, D., & Carpenter, F. L. (1986). Digestive physiology is a determinant of foraging bout frequency in hummingbirds. *Nature, Lond.*, **320**: 62–63.

Dunning, J. B. (Ed.) (1993). *CRC handbook of avian body masses.* CRC Press, Boca Raton, Florida.

Elgar, M. A. & Harvey, P. H. (1987). Basal metabolic rates in mammals: allometry, phylogeny and ecology. *Funct. Ecol.* **1**: 25–36.

Epting, R. J. (1980). Functional dependence of the power for hovering on wing-disc loading in hummingbirds. *Physiol. Zool.* **53**: 347–357.

Ewald, P. W. & Orians, G. H. (1983). Effects of resource depression on use of inexpensive and escalated aggressive behavior: experimental tests using Anna hummingbirds. *Behav. Ecol. Sociobiol.* **12**: 95–101.

Ewald, P. W. & Rohwer, S. (1980). Age coloration and dominance in nonbreeding hummingbirds: a test of the asymmetry hypothesis. *Behav. Ecol. Sociobiol.* **7**: 273–279.

Faegri, K. & van der Pijl, L. (1971). *The principles of pollination ecology.* Pergamon Press, Oxford.

Feinsinger, P. (1976). Organisation of a tropical guild of nectarivorous birds. *Ecol. Monogr.* **46**: 257–291.

Feinsinger, P. (1978). Ecological interactions between plants and hummingbirds in a successional tropical community. *Ecol. Monogr.* **48**: 269–287.

Feinsinger, P. (1983a). Coevolution and pollination. In *Coevolution*: 281–310. (Eds Futuyama, D. J. & Slatkin, M.). Sinauer Associates Inc., Massachusetts.

Feinsinger, P. (1983b). Variable nectar secretion in a *Heliconia* species pollinated by hermit hummingbirds. *Biotropica* **15**: 48–52.

Feinsinger, P. & Colwell, R. K. (1978). Community organization among Neotropical nectar-feeding birds. *Am. Zool.* **18**: 779–795.

Feinsinger, P., Colwell, R. K., Terborgh, J., & Chaplin, S. B. (1979). Elevation and the morphology, flight energetics, and foraging ecology of tropical hummingbirds. *Am. Nat.* **113**: 481–497.

Forshaw, J. M. (1989). *Parrots of the world.* (3rd edn). Blandford Press, London.

French, N. R. & Hodges, R. W. (1959). Torpidity in cave-roosting hummingbirds. *Condor* **61**: 223.

Gass, C. L. (1978). Rufous hummingbird feeding territoriality in a suboptimal habitat. *Can. J. Zool.* **56**: 1535–1539.

Gass, C. L. (1979). Territory regulation, tenure, and migration in rufous hummingbirds. *Can. J. Zool.* 57: 914–923.
Gass, C. L. & Montgomerie, R. D. (1981). Hummingbird foraging behavior: decision-making and energy regulation. In *Foraging behavior: ecological, ethological and psychological approaches*: 159–194. (Eds Kamil, A. C. & Sargent, T. D.). Garland STMP Press, New York.
Gass, C. L. & Roberts, W. M. (1992). The problem of temporal scale in optimization: three contrasting views of hummingbird visits to flowers. *Am. Nat.* **140**: 829–853.
Gill, F. B. (1985). Hummingbird flight speeds. *Auk* **102**: 97–101.
Gill, F. B. (1988). Trapline foraging by hermit hummingbirds: competition for an undefended, renewable resource. *Ecology* **69**: 1933–1942.
Gill, F. B., Mack, A. L., & Ray, R. T. (1982). Competition between hermit hummingbirds Phaethorninae and insects for nectar in a Costa Rica rain forest. *Ibis* **124**: 44–49.
Gill, F. B. & Wolf, L. L. (1975). Economics of feeding territoriality in the golden-winged sunbird. *Ecology* **56**: 333–345.
Gill, F. B. & Wolf, L. L. (1979). Nectar loss by golden-winged sunbirds to competitors. *Auk* **96**: 448–461.
Grant, K. A. & Grant, V. (1968). *Hummingbirds and their flowers.* Columbia University Press, New York.
Greenewalt, C. H. (1975). The flight of birds. *Trans. Am. Phil. Soc.* **65**: 1–67.
Hainsworth, F. R. (1978). Feeding: models of costs and benefits in energy regulation. *Am. Zool.* **18**: 701–714.
Hainsworth, F. R. (1986). Why hummingbirds hover: a commentary. *Auk* **103**: 832–833.
Hainsworth, F. R., Collins, B. G., & Wolf, L. L. (1977). The function of torpor in hummingbirds. *Physiol. Zool.* **50**: 215–222.
Hainsworth, F. R. & Wolf, L. L. (1970). Regulation of oxygen consumption and body temperature during torpor in a hummingbird, *Eulampis jugularis. Science* **168**: 368–369.
Hainsworth, F. R. & Wolf, L. L. (1972a). Crop volume, nectar concentration, and hummingbird energetics. *Comp. Biochem. Physiol. (A)* **42**: 359–366.
Hainsworth, F. R. & Wolf, L. L. (1972b). Power for hovering flight in relation to body-size in hummingbirds. *Am. Nat.* **106**: 589–596.
Hainsworth, F. R. & Wolf, L. L. (1976). Nectar characteristics and food selection by hummingbirds. *Oecologia* **25**: 101–113.
Hanski, I. (1984). Food consumption, assimilation and metabolic rate in six species of shrew (*Sorex* and *Neomys*). *Annls zool. fenn.* **21**: 157–165.
Herrera, C. M. (1984). A study of avian frugivores, bird-dispersed plants, and their interaction in Mediterranean scrublands. *Ecol. Monogr.* **54**: 1–23.
Hiebert, S. (1994). Seasonal changes in body mass and use of torpor in a migratory hummingbird. *Auk* **110**: 787–797.
Hixon, M. A., Carpenter, F. L., & Paton, D. C. (1983). Territory area, flower density, and time budgeting in hummingbirds: an experimental and theoretical analysis. *Am. Nat.* **122**: 336–391.
Houston, A. I. & Krakauer, D. C. (1993). Hummingbirds as net rate maximisers. *Oecologia* **94**: 135–138.
Howe, H. F. & Vande Kerckhove, G. A. (1979). Fecundity and seed dispersal of a tropical tree. *Ecology* **60**: 180–189.
Jones, D. R. & Johansen, K. (1972). The blood vascular system of birds. In *Avian biology*: 157–285. (Eds Farner, D. S., King, J. R., & Parkes, K. C.). Academic Press, New York.
Kendeigh, S. C., Dol'nik, V. R., & Gavrilov, V. M. (1977). Avian energetics. In *Granivorous*

birds in ecosystems: 127–204. (Eds Pinowski, J. & Kendeigh, S. C.). Cambridge University Press, Cambridge. (*Int. biol. Progm.* **12**.)

Kingsolver, J. G. & Daniel, T. L. (1983). Mechanical determinants of nectar feeding strategy in hummingbirds: energetics, tongue morphology, and licking behaviour. *Oecologia* **60**: 214–226.

Kleiber, M. (1932). Body size and metabolism. *Hilgardia* **6**: 315–353.

Kodric-Brown, A. & Brown, J. H. (1978). Influence of economics, interspecific competition and sexual dimorphism on territoriality of migrant rufous hummingbirds. *Ecology* **59**: 285–296.

Kodric-Brown, A., Brown, J. H., Byers, G. S., & Gori, D. F. (1984). Organization of a tropical island community of hummingbirds and flowers. *Ecology* **65**: 1358–1368.

Krebs, J. R. & Harvey, P. H. (1986). Busy doing nothing—efficiently. *Nature, Lond.* **320**: 18–19.

Krüger, K., Prinzinger, R., & Schuchmann, K.-L. (1982). Torpor and metabolism in hummingbirds. *Comp. Biochem. Physiol. (A)* **73**: 679–689.

Laverty, T. M. & Plowright, R. C. (1985). Competition between hummingbirds and bumble bees for nectar in flowers of *Impatiens biflora*. *Oecologia* **66**: 25–32.

Lindstedt, S. L. & Calder, W. A. (1976). Body size and longevity in birds. *Condor* **78**: 91–94.

Miller, R. S. (1985). Why hummingbirds hover. *Auk* **102**: 722–726.

Mitchell, W. A. (1989). Informational constraints on optimally foraging hummingbirds. *Oikos* **55**: 145–154.

Moermond, T. C. & Denslow, J. S. (1985). Neotropical avian frugivores: patterns of behavior, morphology, and nutrition, with consequences for fruit selection. *Orn. Monogr.* No. 36: 867–897.

Montgomerie, R. D. (1979). *The energetics of foraging and competition in some Mexican hummingbirds*. PhD thesis: McGill University.

Montgomerie, R. D. (1984). Nectar extraction by hummingbirds: response to different floral characters. *Oecologia* **63**: 229–236.

Montgomerie, R. D., Eadie, J. M. & Harder, L. D. (1984). What do foraging hummingbirds maximize? *Oecologia* **63**: 357–363.

Montgomerie, R. D. & Gass, C. L. (1981). Energy limitation of hummingbird populations in tropical and temperate communities. *Oecologia* **50**: 162–165.

Montgomerie, R. D. & Redsell, C. A. (1980). A nesting hummingbird feeding solely on arthropods. *Condor* **82**: 463–464.

Nagy, K. A. (1987). Field metabolic rate and food requirement scaling in mammals and birds. *Ecol. Monogr.* **57**: 111–128.

Norberg, V. M. (1990). *Vertebrate flight*. Springer-Verlag, Berlin.

Pennycuick, C. (1972). *Animal flight*. Edward Arnold, London. (*Stud. Biol.* No. 33).

Pennycuick, C. (1989). *Bird flight performance: a practical calculation manual*. Oxford University Press, Oxford.

Pimm, S. L. (1978). An experimental approach to the effects of predictability on community structure. *Am. Zool.* **18**: 797–808.

Pimm, S. L., Rosenzweig, M. L., & Mitchell, W. (1985). Competition and food selection: field tests of a theory. *Ecology* **66**: 798–807.

Powers, D. R. & Conley, T. M. (1994). Field metabolic rate and food consumption of two sympatric hummingbird species in southeastern Arizona. *Condor* **96**: 141–150.

Powers, D. R. & Nagy, K. A. (1988). Field metabolic rate and food consumption by free-living Anna's hummingbirds (*Calypte anna*). *Physiol. Zool.* **61**: 500–506.

Proctor, M. & Yeo, P. (1973). *The pollination of plants*. Collins, London.
Pyke, G. H. (1979). The economics of territory size and time budget in the golden-winged sunbird. *Am. Nat.* **114**: 131–145.
Pyke, G. H. (1981). Why hummingbirds hover and honeyeaters perch. *Anim. Behav.* **29**: 861–867.
Rahn, H. & Ar, A. (1974). The avian egg: incubation time and water loss. *Condor* **76**: 147–152.
Rahn, H., Paganelli, C. V. & Ar, A. (1975). Relation of avian egg weight to body weight. *Auk* **92**: 750–765.
Raths, P. & Kulzer, E. (1976). Physiology of hibernation and related lethargic states in mammals and birds. *Bonn. zool. Monogr.* No. 9: 1–93.
Raven, P. H. (1972). Why are bird-visited flowers predominantly red? *Evolution* **26**: 674.
Rayner, J. M. V. (1988). Form and function in avian flight. *Curr. Orn.* **5**: 1–66.
Remsen, J. V., Stiles, F. G., & Scott, P. E. (1986). Frequency of arthropods in stomachs of tropical hummingbirds. *Auk* **103**: 436–441.
Rosenzweig, M. L. (1986). Hummingbird isolegs in an experimental system. *Behav. Ecol. Sociobiol.* **19**: 313–322.
Schmidt-Marloh, D. & Schuchmann, K.-L. (1980). Zur Biologie des Blauen Veilchenohr-Kolibris (*Colibri coruscans*). *Bonn. zool. Beitr.* **31**: 61–77.
Schoener, T. W. (1968). Sizes of feeding territories among birds. *Ecology* **49**: 123–141.
Schuchmann, K.-L. (1979). Energieumstatz in Abhängigheit von der Umgebungstemperatur beim Kolibri *Ocreatus underwoodii underwoodii*. *J., Orn., Berl.* **120**: 311–315.
Schuchmann, K.-L. & Jacob, H. (1981). Energy expenditure of an incubating tropical hummingbird under laboratory conditions. *Gerfaut* **71**: 227–233.
Schuchmann, K. L., Krüger, K. & Prinzinger, R. (1983). Torpor in hummingbirds. *Bonn. zool. Beitr.* **34**: 273–277.
Schuchmann, K. L. & Schmidt-Marloh, D. (1979). Temperature regulation in non-torpid hummingbirds. *Ibis* **121**: 354–356.
Skutch, A. F. (1974). *The life of the hummingbird*. Octopus Books Ltd., London.
Snow, B. K. & Snow, D. W. (1972). Feeding niches of hummingbirds in a Trinidad valley. *J. Anim. Ecol.* **41**: 471–485.
Snow, D. W. (1981). Coevolution of birds and plants. In *The evolving biosphere*: 169–178. (Ed. Forey, P. L.). British Museum (Natural History), London & Cambridge University Press, Cambridge.
Snow, D. W. & Snow, B. K. (1973). The breeding biology of the hairy hermit *Glaucis hirsuta* in Trinidad. *Ardea* **61**: 106–122.
Snow, D. W. & Snow, B. K. (1980). Relationships between hummingbirds and flowers in the Andes of Colombia. *Bull. Br. Mus. nat. Hist. (Zool.)* **38**: 105–139.
Stiles, F. G. (1971). Time, energy and territoriality of the anna hummingbird (*Calypte anna*). *Science* **173**: 818–821.
Stiles, F. G. (1975). Ecology, flowering phenology, and hummingbird pollination of some Costa Rican *Heliconia* species. *Ecology* **56**: 285–301.
Stiles, F. G. (1976). Taste preferences, color preferences, and flower choice in hummingbirds. *Condor* **78**: 10–26.
Stiles, F. G. (1978). Ecological and evolutionary implications of bird pollination. *Am. Zool.* **18**: 715–727.
Stiles, F. G. (1979). El ciclo anual en una comunidad coadaptada de colibríes y flores en el bosque tropical muy húmedo de Costa Rica. *Revta Biol. trop.* **27**: 75–101.

Stiles, F. G. (1980). Ecological and evolutionary aspects of bird–flower coadaptations. In *Acta XVII congressus internationalis ornithologici* 2: 1173–1178. (Ed. Nöhring, R.). Verlag der Deutschen Ornithologen-Gesellschaft, Berlin.

Stiles, F. G. (1985). Seasonal patterns and coevolution in the hummingbird–flower community of a Costa Rican subtropical forest. *Orn. Monogr.* No. 36: 757–785.

Stiles, F. G. & Wolf, L. L. (1970). Hummingbird territoriality at a tropical flowering tree. *Auk* **87**: 467–491.

Stiles, F. G. & Wolf, L. L. (1979). Ecology and evolution of lek mating behaviour in the long-tailed hermit hummingbird. *Orn. Monogr.* No. 27: 1–78.

Tamm, S. (1987). Tracking varying environments: sampling by hummingbirds. *Anim. Behav.* **35**: 1725–1734.

Tiebout, H. M. (1991). Day-time energy management by tropical hummingbirds: responses to foraging constraint. *Ecology* **72**: 839–851.

Trombulak, S. C. (1990). Assessment of territory value by a tropical hummingbird (*Amazilia saucerottei*). *Biotropica* **22**: 9–15.

Valone, T. J. (1992). Information for patch assessment: a field investigation with black-chinned hummingbirds. *Behav. Ecol.* **3**: 211–222.

Vleck, C. M. & Vleck, D. (1987). Metabolism and energetics of avian embryos. *J. exp. Zool. Suppl.* No. 1: 111–125.

Weathers, W. W. & Stiles, F. G. (1989). Energetics and water balance in free-living tropical hummingbirds. *Condor* **91**: 324–331.

Wells, D. J. (1993a). Ecological correlates of hovering flight of hummingbirds. *J. exp. Biol.* **178**: 59–70.

Wells, D. J. (1993b). Muscle performance in hovering hummingbirds. *J. exp. Biol.* **178**: 39–57.

Wheelwright, N. T. (1985). Competition for dispersers, and the timing of flowering and fruiting in a guild of tropical trees. *Oikos* **44**: 465–477.

Wheelwright, N. T. (1988a). Four constraints on coevolution between fruit-eating birds and fruiting plants: a tropical case history. In *Acta XIX congressus internationalis ornithologici* 1: 827–845. (Ed. Ouellet, H.). University of Ottawa Press, Ottawa.

Wheelwright, N. T. (1988b). Fruit-eating birds and bird-dispersed plants in the tropics and temperate zone. *Trends Ecol. Evol.* **3**: 270–274.

Willmer, P. G. & Corbet, S. A. (1981). Temporal and microclimatic partitioning of the floral resources of *Justicia aurea* amongst a concourse of pollen vectors and nectar robbers. *Oecologia* **51**: 67–78.

Wolf, L. L. (1978). Aggressive social organization in nectarivorous birds. *Am. Nat.* **18**: 765–778.

Wolf, L. L. & Gill, F. B. (1986). *Physiological and ecological adaptations of high montane sunbirds and hummingbirds.* Oxford University Press, Oxford.

Wolf, L. L. & Hainsworth, F. R. (1972). Environmental influence on regulated body temperature in torpid hummingbirds. *Comp. Biochem. Physiol. (A)* **41**: 167–173.

Wolf, L. L. & Hainsworth, F. R. (1983). Economics of foraging strategies in sunbirds and hummingbirds. In *Behavioral energetics: the cost of survival in vertebrates*: 223–264. (Eds Aspey, W. P. & Lustick, S. I.). Ohio State University Press, Columbus, Ohio.

Wolf, L. L. & Hainsworth, F. R. (1986). Information and hummingbird foraging at individual inflorescences of *Ipomopsis aggregata*. *Oikos* **46**: 15–22.

Wolf, L. L., Stiles, F. G., & Hainsworth, F. R. (1976). Ecological organisation of a tropical, highland hummingbird community. *J. Anim. Ecol.* **45**: 349–379.

Symp. zool. Soc. Lond. (1996) No. 69: 259–276

Ecology of very small terrestrial mammals

SARA CHURCHFIELD

Division of Life Sciences
King's College
Campden Hill Road
London W8 7AH

Synopsis

Miniaturism among terrestrial mammals is a rare phenomenon, with only 8% of all species being < 15 g in body mass and < 100 mm in body length. It is confined to just three orders (Marsupialia, Insectivora, and Rodentia) and is the major preserve of four families (Dasyuridae, Tenrecidae, Soricidae, and Muridae). The smallest representatives have body masses of 1.5–3.0 g. Most species have a tropical and subtropical distribution, and numbers of species (and their mean body mass) decline with increasing latitude. The predominant feeding modes are insectivory and granivory. Most miniature mammals occur in forest and grassland/steppe habitats and are mainly ground-surface foragers. Very small mammals have lower absolute food requirements than do larger counterparts, and miniature insectivores subsist on small arthropods with low biomass but high abundance. This permits them to inhabit less productive areas at higher latitudes and altitudes compared with larger congeners which rely on richer habitats supporting larger prey such as earthworms. The occurrence of insectivorous shrews, but not miniature granivorous rodents, in Arctic regions probably reflects greater availability of small arthropods than of seeds in winter. Exceedingly small mammals have low population densities and there is evidence of reduced population numbers, body size, and niche breadth when they are sympatric with larger congeners. But differences in body size and niche occupancy help to reduce overlap and competition in communities of similar species. Very small species tend to be more specialized with narrower niches, and have lower levels of overlap, than larger congeners.

The occurrence of miniaturism amongst terrestrial mammals

For the purpose of this paper a miniature mammal is defined as one whose adult body mass does not exceed 15 g and whose head and body length does not exceed 100 mm. Although the Chiroptera (bats) contain a number of species which satisfy this definition, they have been excluded from this survey because the remit was to examine miniaturism only among terrestrial mammals. It is noteworthy that,

ZOOLOGICAL SYMPOSIUM No. 69
ISBN 0-19-857787-7

while there are several miniature semi-aquatic species which occupy the terrestrial/freshwater interface and forage in water (and have been included in the survey), none are marine.

Of the 15 orders of terrestrial mammals, only three contain miniature species as defined above. These are the Marsupialia, the Insectivora, and the Rodentia. Of approximately 266 species of marsupials, only 7.1% demonstrate miniaturism, and these are distributed among just four families (Table 1). Small size is found predominantly in the Dasyuridae of Australia where 25% of species are miniatures. They mostly comprise planigales, ningauis, and dunnarts of the genera *Planigale, Ningaui*, and *Sminthopsis*, respectively.

Table 1. The occurrence of miniature species amongst the terrestrial Mammalia.

Order	Family	Approximate number of species	Number of miniature species	Common names of miniature species
Marsupialia	Burramyidae	7	3	Pygmy possums
	Tarsipedidae	1	1	Honey possum
	Didelphidae	75	2	Mouse opossums
	Dasyuridae	51	13	Dunnarts, planigales, ningauis
	Total	141	19	
Insectivora	Chrysochloridae	18	1	Desert mole
	Talpidae	29	1	Shrew mole
	Tenrecidae	34	13	Shrew tenrecs
	Soricidae	246	228	Shrews
	Total	327	243	
Rodentia	Anomaluridae	7	1	Scaly-tailed squirrel
	Sciuridae	267	1	Pygmy squirrel
	Heteromyidae	65	7	Pocket mice
	Dipodidae	31	6	Pygmy jerboas
	Zapodidae	14	6	Birch mice
	Muridae	1082	53	Small mice
	Total	1466	74	

Main sources: van den Brink (1977); Corbet & Hill (1980); Whitaker (1980); Nowak & Paradiso (1983); Smithers (1990); Macdonald (1985); Strahan (1991); Redford & Eisenberg (1992); Wilson & Reeder (1993).

It is the order Insectivora which contains the greatest number of miniature species. Again, these are confined to just four families (see Table 1), only two of which contain significant numbers of very small species. These include the shrew tenrecs of the genus *Microgale* (Tenrecidae) and the true shrews of the family Soricidae. Of approximately 345 species of insectivores, 70.4% are miniatures but such small size is found predominantly amongst the Soricidae in which 92.7% of species have this status.

Miniaturism is confined to six families of the Rodentia (Table 1) and, despite the abundance of rodent species, it is a relatively rare phenomenon, with only 4.3% possessing this status. Several species of pocket mice (Heteromyidae), jerboas (Dipodidae), and birch mice (Zapodidae) satisfy the size criteria, but such small size is found predominantly amongst the Muridae. Even so, only approximately 4.9% of murid species are miniature.

Miniaturism, then, is a relatively rare feature among terrestrial mammals, encompassing only 8% of all species, and it is the major preserve of just four families: the Dasyuridae, Tenrecidae, Soricidae, and Muridae. The purpose of this paper is to explore aspects of the ecology of very small mammals, and elucidate the advantages and disadvantages that small size confers in terms of distribution, habitat utilization, food exploitation, and niche occupancy and competition.

Examples of very small size in terrestrial mammals

Examples of species exhibiting the extremes of small size amongst terrestrial mammals are shown in Table 2. The ranges given for body mass and length include immature (but weaned, independent) individuals and sexually mature ones. The list is by no means exclusive, but comprises those species which are best documented. Sixteen of those listed have an adult body mass of less than 7 g. The smallest known terrestrial mammals are shrews, exemplified by *Suncus etruscus*, with a body mass of 1.5–2.0 g, and *Sorex minutissimus* (1.5–3.0 g).

Table 2. Examples of the smallest terrestrial mammals.

	Family	Body mass (g)	Head + body length (mm)	Source
Suncus etruscus	Soricidae	1.5–2.0	36–52	van den Brink (1977)
Sorex minutissimus	Soricidae	1.5–3.0	35–53	van den Brink (1977)
Crocidura luna	Soricidae	*c.* 2.0	35–45	Smithers (1990)
Microsorex hoyi	Soricidae	2.2–3.8	51–63	Whitaker (1980)
Sorex longirostris	Soricidae	2.0–4.0	46–48	Whitaker (1980)
Notiosorex crawfordi	Soricidae	2.9–5.0	55–61	Whitaker (1980)
Salpingotulus michaelis	Dipodidae	*c.* 3.0	36–47	Nowak & Paradiso (1983)
Suncus infinitesimus	Soricidae	3.0–4.0	42–62	Smithers (1990)
Sorex minutus	Soricidae	2.5–5.5	43–63	Corbet & Harris (1991)
Planigale ingrami	Dasyuridae	3.9–4.5	55–65	Strahan (1991)
Salpingotus crassicauda	Dipodidae	*c.* 4.0	41–57	Nowak & Paradiso (1983)
Cryptotis parva	Soricidae	4.0–6.5	57–67	Whitaker (1980)
Sorex merriami	Soricidae	4.4–6.5	55–65	Whitaker (1980)
Sorex dispar	Soricidae	5.0–6.0	55–73	Whitaker (1980)
Crocidura bicolor	Soricidae	4.2–6.8	48–75	Smithers (1990)
Suncus varilla	Soricidae	*c.* 6.5	44–68	Smithers (1990)
Planigale tenuirostris	Dasyuridae	4.0–9.0	50–75	Strahan (1991)
Micromys minutus	Muridae	5.0–9.0	58–76	Corbet & Harris (1991)
Mus minutoides	Muridae	2.5–12.0	45–82	Smithers (1990)

Geographical distribution of miniature terrestrial mammals

Very small mammals have high ratios of surface area to volume, and those as small as shrews approach a critical mass for maintaining endothermy at low temperatures, losing much more energy as heat than larger mammals do (McNab 1983; Genoud 1988). As ambient temperature falls below their thermoneutral zone, metabolic rate increases to generate heat and their energy requirements rise. Shrews show an inverse relationship between body mass and (1) metabolic rate (Fig. 1) and (2) food consumption as a proportion of body mass (Fig. 2). While *Sorex araneus* consumes, on average, 85% of its body weight in food daily, the tiny *Sorex minutus* consumes a massive 125%. These characteristics should not be conducive to life in cold climates but are suited to hot climates requiring high rates of heat loss. It could be predicted, therefore, that the distribution of miniature mammals favours regions of hot, equable climate and discriminates against regions of cold climate with severe winters.

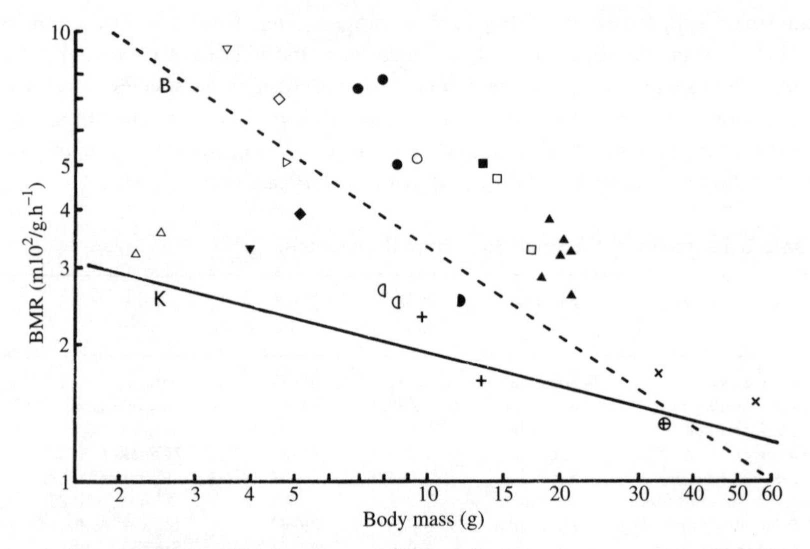

Fig. 1. The relationship between body mass and basal metabolic rate of 16 species of shrews (each depicted by a different symbol). Line K is the Kleiber curve and line B is the boundary curve for continuous maintenance of endothermy (after Genoud 1988).

This proves to be the case, for, while very small mammals are widely distributed in the Old and New Worlds, the majority of species have a tropical and subtropical distribution (Fig. 3). As with all the data provided in the accompanying tables, the percentage values express species occurrences which take account of the fact that some species may occur in more than one category. Certain species, principally among the shrews, have very large north–south geographic ranges which incorporate different climatic zones and habitats (e.g. *Sorex cinereus* and *Sorex monticolus* found from subtropical to Arctic latitudes). Despite this, the number of species occurrences decreases with increasing latitude.

Today's distribution of marsupials in Australia and the Americas is the result of biogeographical factors, geological history, isolation, and competition with

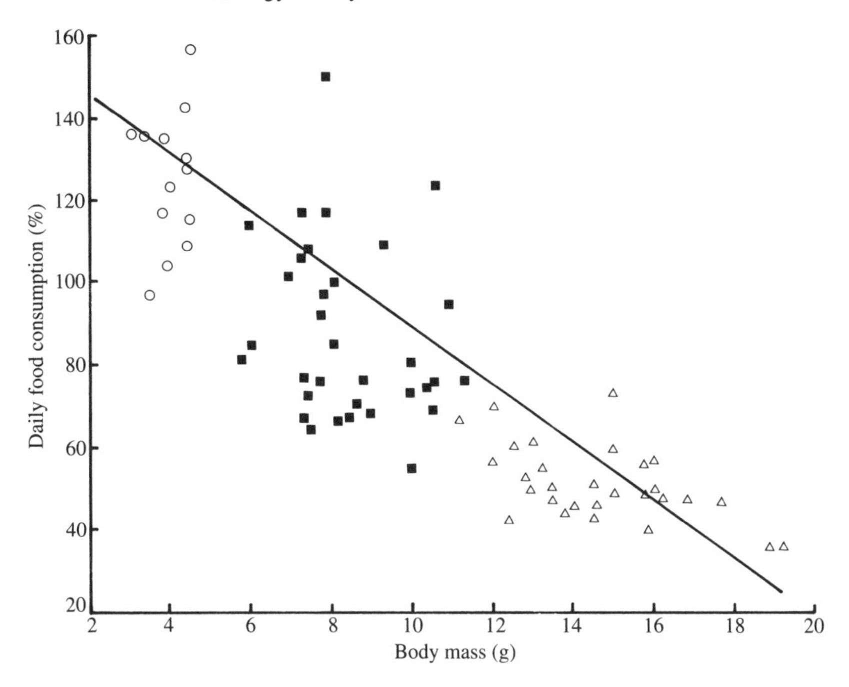

Fig. 2. The relationship between body mass and daily food consumption as a percentage of body mass for three species of soricine shrews: *S. minutus* (open circles), *S. araneus* (solid squares), and *N. fodiens* (open triangles); $r = 0.689$, $n = 75$, $P < 0.001$. (After Churchfield 1990.)

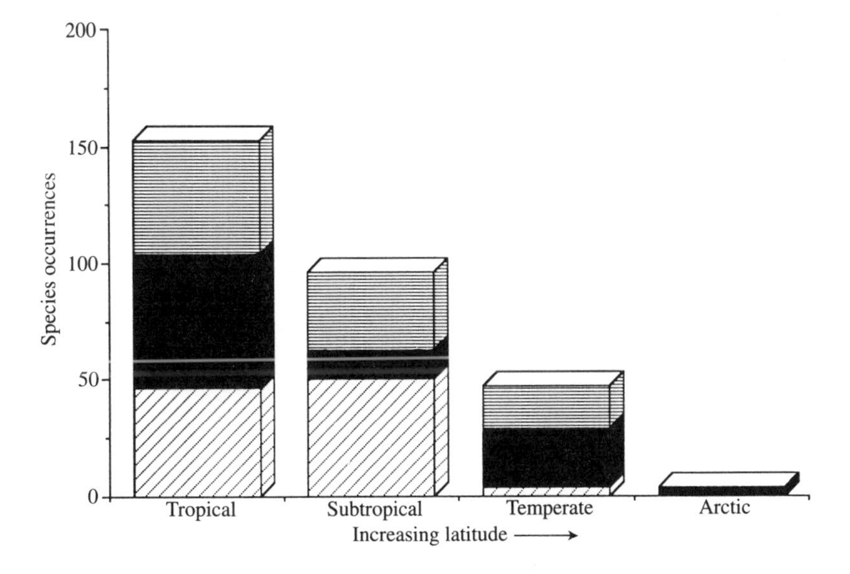

Fig. 3. The occurrence of miniature mammal species in different latitudinal regions: marsupials (cross hatched), insectivores (black), rodents (horizontal stripes). Tropical (0–20°N/S; Subtropical (20–35°N/S); Temperate (35–68°N); Arctic (>68°N).

placental mammals. Nevertheless, it is worth noting that the smallest marsupials are confined to Australasia, almost exclusively to tropical and subtropical areas. Rodents, with their world-wide distribution, provide a better example of the relationship between latitude, climate, and occurrence, with a significant negative correlation between species occurrence and increasing latitude ($r = 0.998$, $P < 0.001$, see Fig. 3). Although some 18% of miniature species occurrences are in temperate climatic zones, rodents maintain their smallest representatives in tropical and subtropical areas. While larger species of rodents are found in Arctic regions, miniature ones are not. Most species of miniature insectivores also have a tropical/subtropical distribution, but some 25% are temperate in occurrence. Most remarkable is the observation that, unlike the rodents, 4% (some 12 species) of miniature insectivores are found in Arctic regions. All of these are shrews. In Eurasia (Scandinavia and Siberia) they include the tiny *Sorex minutissimus* (1.5–3.0 g) plus the larger *S. minutus, S. caecutiens, S. isodon, S. araneus*, and *Neomys fodiens*, whose ranges extend up beyond the Arctic Circle (Hanski & Kaikusalo 1989). In North America they include the tiny *Microsorex hoyi* (2.2–3.8 g) and slightly larger *Sorex cinereus* and *S. arcticus* found in Alaska and northern Canada in boreal forest and tundra (Aitchison 1987a). Shrew distribution northwards in Russia is apparently limited by the −30° isotherm of mean January temperatures (Yudin 1964). Remarkably, mean body mass of species in small mammal assemblages decreases with increasing latitude (Cotgreave & Stockley 1994), as Fig. 4 demonstrates. Areas with the lowest winter temperatures have the smallest species of shrews (Mezhzherin 1964).

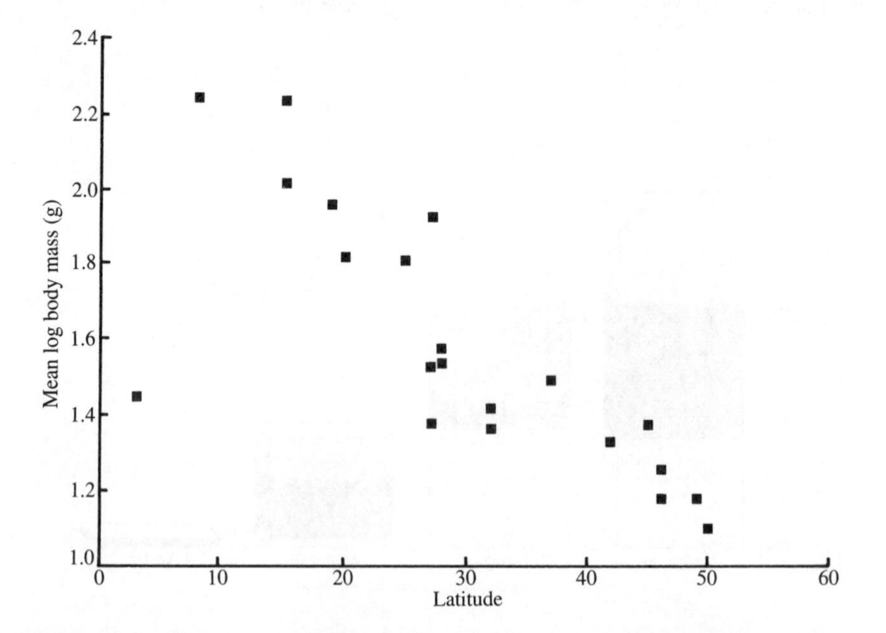

Fig. 4. The relationship between the average logarithmically transformed mass of the species in an assemblage of small mammals and the latitude of the assemblage ($r = -0.79$, $P = 0.001$) (after Cotgreave & Stockley 1994).

The occurrence of miniature terrestrial mammals in regions of severe winter climate seems paradoxical, not only because of their unfavourable surface-area to volume ratios but also because they are unable to escape these conditions by migration or hibernation, and must remain active all year round. Wild soricine shrews and murid rodents are unable to lower their metabolic rates and body temperatures sufficiently to hibernate, neither can they store adequate body fat to last them through a period of hibernation. For example, *Eptesicus fuscus*, a typical northern temperate bat, enters hibernation with a body mass of approximately 21 g of which some 7 g is fat, enough to last it for some 170 days (Wimsatt 1970). In marked contrast, a young *Apodemus flavicollis* entering winter at 27 g has only 1.1 g of fat (M. Pucek 1973) and *S. araneus* enters winter weighing approximately 7 g, of which only 0.1 g is fat (Churchfield 1981).

A further paradox is the winter decrease in body size in northern temperate shrews when they lose 25–40% of their body weight, these losses being greatest at most northerly latitudes (Dehnel 1949; Z. Pucek 1965; Hyvärinen 1969; Churchfield 1981). This would seem to disadvantage them even further, by increasing the problems of surface-area to volume ratio. However, the metabolic rate of shrews caught in winter is, weight for weight, actually lower than that of summer-caught ones (Gebczyński 1965). This is assisted by decreased activity of the endocrine glands (Hyvärinen 1984). Heat loss and energy requirements are minimized by greater insulation of the winter pelage, reduced activity on the cold ground surface, and more time spent in insulated nests (Churchfield 1990).

Small shrews require less food, in absolute terms, than large ones (Fig. 5) and the decrease in body size in winter is suggested to be a strategy for reducing absolute food requirements at a time of low prey abundance (Z. Pucek 1970). It is estimated that a 30% decrease in body mass could reduce daily foraging time of *S. araneus* by some 5 h (Churchfield 1990). Lower food requirements may also explain the predominance of very small species at high latitudes where prey numbers and biomass may be low (see below).

A critical factor for the survival of small mammals during northern winters is the protective snow cover. Under a mere 15–20 cm of snow, nearly constant conditions of temperature, air flow, and relative humidity prevail and temperatures are significantly higher than on the ground surface: while ground surface temperature reads −30°C, the subnivean space is −4°C (Aitchison 1987b). Confining activity to subterranean burrows and the subnivean space minimizes heat loss (and predation).

The reason why shrews but not miniature rodents can inhabit such cold Arctic regions may reflect their feeding habits: shrews feed predominantly on invertebrates which are temporally more predictable in supply (being available all year) than small seeds on which rodents feed. Examination of the distribution of small mammal assemblages reveals that the number and proportion of insectivorous species in the assemblage increases with increasing latitude (Cotgreave & Stockley 1994). Small insectivores feed predominantly on small arthropods with relatively high lipid and energy contents and low water and sclerotin contents. Winter-active invertebrates are predominantly small arthropods such as spiders, beetles, and collembolans of 1–5 mm in length, and these are surprisingly abundant in midwinter (Yalden 1981; Churchfield 1982; Aitchison 1984). This may be sufficient to sustain small species, but not large ones with their higher absolute food requirements. Nevertheless, the demands of even small shrews must not be underestimated. *Sorex minutus* of 2 g body mass requires some 125% of its body weight in food daily. If it subsists on spiders with an average individual mass of 1 mg (Aitchison 1987a), it must catch and eat 2500 of these prey each day.

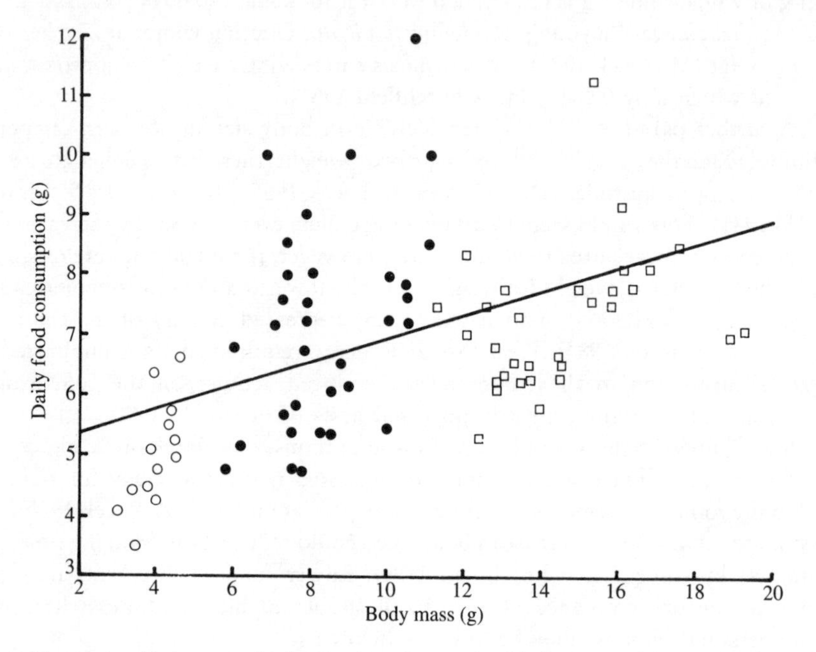

Fig. 5. The relationship between absolute food requirements and body mass in three species of soricine shrew: *S. minutus* (open circles), *S. araneus* (solid circles), and *N. fodiens* (open squares); $r=0.209$, $n=77$, $P < 0.1$. (After Churchfield 1990.)

Feeding habits

Miniature mammals exhibit a range of feeding habits, from strictly vegetarian to subsistence on live prey (see Table 3). The predominant feeding mode amongst marsupials and insectivores is insectivory, comprising a diet of insects and other invertebrates. Despite their small size, some species (mostly of the Insectivora) also eat vertebrate material, in the form of small rodents, birds, lizards and amphibians, either as carrion or as live prey. The major feeding mode for rodents is granivory, although inclusion of green plant material and invertebrates is important for many miniature species. No miniature species are strictly herbivorous (feeding only on leaves, stems, or roots of plants). Coping with indigestible, fibrous, herbaceous vegetation requires specialist feeding apparatus with large teeth and long guts, and large quantities must be ingested. Very small mammals are better suited to feeding on small quantities of easily-digested items rich in energy and nutrients, such as seeds. Specialist feeders subsisting primarily on nectar, pollen, or flowers are found only among the marsupials.

Table 3. Percentage species occurrences in different dietary modes.

	Vertebrates	Invertebrates	Green Vegetation	Seeds/fruits	Flowers	Nectar/pollen
Marsupials	5.6	66.7	0.0	5.6	11.1	11.1
Insectivores	6.1	83.1	5.0	5.7	0.0	0.0
Rodents	0.0	19.1	27.3	53.7	0.0	0.0

Main sources: van den Brink (1977); Whitaker (1980); Nowak & Paradiso (1983); Smithers (1990); Macdonald (1985); Corbet & Harris (1991); Strahan (1991).

Small size may convey certain feeding advantages. In particular, small species can exploit small food items which may be overlooked by larger counterparts, and they may be more efficient at locating and handling very small items. They may also subsist on food items which could not sustain larger counterparts. Certainly, small species tend to consume greater proportions of small food items than larger congeners. In a study of 18 species of dasyurid marsupial ranging in body mass from 5 to 200 g, a strong positive correlation was found between body size of dasyurid marsupials and the mean sizes of their invertebrate prey (Fisher & Dickman 1993). Amongst six species of *Sorex* shrews exhibiting a narrower range of body size (3–10 g) and co-existing in the Siberian taiga (Churchfield & Sheftel 1994), small species of 3–6 g fed predominantly, though not exclusively, on prey of 3–10 mm in length while larger species fed on increasing quantities of prey of > 16 mm. A clear relationship existed between the incidence of very small prey in the diet and the shrews' body mass (Fig. 6). Similar findings have been reported for desert rodents, with a positive correlation between body mass and seed size taken, small species selecting predominantly small seeds (Brown 1975).

There is little evidence to show that very small species are more efficient foragers on small food items than large species but studies of insectivorous dasyurids revealed that smaller species (5–16 g) maximized rates of energy gain by feeding preferentially on small prey whereas larger species (> 20 g) maximized gain with large prey (Fisher & Dickman 1993). In laboratory tests, larger species (> 10 g) also exhibited a preference for the largest prey and ignored the smallest.

The great advantage of subsisting on small food items is that they are very abundant. Amongst invertebrates, for example, large taxa such as earthworms and insect larvae may provide a good return on foraging effort in terms of energy content per item and their populations may have a high biomass relative to small taxa, but, as Fig. 7 shows, they are far exceeded by small taxa in terms of abundance and hence the rate at which they are encountered by a hungry forager (Dickman 1988; Churchfield 1991).

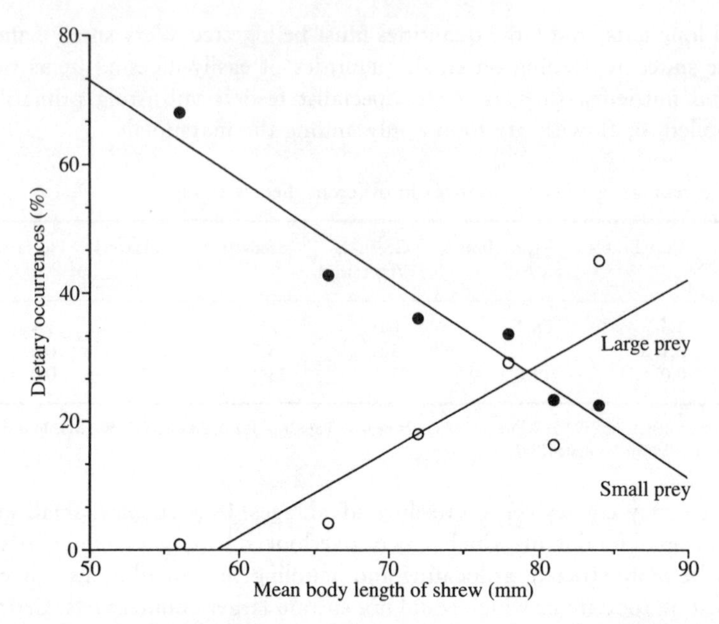

Fig. 6. The relationship between body size of shrews and the incidence of small prey of < 5 mm in length (closed circles) and large prey of > 16 mm in length (open circles). $r=0.877$, $P < 0.01$ and $r=0.725$, $P < 0.10$, respectively.

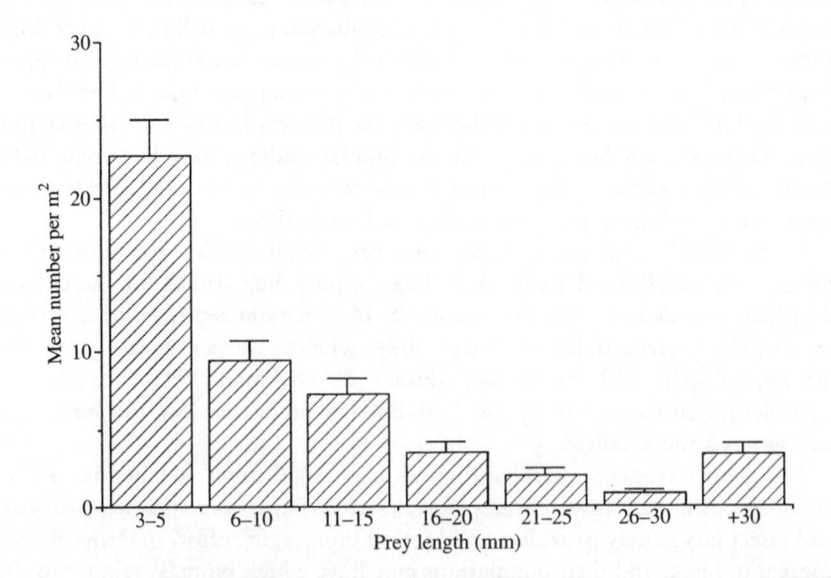

Fig. 7. The abundance of invertebrate prey of different size classes known to be utilized by shrews in taiga habitats in central Siberia, with standard errors.

As to exploiting a food type which is not utilized by other species, this is only partly true. Certainly in the case of shrews, dunnarts, and planigales, large congeners tend to take predominantly larger prey than small congeners, but all species have a wide range of prey sizes which may reflect the relative abundance, availability, and encounter rate of different prey types. If large prey are scarce, a hungry predator of whatever size is unlikely to reject small, abundant prey, despite their low individual profitability and low rates of energy gain, and this is reflected in diet studies of wild species. There are also many relatively large predators which are specialists (e.g. anteaters, *Myrmecophaga*) or opportunists (e.g. elephant shrew *Elephantulus myurus*, and large shrew *Crocidura viaria* of > 14 g) on very small prey such as ants and termites. So, exploiting small food items may have advantages in reducing niche overlap, but does not provide an exclusive niche for small species.

Small size may convey feeding disadvantages by restricting the range of food items which fall within the physical capabilities of the predator. Some invertebrate prey may be too large or too strong to handle and this may explain why very small species of shrew such as *Sorex minutissimus*, *S. minutus*, and *S. caecutiens* do not forage on earthworms as do their congeners *S. araneus*, *S. tundrensis*, and *S. isodon* (Churchfield & Sheftel 1994). Nevertheless, even very small predators feed on some large invertebrates: some 20% of the diet of small shrew species (1.5–6 g) comprises prey of > 10 mm in length, and these include large insect larvae of 30 mm or more (Churchfield 1991; Churchfield & Sheftel 1994). The water shrew *Neomys fodiens* (10–18 g) feeds on frogs, newts, and small fish (Wolk 1976), aided by the mild neurotoxin in its saliva which helps to subdue such prey (M. Pucek 1959), a rare occurrence amongst mammals. In fact, anatomical constraints seem to be minimal: most large insects of 25–80 mm long can be killed and at least partially consumed by dasyurid species of all sizes; even very small dasyurids (< 10 g) can handle these prey and puncture the cuticle, and may show great tenacity in their attacks (Van Dyck 1979; Fisher & Dickman 1993). However, these small species take longer to kill large prey than their larger counterparts, and this may ultimately influence their selection of prey.

Miniature rodents such as *Micromys minutus* feed on small seeds of grasses and herbaceous forbs while larger co-existing seed-eaters, such as *Apodemus sylvaticus*, are not so physically constrained and can also tackle and consume acorns and other large tree seeds with a thicker, harder testa (Churchfield & Brown 1987; Flowerdew 1991; Harris & Trout 1991). Unlike small insectivorous mammals whose food resources show only limited temporal changes in availability (and represent a fairly predictable and dependable food resource), granivorous species rely on grass and forb seeds which are seasonal in supply, with periods of glut and of scarcity. For both seed-eaters and insectivores the availability of food resources is probably a major influence on which habitats they occupy and how they make use of them.

Habitat occurrence

While miniature mammals are distributed amongst a wide range of terrestrial and semi-aquatic habitats, the majority are found in forest and grassland or steppe (see Table 4). Some 67% of marsupials, 73% of rodents, and 74% of insectivores occur in these habitats. Scrub and bush habitats also contain many small mammals. Desert, rocky, and sandy habitats support some species, notably of rodents, but tundra, heath, and marsh or bog support very few. Again, many species occur in more than one habitat and so the results have been scored in terms of species occurrences, which permits any named species to be included in more than one category of habitat. Species were assigned to habitats according to the level and frequency of use, as indicated in the literature. The aim was to discern the habitats most preferred and occupied by each species, and rare occurrences were not scored.

Table 4. Percentage species occurrences in different habitats. (Main sources as in Table 3).

	Forest	Scrub	Heath	Grassland/ steppe	Tundra	Rocky/ sandy	Desert	Semi-aquatic
Marsupials	33.4	14.3	9.5	33.4	0.0	4.7	4.7	0.0
Insectivores	63.3	17.6	0.0	11.0	2.4	8.0	2.0	2.9
Rodents	27.0	10.0	0.0	46.0	0.0	9.0	8.0	0.0

Bearing in mind the feeding habits of small rodents which subsist primarily on small seeds of grasses and forbs, it is not surprising that the majority (46%) of these species inhabit grassland or steppe habitats. Second in importance was forest (27% of species occurrences), with other habitats supporting relatively few species. In contrast to marsupials and insectivores, many rodents (some 19 species) were associated with arid habitats such as desert, rocky and sandy areas. No miniature rodent species were found to occur in heathland or wet habitats such as marsh and bog. Their absence from tundra habitats reflects their geographical distribution which is restricted to low and mid latitudes.

Marsupials are equally numerous in forest and grassland habitats (33% in each), which does not appear to reflect dietary trends (few species are seed eaters). Habitat occurrence in these mammals tends towards dryness, including seasonally dry scrub, heath and sclerophyll forest.

Insectivores show a high degree of habitat bias in favour of forest, including wet, tropical evergreen, temperate deciduous and boreal forests (63% of species occurrences). Only 11% were found in grassland, and rather more in scrub habitats. Of all three taxonomic groups, insectivores have the widest habitat occurrence (see Table 4). Although several species inhabit arid areas, the majority occur in moist habitats, and some seven species (mostly shrews) are found in semi-aquatic habitats such as marshes and stream-sides. Again, this may reflect feeding habits, since such habitats offer the greatest abundance and variety of invertebrates and less temporal variation in availability of prey than arid habitats subject to marked dry seasons.

The distribution of insectivore species such as shrews shows close links with the occurrence of invertebrate prey, both on a local habitat scale and on a wider geographical scale. Earthworms and other large invertebrates are mostly confined to productive, base-rich soils (grasslands and deciduous forests) and are rare or absent from acid soils in coniferous forests, tundra, and Alpine areas where microarthropods predominate (Kühnelt 1961; Wallwork 1970). Large, earthworm-eating species predominate in the more productive habitats while small species, with their lower per capita food requirements and subsistence on small arthropod prey, predominate in less productive habitats and at high latitudes. For example, *S. araneus* greatly outnumbers the smaller *S. minutus* in lowland habitats in Britain but the latter is numerically dominant on upland acid moorland, which supports a wealth of small arthropod prey for *S. minutus*, but not the earthworms on which *S. araneus* mostly feeds (Butterfield, Coulson & Wanless 1981). The small, arthropod-eating *S. caecutiens* is numerically dominant in the most barren, northern areas of Lapland where the worm-eating *S. isodon*, the largest shrew in the region, is absent (Hanski & Kaikusalo 1989) and where the density of lumbricids is low (Terhivuo 1988).

Habitat utilization

Body size is predicted to influence the way in which small mammals utilize their habitats and the mode of life they adopt. Very small mammals, for example, may be able to utilize an arboreal or scansorial mode of life which is not available to their larger and heavier counterparts. However, their small size and slight physique may limit burrowing activity and affect the use of the subterranean mode of life for foraging and shelter.

The majority of miniature mammals are active predominantly on the ground surface (see Table 5). This may reflect the absence of specific anatomical adaptations for alternative modes of life and a more favourable distribution, availability and predictability of their required food resources at ground level then above or below it. Very few miniature mammals are arboreal, although there are many examples from all three taxonomic groups which exhibit scansorial behaviour and climb amongst vegetation just above the ground surface. Several species, such as the shrews *Sylvisorex megalura* and *Soriculus leucops* and the harvest mouse *Micromys minutus*, possess anatomical adaptations for such an existence, including some prehensility with longer, more naked tails and feet than their more terrestrial counterparts (Hutterer 1985; Harris & Trout 1991).

Table 5. Percentage species occurrences in different habitat modes. (Main sources as in Table 3).

	Arboreal	Scansorial	Ground surface	Subterranean	Semi-aquatic
Marsupials	15.8	26.3	42.1	15.8	0.0
Insectivores	0.0	6.2	79.6	12.7	1.6
Rodents	3.3	20.3	56.9	19.5	0.0

Contrary to expectation, a significant proportion of species were found to be active below ground level, where they foraged and nested, although none was exclusively subterranean. They do not, however, possess adaptations for burrowing and are not observed to construct extensive burrows of their own. Instead, they utilize and modify the burrows of other species and naturally existing cracks and crevices in the ground. The small dasyurids *Planigale gilesi* and *Planigale tenuirostris* even have flattened heads enabling them better to exploit crevices in the soil, and they occur only in such areas (Read 1987).

Several insectivores are semi-aquatic, living on land but foraging for prey in fresh water. They include the water shrews, notably *Neomys fodiens* and *Sorex palustris*.

It would be a reasonable assumption that small-bodied species require smaller home-ranges or territories than their larger counterparts do (principally because food requirements are lower). This may be true for certain species: the harvest mouse *Micromys minutus* has a mean home-range size of approximately 400 m^2 (Trout 1978) compared with 2000 m^2 or more for the larger wood mouse *Apodemus sylvaticus* (Flowerdew 1991), both principally seed eaters. But there are other examples of very small species having considerably bigger home ranges than their larger counterparts: live trapping and tracking reveal that *S. minutus*, both in allopatry and in sympatry with *S. araneus*, maintains home ranges 1.2–2.2 times bigger than those of the larger *S. araneus* (Michielsen 1966; Ellenbroek 1980; Churchfield 1990). Home-range size reflects the resource requirements of species, and food distribution and availability are major factors. The feeding ecology of these two shrew species may provide the explanation here. *S. minutus* is mainly a ground-surface forager on small arthropods which, though abundant, have a patchy occurrence and low biomass. It may have to cover a larger area in its essentially two-dimensional habitat to satisfy its daily needs than *S. araneus*, which feeds extensively on subterranean earthworms as well as surface-active prey and so occupies a three-dimensional foraging niche. Home-range size, therefore, is not determined solely by body size.

Competition and niche separation

As we have seen, small body size can expand the ecological opportunities available and permit exploitation of resource-poor habitats. But, in most habitats, small species, whether rodents, marsupials or insectivores, co-exist with larger congeners and the problem of niche overlap and competition arises. It is notable that in many small mammal communities where congeners of different body sizes are sympatric and syntopic, populations of small species tend to be outnumbered by larger ones. This is clearly demonstrated by shrews which have particularly species-rich communities. Where *S. araneus* and *S. minutus* are syntopic, the larger *S. araneus* outnumbers the smaller *S. minutus* by at least 2.3:1 (Michielsen 1966; Churchfield 1990). Very small species may be classed as 'rare': in a nine-species community in the central Siberian taiga, the smallest species (*S. minu-*

tissimus) formed only 0.8% of pitfall captures over a 13 year period, in contrast to the numerically dominant, larger *S. caecutiens* (33% of captures) and *S. araneus* (48%) (Sheftel 1989). Similarly, amongst six species of North American shrews in various habitats, the tiny *Microsorex hoyi* formed only 1.1% of captures compared with 70% for the larger *S. cinereus* (Wrigley, Dubois, & Copland 1979).

The foraging mode and large home range/territory typical of some very small species (see above) may partly explain low population densities. But, these examples also suggest that competition may influence population numbers and the niche occupancy of small species, with larger species being competitively superior. Differential competitive effects have been demonstrated by several removal experiments. Removal of the dasyurid *Antechinus swainsonii* from enclosures inhabited by *Antechinus stuartii* resulted in increased numbers, increased survival of young, and increased activity and home-range area in the latter, smaller species, but no reciprocal effect was found when *A. stuartii* was removed from *A. swainsonii* enclosures (Dickman 1986). Removal of the larger species in two-species communities typically results in an increase in niche breadth of the smaller: removal of *A. swainsonii* and *S. araneus* leads to the smaller species (*A. stuartii* and *S. minutus*, respectively) exploiting greater proportions of large-sized invertebrate prey (Dickman 1988). Morphological studies of the skulls of shrews showed that jaw size of *S. minutus* was greater in allopatry than in sympatry with the larger *S. araneus* (Malmquist 1985), again suggesting competitive release in the absence of the larger species and an increase in niche breadth to encompass a wider range of prey sizes. No such effect was found in *S. araneus*.

So, small species appear to have a competitive disadvantage, with reduced niches, when co-existing with larger congeners, although investigations of direct competition have produced some equivocal results. For example, there is no evidence of a change in habitat use or foraging mode by *S. minutus* in the absence of *S. araneus* (Ellenbroek 1980). These two species appear to have evolved slightly different modes of life which are entrenched and not modified by the presence or absence of potential competitors. As we have seen, small species (such as *S. minutus*) forage mostly on the ground surface for small prey while bigger ones (such as *S. araneus*) forage mostly in the soil for larger prey such as earthworms, providing a form of vertical segregation (Michielsen 1966; Yoshino & Abe 1984; Churchfield 1991; Ellenbroek & Hamburger 1991; Churchfield & Sheftel 1994). Dietary overlap may be reduced solely by exploitation of prey of different sizes, as in the dasyurid *Planigale gilesi* and the smaller *P. tenuirostris* (Read 1987).

Differences in body size, therefore, offer a clue to the ecological roles and niche occupancy of species, and provide a means of reducing (but not eradicating) overlap and competition. There is a trend for reduced niche overlap with increasing divergence of body sizes within small mammal communities such as shrews (e.g. Churchfield & Sheftel 1994). Very small species tend to be more specialized with narrower niches but lower levels of overlap than their larger congeners.

References

Aitchison, C. W. (1984). A possible subnivean food chain. *Spec. Publs Carnegie Mus. nat. Hist.* No. 10: 363–372.

Aitchison, C. W. (1987a). Review of winter trophic relations of soricine shrews. *Mammal Rev.* **17**: 1–24.

Aitchison, C. W. (1987b). Winter energy requirements of soricine shrews. *Mammal Rev.* **17**: 25–38.

Brown, J. H. (1975). Geographical ecology of desert rodents. In *Ecology and evolution of communities*: 315–341. (Eds Cody, M. L. & Diamond, J. M.). Bellknap Press of Harvard University Press, Cambridge, Ma.

Butterfield, J., Coulson, J. C., & Wanless, S. (1981). Studies on the distribution, food, breeding biology and relative abundance of the pygmy and common shrews (*Sorex minutus* and *S. araneus*) in upland areas of northern England. *J. Zool., Lond.* **195**: 169–180.

Churchfield, S. (1981). Water and fat contents of British shrews and their role in the seasonal changes in body weight. *J. Zool., Lond.* **194**: 165–173.

Churchfield, S. (1982). Food availability and the diet of the common shrew, *Sorex araneus*, in Britain. *J. Anim. Ecol.* **51**: 15–28.

Churchfield, S. (1990). *The natural history of shrews.* Christopher Helm, Bromley.

Churchfield, S. (1991). Niche dynamics, food resources, and feeding strategies in multi-species communities of shrews. In *The biology of the Soricidae*: 23–34. (Eds Findley, J. S. & Yates, T. L.). University of New Mexico, Albuquerque. (*Spec. Publs Mus. southwest. Biol.* No. 1.)

Churchfield, S. & Brown, V. K. (1987). The trophic impact of small mammals in successional grasslands. *Biol. J. Linn. Soc.* **31**: 273–290.

Churchfield, S. & Sheftel, B. I. (1994). Food niche overlap and ecological separation in a multi-species community of shrews in the Siberian taiga. *J. Zool., Lond.* **234**: 105–124.

Corbet, G. B. & Harris, S. (Eds) (1991). *The handbook of British mammals.* (3rd edn). Blackwell Scientific Publications, Oxford.

Corbet, G. B. & Hill, J. E. (1980). *A world list of mammalian species.* British Museum (Natural History), London; Comstock Publishing Associates, Ithaca.

Cotgreave, P. & Stockley, P. (1994). Body size, insectivory and abundance in assemblages of small mammals. *Oikos* **71**: 89–96.

Dehnel, A. (1949). [Studies on the genus *Sorex* L.]. *Annls Univ. Mariae Curie-Sklodowska (C)* **4**: 17–102. [In Polish.]

Dickman, C. R. (1986). An experimental study of competition between two species of dasyurid marsupials. *Ecol. Monogr.* **56**: 221–241.

Dickman, C. R. (1988). Body size, prey size and community structure in insectivorous mammals. *Ecology* **69**: 569–580.

Ellenbroek, F. J. M. (1980). Interspecific competition in the shrews *Sorex araneus* and *Sorex minutus* (Soricidae, Insectivora): a population study of the Irish pygmy shrew. *J. Zool., Lond.* **192**: 119–136.

Ellenbroek, F. J. M. & Hamburger, J. (1991). Interspecific interactions between the shrews *Sorex araneus* L. and *S. minutus* L. (Soricidae, Insectivora) and the use of habitat: a laboratory study. *Neth. J. Zool.* **41**: 32–62.

Fisher, D. O. & Dickman, C. R. (1993). Body size–prey size relationships in insectivorous marsupials: tests of three hypotheses. *Ecology* **74**: 1871–1883.

Flowerdew, J. R. (1991). Wood mouse. In *The handbook of British mammals* (3rd edn): 220–229. (Eds Corbet, G. B. & Harris, S.). Blackwell Scientific Publications, Oxford.

Gebczyński, M. (1965). Seasonal age changes in the metabolism and activity of *Sorex araneus* Linnaeus 1758. *Acta theriol.* **10**: 303–331.

Genoud, M. (1988). Energetic strategies of shrews: ecological constraints and evolutionary implications. *Mammal Rev.* **18**: 173–193.

Hanski, I. & Kaikusalo, A. (1989). Distribution and habitat selection of shrews in Finland. *Annls zool. fenn.* **26**: 339–348.

Harris, S. & Trout, R. C. (1991). Harvest mouse. In *The handbook of British mammals* (3rd edn): 233–239. (Eds Corbet, G. B. & Harris, S.). Blackwell Scientific Publications, Oxford.

Hutterer, R. (1985). Anatomical adaptations of shrews. *Mammal Rev.* **15**: 43–55.

Hyvärinen, H. (1969). On the seasonal changes in the skeleton of the common shrew (*Sorex araneus* L.) and their physiological background. *Aquilo (Ser. zool.)* 7: 1–32.

Hyvärinen, H. (1984). Wintering strategy of voles and shrews in Finland. *Spec. Publs Carnegie Mus. nat. Hist.* No. 10: 139–148.

Kühnelt, W. (1961). *Soil biology, with special reference to the animal kingdom.* Faber & Faber, London.

Macdonald, D. (Ed.) (1985). *The encyclopaedia of mammals.* **2**. George Allen & Unwin, London.

McNab, B. K. (1983). Energetics, body size, and the limits to endothermy. *J. Zool., Lond.* **199**: 1–29.

Malmquist, M. G. (1985). Character displacement and biogeography of the pygmy shrew in northern Europe. *Ecology* 77: 372–377.

Mezhzherin, V. A. (1964). Dehnel's phenomenon and its possible explanation. *Acta theriol.* **8**: 95–114.

Michielsen, N. C. (1966). Intraspecific and interspecific competition in the shrews *Sorex araneus* L. and *Sorex minutus* L. *Archs néerl. Zool.* **17**: 73–174.

Nowak, R. M. & Paradiso, J. L. (1983). *Walker's mammals of the world.* (4th edn). Johns Hopkins University Press, Baltimore & London.

Pucek, M. (1959). The effect of the venom of the European water shrew (*Neomys fodiens fodiens* Pennant) on certain experimental animals. *Acta theriol.* **3**: 93–104.

Pucek, M. (1973). Variability of fat and water content in two rodent species. *Acta theriol.* **18**: 57–80.

Pucek, Z. (1965). Seasonal and age changes in the weight of internal organs of shrews. *Acta theriol.* **10**: 369–438.

Pucek, Z. (1970). Seasonal and age change in shrews as an adaptive process. *Symp. zool. Soc. Lond.* No. 26: 189–207.

Read, D. G. (1987). Diets of sympatric *Planigale gilesi* and *P. tenuirostris* (Marsupialia: Dasyuridae): relationships of season and body size. *Aust. Mammal.* **10**: 11–21.

Redford, K. H. & Eisenberg, J. F. (1992). *Mammals of the Neotropics. The southern cone.* **2**. *Chile, Argentina, Uruguay, Paraguay*. University of Chicago Press, Chicago & London.

Sheftel, B. I. (1989). Long-term and seasonal dynamics of shrews in Central Siberia. *Annls zool. fenn.* **26**: 357–369.

Smithers, R. H. N. (1990). *The mammals of the southern African subregion.* (2nd edn). University of Pretoria, Pretoria.

Strahan, R. (Ed.). (1991). *The Australian Museum complete book of Australian mammals.* Cornstalk edn. Collins, Angus & Robertson, Australia.

Terhivuo, J. (1988). The Finnish Lumbricidae (Oligochaeta) fauna and its formation. *Annls zool. fenn.* **25**: 229–247.

Trout, R. C. (1978). A review of studies on populations of wild harvest mice (*Micromys minutus* (Pallas)). *Mammal Rev.* **8**: 143–158.

Van den Brink, F. H. (1977). *A field guide to the mammals of Britain and Europe*. (5th edn.) Collins, London.

Van Dyck, S. (1979). Behaviour in captive individuals of the dasyurid marsupial *Planigale maculata* (Gould, 1851). *Mems Qd Mus.* **19**: 413–439.

Wallwork, J. A. (1970). *Ecology of soil animals*. McGraw-Hill, London.

Whitaker, J. O., Jr (1980). *The Audubon Society field guide to North American mammals*. Alfred A. Knopf Inc., New York.

Wilson, D. E. & Reeder, D. M. (Eds). (1993). *Mammal species of the world. A taxonomic and geographic reference*. (2nd edn). Smithsonian Institution Press, Washington & London.

Wimsatt, W. A. (1970). *Biology of bats*. **1**. Academic Press, New York.

Wolk, K. (1976). The winter food of the European water-shrew. *Acta theriol.* **21**: 117–129.

Wrigley, R. E., Dubois, J. E., & Copland, H. W. R. (1979). Habitat, abundance, and distribution of six species of shrews in Manitoba. *J. Mammal.* **60**: 505–520.

Yalden, D. W. (1981). The occurrence of the pigmy shrew *Sorex minutus* on moorland, and the implications for its presence in Ireland. *J. Zool., Lond.* **195**: 147–156.

Yoshino, H. & Abe, H. (1984). Comparative study on the foraging habits of two species of soricine shrews. *Acta theriol.* **29**: 35–43.

Yudin, B. S. (1964). [The geographic distribution and interspecific taxonomy of *Sorex minutissimus* Zimmerman, 1780 in west Siberia]. *Acta theriol.* **8**: 167–179. [In Russian.]

Symp. zool. Soc. Lond. (1996) No. 69: 277–292

Dwarfing in island elephants and deer: processes in relation to time of isolation

A. M. LISTER

Department of Biology
University College
Gower Street
London WC1E 6BT, UK

Synopsis

Dwarfed populations of large mammal species evolved on many islands in the Pleistocene. Recent discoveries and advances in dating allow us to take the first steps in observing the time scale of dwarfing and the stages by which size reduction and associated morphological changes occurred. Dwarfed red deer on Jersey, and dwarfed woolly mammoths on Wrangel Island, evolved in a few thousand years or less. Their diminution probably occurred by an interplay of ecophenotypic and genetic effects, and was associated with some changes in proportion and morphology. These can be attributed to direct allometric and developmental effects of size reduction, and may in some cases have been adaptively suboptimal. By comparison, dwarfed deer from Crete, elephants on other Mediterranean islands, and Columbian mammoths from islands off California, had probably been isolated for at least an order of magnitude longer than the Jersey and Wrangel animals. In consequence, they show modifications of cranial, dental, and limb morphology which can be interpreted as adaptive to small body size or the island habitat. Other proportions have readjusted to counter the initial developmental effects of dwarfing. Some forms evolved unique display organs implying biological speciation. In several cases, the morphology appears paedomorphic relative to the full-sized ancestor.

Introduction

The dwarfing of large mammals on islands was a widespread phenomenon in the Pleistocene. As pointed out by Sondaar (1977), taxa which are good swimmers (and therefore colonizers) most commonly produced island dwarfs. Thus, fossils of elephantids (Elephantidae) and deer (Cervidae) are among the most frequently encountered as dwarfed forms. Different species within these families became dwarfed in various parts of the world, and the same species often became repeatedly dwarfed as a result of isolation on different islands. In the majority of cases (though not all), it has been possible to identify with high probability the

ZOOLOGICAL SYMPOSIUM No. 69
ISBN 0–19–857787–7

mainland species ancestral to a given island dwarf. This identification is based on morphology, as well as a knowledge of the available contemporary mainland fauna. The adaptive reasons for size reduction in island mammals have been widely discussed (Foster 1964; Thaler 1973; Sondaar 1977; Case 1978; Heaney 1978; Lomolino 1985; Roth 1990, 1992) and will not be reiterated here. Most of the explanations centre around the need to reduce absolute food requirement in an area of finite foraging, coupled with the absence or rarity of predators.

The island dwarfs frequently differ from their full-sized forbears in features apart from size. As shown by Sondaar (1977), Roth (1992), and others, similar patterns are repeated in different examples. Common features of dwarfed ungulates are shorter limb bones (especially the more distal elements), larger teeth and/or skull, simplified dental morphology, and fused foot bones. Display organs (e.g. horns or antlers) may also be simplified or modified. The processes leading to dwarfing probably included ecophenotypic effects as well as genetic change through drift or selection, the two perhaps linked by genetic assimilation (Bonner 1968; Roth 1992; Lister 1995). The causes of more detailed morphological changes include the allometric effects of reduced body size, developmental readjustments, and adaptive shifts brought about by natural selection.

The population biology and genetics of the dwarfing process have been modelled by Roth & Mercer (in prep; V. L. Roth, pers. comm.). In the present paper, the main question posed is to what extent the progress of change can be followed through time in the fossil record. At present, only very few fossil examples of island dwarfs allow any clear deductions on this question. The geological age of the dwarfs is in most cases known only within very broad limits if at all. Even more poorly known is the length of time the animals had been on the island, and how long the dwarfing process itself took. Furthermore, in few cases is there any sequence of fossils showing the morphological transitions from the large ancestor to the dwarf.

Recently, however, advances in dating and the excavation of some new finds make possible some tentative steps in this direction. In this paper, I will take examples from the deer and elephants. Within each of these families, I will compare recently-isolated dwarfs in an early stage of the adaptive process with longer-isolated examples which show more profound changes.

First case study: deer from Jersey and Crete

In this section, red deer (*Cervus elaphus* L.), dwarfed for a relatively short time on Jersey in the last interglacial, will be compared with a series of extinct deer (species of *Candiacervus* de Vos) recorded over a longer Pleistocene time interval on Crete.

Jersey red deer, *Cervus elaphus*

The dwarf red deer of Jersey, and their geological setting, have been described by Zeuner (1946) and Lister (1989, 1993a, 1995). Jersey, an island of some 130 km^2, is situated approximately 25 km from the coast of north-west France, but has been part of the mainland for much of the Pleistocene. During the penultimate cold stage, *c.* 200 000–127 000 years ago, Jersey was connected by a broad land mass to France and Britain, and was populated by 'normal', large-sized red deer indistinguishable from contemporary fossils found in those countries.

As the sea level rose at the beginning of the last interglacial, Jersey became an island, and remained thus for around 11 000 years, between approximately 126 000 and 115 000 years ago. Dwarf deer occur in deposits which appear to date no later than 6000 years into the period of isolation. Assuming, therefore, that the dwarfing process did not begin until the deer were largely or wholly geographically and genetically isolated, 6000 years is a minimum period during which the size reduction occurred. During this period, Jersey was presumably, as today, approximately 25 km from the French coast, and since records of red deer swimming do not exceed 7 km, genetic isolation seems probable. Unfortunately, there is no fossil record of the stages in the dwarfing process, but a comparison of the large ancestral form with the Jersey dwarf indicates what was achieved in no more than 6000 years.

In body size, the ancestral mainland red deer of the Pleistocene weighed approximately 200 kg in males, based on size comparison of fossil limb bones with those of present-day subspecies (Lister 1989, 1995). In contrast, limb bone diameters of the Jersey dwarfs were reduced to approximately 56% of those of their ancestors (Fig. 1). This measurement correlates well with body mass among deer according to the cube law (Scott 1987), so the dwarf males had suffered a weight reduction to 0.56^3 or approximately one-sixth. They therefore weighed around 36 kg. In shoulder height, they would have stood about 70 cm, compared to 125 cm on the mainland (Lister 1989, 1995).

In tooth size, the Jersey deer were also reduced compared to their ancestors, but to a considerably lesser extent than in the limb bones. The average molar length is 77% of that on the mainland (Lister 1989; Fig. 1). This indicates a higher ratio of tooth size to body size, a phenomenon commonly observed in small-bodied populations of mammals (Gould 1975; Fortelius 1985).

A common occurrence in dwarfed island herbivores was a great shortening of the limbs, especially the distal limb elements, the metacarpals and metatarsals. This has been explained as an adaptation for 'low gear locomotion' (Sondaar 1977). On the one hand, the need to run away from predators had disappeared, and on the other, there was often a need to scramble over rocky areas to reach patches of food in the island habitat. Among fossils of the Jersey dwarf, only one metacarpal and one metatarsal are preserved complete. These are long and thin, not differing significantly in proportion from 'normal', mainland bones (Lister 1989). Either the adaptive pressure for shortened metapodials was not present in the Jersey topography or else the deer had not been dwarfed long enough for this feature to evolve.

Finally, several antlers of the dwarf deer have been recovered. These are preserved only in their lower parts, and show there a single tine, sprouting a few centimetres above the base. This differs from 'normal', large red deer, in

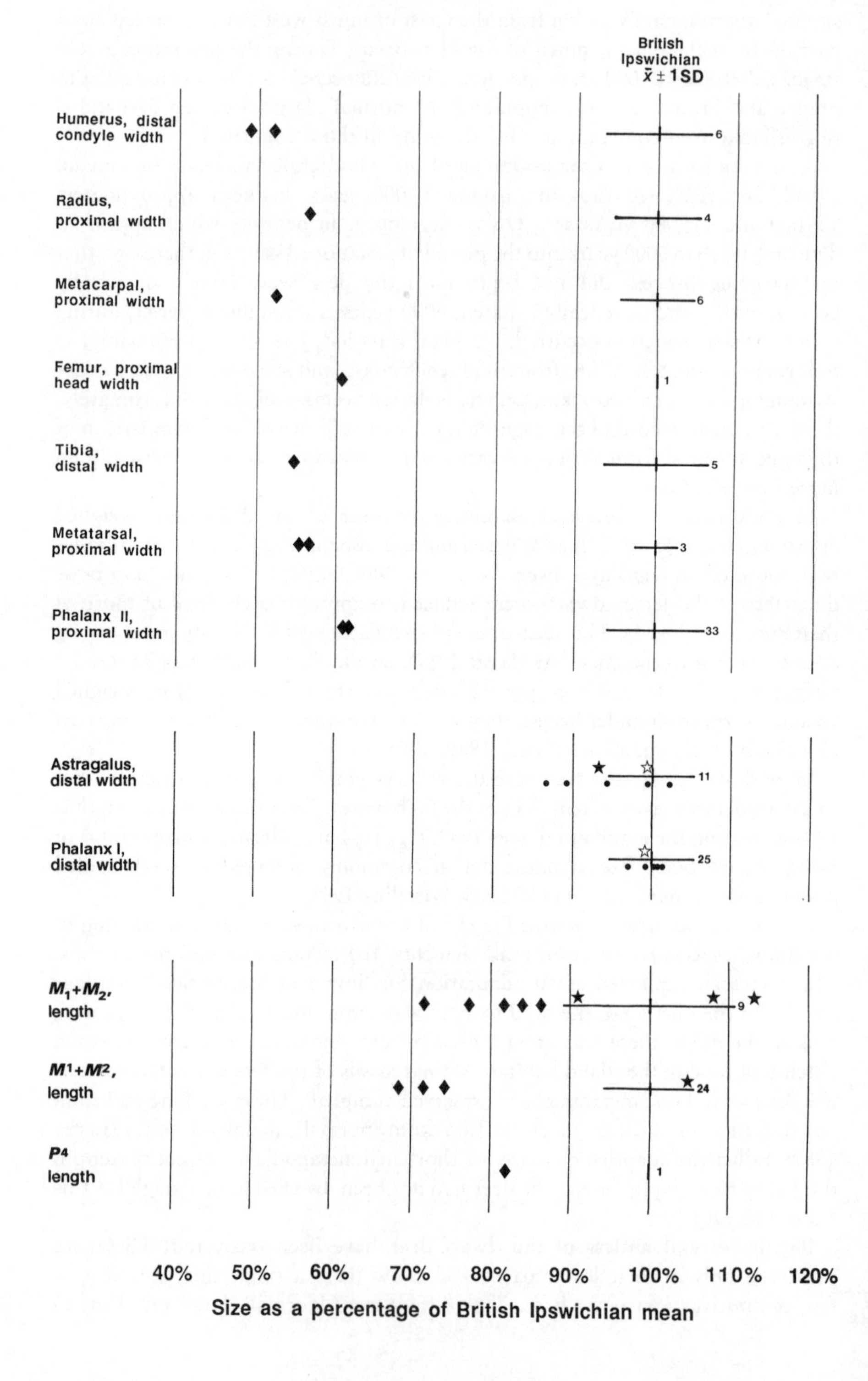
British
Ipswichian
$\bar{x} \pm 1$ SD
Humerus, distal condyle width
Radius, proximal width
Metacarpal, proximal width
Femur, proximal head width
Tibia, distal width
Metatarsal, proximal width
Phalanx II, proximal width
Astragalus, distal width
Phalanx I, distal width
M_1+M_2, length
M^1+M^2, length
P^4 length
6
4
6
1
5
3
33
11
25
9
24
1
40%
50%
60%
70%
80%
90%
100%
110%
120%
Size as a percentage of British Ipswichian mean

which two adjacent lower tines spring close to the antler base. However, as discussed by Zeuner (1946), the Jersey form is similar to that commonly seen in small-bodied populations of red deer at the present day, such as those at the margins of the modern range in Corsica and North Africa, even though these do not reach the truly dwarfed dimensions of the Jersey animals. A stressed red deer population of small body size from the Po delta, northern Italy, also consistently lacks the second basal tine as well as the distal crown (Mattioli 1993), and the remaining basal tine sometimes occurs in a high position (S. Mattioli pers. comm.). This reduction in antler complexity seems therefore to be a common and direct effect of size reduction in red deer.

Cretan deer, *Candiacervus*

Pleistocene faunal assemblages have been excavated from many localities on Crete. Among them are numerous remains of deer, which have been divided into at least seven species in a thorough study by de Vos (1979, 1984). When the localities are placed in stratigraphic order, it can be seen that the earliest deposits contain only a single species of deer, but this number progressively increases to a maximum of four at some sites, before declining again until the last one became extinct (Table 1). Although there is no absolute dating, sedimentary lithology and the changing faunal assemblage indicate that the sequence covers several hundred thousand years, probably beginning in the Middle Pleistocene and extending into the Holocene. It seems likely, therefore, that Crete was seeded by deer from the mainland on one or more occasions, and that an endemic fauna evolved from these founders. Geological evidence indicates that Crete has been an island since at least the Miocene, so the founders arrived by swimming, probably at a time of lowered sea level when the distance from the mainland was reduced.

In contrast to the Jersey deer, therefore, the Cretan deer had very much longer to evolve on their island habitat. The fact that the deer cannot, from their morphology, be linked easily to any putative mainland ancestor is one consequence of this. A possible ancestral form may be found among early megalocerine deer, perhaps *Praemegaceros verticornis* (as discussed by de Vos 1984).

Fig. 1. Comparison of skeletal and dental size in red deer. A composite sample from sites of last interglacial (Ipswichian) age on the British mainland is used as a standard of comparison. The mainland mean for each measurement is plotted as 100%, and shown plus or minus one standard deviation, with sample size. Ten limb bones from the last interglacial of Belle Hougue Cave, Jersey (diamonds) average 55.9% the diameter of their mainland contemporaries. Nine cheek teeth from Belle Hougue (diamonds) are relatively larger, averaging 77.4% the length of their mainland counterparts. First and second molars are often indistinguishable, so have been pooled to increase sample size. By contrast, two bones from the penultimate cold stage (preceding the last interglacial) of La Cotte, Jersey (open stars) are very close in size to the mainland mean, as is a similar-aged sample from La Fage, France (small closed circles). One limb bone and four teeth from the last cold stage (following the last interglacial) of La Cotte (closed stars) show the reversion to large size, again within the range of mainland deer. Jersey was connected to the mainland during both of these cold stages. All data original except La Fage, from Bouchud (1978).

Table 1. Pleistocene fossil localities on Crete and their cervid faunas. The sites are arranged in approximate stratigraphic order with the oldest at the base. Modified after de Vos (1984).

Locality	Species					
	Candiacervus sp. indet.	*Candiacervus* sp. II	*Candiacervus cretensis*	*Candiacervus ropalophorus*	*Candiacervus rethymnensis*	*Candiacervus* sp. VI	*Candiacervus* sp. V
Gerani 5				+			
Gerani 6				+			
Gerani 2_3				+			
Gerani 4		+		+			
Gerani 2_4				+	+		
Bate Cave	+					+	+
Liko		+	+		+	+	
Mavro Mouri 4c		+	+	+	+		
Sourida		+		+	+		
Rethymnon Fissure		+	+	+	+		
Kalo Chorafi		+		+			
Simonelli Cave		+		+	?		
Kharoumes 3	+						
Kharoumes 2		+	+				
Milatos 3 stal. column	+						
Stavros macro		+					
Kharoumes 5	+						
Kharoumes 4	+						

The Cretan species are of varying body size, limb length, hypsodonty, and antler form, reflecting an adaptive radiation on the island (de Vos 1984). Here I will concentrate on one relatively late species, *Candiacervus ropalophorus*, which is of dwarfed size and very well represented in the fossil record, especially at the site of Gerani 4. By roughly piecing together a skeleton from preserved bones at the Natural History Museum, London, it is clear that the animal when fully grown did not exceed 60 cm in height.

De Vos (1979) compared limb bone dimensions of *C. ropalophorus* with an array of living deer species, and indicated that its bones were relatively short, with a lower ratio of length to diameter than in the comparative species. Some caution attaches to this comparison, because if *Candiacervus* evolved from *Praemegaceros*, the relatively stout bones in this mainland form would mean that the effect attributable to island dwarfing was less.

In addition, de Vos (1979) found that the metacarpal of *C. ropalophorus* was short even in relation to the rest of the skeleton, including the metatarsal, its analogue in the hind limb. The mean ratio of metacarpal to metatarsal length was 0.81 in a large sample of the Cretan species, compared to 0.89 in a sample of red deer, *Cervus elaphus*. This has been corroborated by measurements on six skeletons of different subspecies of *C. elaphus* in the Natural History Museum, London, where the range is 0.87–0.91, with the mean 0.89. In *Praemegaceros*, the possible ancestor of *Candiacervus*, a skeleton from the Middle Pleistocene of Voigtstedt, Germany, also has a metacarpal/metatarsal length ratio of 0.89 (from data in Kahlke 1965), significantly higher than *C. ropalophorus*. Unfortunately, there is insufficient fossil material from Jersey (one metacarpal and one metatarsal, not associated) to quantify this proportion.

The shortening of the metacarpal in relation to the metatarsal in the Cretan dwarf may be significant. Since the two bones may to a considerable degree be regarded as serially homologous, the simplest developmental change would presumably have been to shorten each to the same degree. Shortening the front element more than the back implies a more profound and specific effect, perhaps selected to produce low-gear 'front-wheel-drive' locomotion.

Another feature common among dwarfed island populations of herbivorous mammals is the presence of fusions between bones of the carpal or tarsal series which normally remain separate. According to Sondaar (1977), this may be an adaptation conferring added rigidity for locomotion over rugged terrain. In a sample ($n = 28$) of the Cretan dwarf *C. ropalophorus*, 68% showed fusion between the naviculo-cuboid and the central cuneiform, bones of the proximal and distal tarsal rows respectively (de Vos 1979).

The most interesting difference between the Cretan and Jersey deer, however, is in the relative sizes of teeth and bones (Table 2). As indicated above and in Table 1, bone diameters in the Jersey dwarfs had reduced to approximately 56% in comparison with their mainland ancestors, but linear tooth dimensions were reduced by less, to only 77%. In Table 2 the size of the Cretan remains has been given relative to the same standard, by expressing the dimensions given in de Vos (1979) as a percentage of the same sample of British last interglacial red deer as was used in the Jersey comparisons. It can be seen that bone diameter is 49% of the standard, slightly smaller than the Jersey dwarf, but, interestingly, tooth size has come down to practically the same relative size, 51% of the standard. So the Cretan deer had the normal ratio of tooth to body size, as in mainland deer species.

Table 2. The relative dimensions of teeth and limb bones in dwarf deer from Jersey (*Cervus elaphus*) and Gerani 4, Crete (*Candiacervus ropalophorus*).

	Molar length	Metatarsal proximal width
Gerani 4, Crete	51% ($n=82$)	49% ($n=63$)
Belle Hougue, Jersey	76.5% ($n=5$)	56% ($n=2$)

The mean dimensions for each sample are expressed as a percentage of a standard sample ($n=33$) of last interglacial red deer from Britain (details in Lister 1989). The Gerani teeth are lower M3s, but because of small sample availability, the Belle Hougue tooth sample is pooled from various upper and lower molars; however, the different teeth give similar percentage differences. For limb bones, the metatarsal is taken as representative; other limb bones give similar percentage differences. *Candiacervus* data from de Vos (1979, 1984); *Cervus* data from Lister (1989).

It is as though the Jersey deer, isolated for a relatively short time, are showing the immediate, automatic developmental or allometric effects of dwarfing. Their relatively large teeth parallel those seen among small-bodied populations, or domesticated forms, of mammalian species at the present day. The Cretan deer, on the other hand, have had time to readjust their development, resulting in isometry between teeth and bones, as seen generally between species of deer at the present day, and indeed across mammals generally (Fortelius 1985).

Finally, it is instructive to consider the antlers of the Cretan deer (de Vos 1984). Some species bore very simple antlers with only two points, which could be regarded as a typically reduced or 'stunted' version of an ancestral type, as in the Jersey *C. elaphus*. Others, however, show more complex and quite particular morphologies unlike any other known deer species. Give the role of antlers as species-specific secondary sexual characters, this may well indicate that the Cretan endemics had become good biological species, a suggestion corroborated by the finding of several of the forms in a single fossil horizon (Table 1), indicating probable sympatry. For the Jersey dwarf, in contrast, it would be hard to make the case for regarding it as more than a subspecies of its ancestor, the red deer.

Second case study: elephants and mammoths from Siberia, California, and the Mediterranean

In this section, a relatively short-term dwarfing experiment is provided by the woolly mammoths from Wrangel Island in the Siberian Arctic. This is compared with two examples of more 'entrenched' dwarfing: dwarf Columbia mammoths from Santa Rosa Island off California, and the most extreme example of all, dwarfed straight-tusked elephants from Sicily, Malta, and Cyprus.

Woolly mammoths from Wrangel Island, Siberia

Wrangel Island, approximately 150 km from the coast of north-east Siberia, has yielded numerous fossils of very small woolly mammoth, *Mammuthus primigenius* (Vartanyan, Garutt, & Sher 1993). These animals are estimated to have been approximately 180 cm in shoulder height, compared to 320 cm for typical mainland Eurasian woolly mammoths (Lister 1993b). Throughout the last cold stage (*c.*100 000–10 000 radiocarbon years ago), the sea level was lower and Wrangel was part of mainland Siberia. Fossils of mammoth from this period have also been recovered from the island, and these are of the 'normal', large size, indicating that at that time the Wrangel mammoths were merely part of the large-sized mainland population. The latest such remains are dated to approximately 12 000 years ago. The dwarfs themselves range in radiocarbon age from approximately 7000 to 3700 years old, making them the latest mammoths to survive anywhere on Earth (Vartanyan *et al.* 1993). Vartanyan *et al.* (1993) estimate the time of isolation of the island as 12 000 years ago, although elsewhere in the world (e.g. the Straits of Dover), the interglacial warming did not flood land-bridges until approximately 8000 years ago. The maximum time available for the evolution of the dwarf mammoths is therefore 5000 years (12 000–7000), although it may have been much shorter.

Because of the difficulty of collecting and transporting fossils from such a remote area, no detailed information on the dimensions of the skull or skeleton of the Wrangel dwarfs is yet available. Almost all of the fossils collected to date are molar teeth, but these can be examined for an interesting correlate of size reduction. The evolution of the mammoth entailed a great increase in the number of enamel loops (lamellae) in its molar teeth over a period of 3 million years or more (Lister 1993a,c). This was associated with a shift from temperate, wooded habitats to the open grasslands of the Pleistocene steppe–tundra. The Late Pleistocene woolly mammoth typically had up to 26 lamellae in its last (third) molar, the highest number to evolve in any proboscidean, living or extinct. As shown by Maglio (1973), this was probably at the functional limit for effective chewing. Twenty-six lamellae means 52 enamel bands in a molar length of approximately 260 mm, or one band every 5 mm. If the bands were any closer together, they would not be able to shear the food effectively when ground together. To fit in more lamellae would also require a thinning of the enamel below its functional limit, or an increase in tooth length beyond the capacity of the skull or mandible to house it (Maglio 1973). The implications of lamellar compression in dwarf elephants have been noted by Sondaar (1977) and Roth (1992). Here some attempt is made to quantify the phenomenon and compare it between taxa.

If we imagine taking a full-sized woolly mammoth tooth as described above, and shrinking it isometrically to 60% of its length as in the Wrangel dwarfs, without reducing the total number of lamellae, then the enamel bands would be pressed even closer together, perhaps to the point of losing function (Lister & Joysey 1992). It is therefore interesting to examine the Wrangel teeth to see if they have reduced the number of lamellae to preserve their separation for effective chewing. Lamellar counts of Wrangel teeth indicate that this is not the case: up to 26 lamellae are present in M3, as in the full-sized mainland forms (Vartanyan *et al.* 1993).

This phenomenon can be explored further by calculating lamellar frequency, the number of lamellae per 10 cm length of tooth, which provides an index of lamellar packing. In Fig. 2, this metric is plotted against tooth width as an index of size, for four dwarfed last molars from Wrangel in comparison with a sample of larger, mainland woolly mammoths, including three Wrangel fossils dating from before isolation and dwarfing (graph based on Vartanyan *et al.* 1993). Note first the reduction in size of the island dwarfs. In lamellar frequency, mainland teeth average around nine to ten, with a maximum value of 11. This represents the degree of lamellar packing which may be regarded as the functional limit (cf. Maglio 1973). In the dwarfs, however, because the same number of lamellae are packed into a smaller tooth, lamellar frequency has risen to between 11 and 13. Conceivably, this suboptimal degree of lamellar packing is retained because the dwarfs had evolved only a short while previously, and had not had time to readapt by losing molar lamellae. The dwarf mammoths had formed in less than 5000 years, and possibly in less than 1000 years.

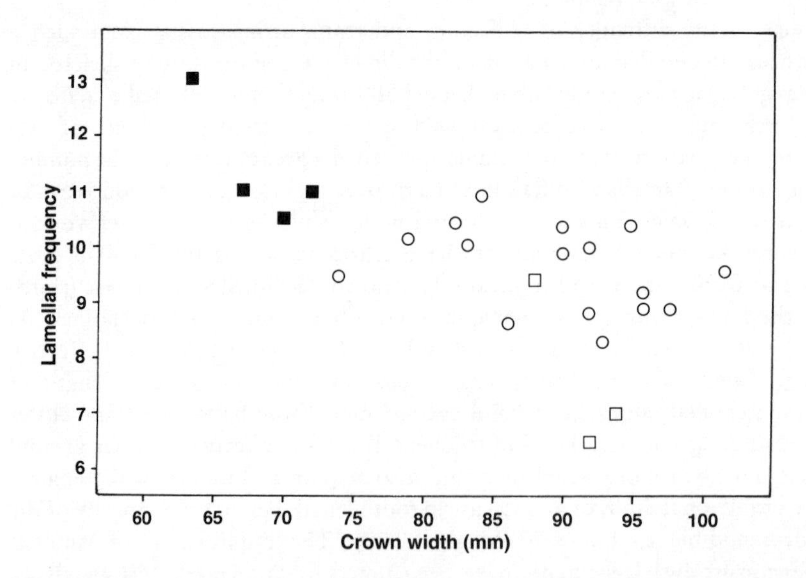

Fig. 2. Crown width (an index of tooth size) plotted against lamellar frequency (the number of enamel loops per 10 cm length of tooth) for upper third molars of *M. primigenius*. Circles: typical mainland Siberian *M. primigenius*, last cold stage. Open squares: teeth from Wrangel Island dated to 12 000 BP or older (Wrangel connected to the mainland). Closed squares: 'dwarf' teeth from Wrangel island dated to approximately 7000–4000 BP. Modified after Vartanyan *et al.* (1993); measurements taken according to the method of Maglio (1973).

Columbian mammoths from Santa Rosa Island, California

Remains of dwarfed mammoths have been found on several of the California Channel islands, but most notably on Santa Rosa (Roth 1982, 1992). The age of this material is not known accurately, but radiocarbon dates on mammoth fossils range from approximately 11 000 years old, to beyond the range of effective radiocarbon dating (i.e. greater than approximately 40 000 years old). Estimates of when the mammoths first arrived on the mainland range from as much as 500 000 years ago, to between 200 000 and 100 000 years ago (Orr 1967; Madden 1981; Roth 1982). In any event, it seems very likely that the dwarfs had a much longer history than those of Wrangel Island.

Data on the molars of the Californian dwarf mammoths, *Mammuthus exilis*, and their large mainland ancestor, *Mammuthus columbi*, are given by Madden (1981). *Mammuthus columbi*, a species less adapted for Arctic conditions than the woolly mammoth, had 18–24 lamellae on the third molars ($n = 14$), with an average of 21. With molar lengths averaging around 296 mm, this gave a lamellar frequency (number of lamellae per 10 cm) of around seven (Table 3). In *M. exilis*, the teeth were reduced in length by approximately 50% compared to its ancestor and, interestingly, the number of lamellae had become reduced to an average of only 15 (range 12–18, $n = 10$). As a result, lamellar frequency, though slightly higher than in *M. columbi*, was held down to eight (Table 3).

Table 3. Comparison of third upper molar morphology in mainland *M. columbi* and Santa Rosa *M. exilis*.

	M. columbi ($n = 13$)	*M. exilis* ($n = 10$)
Length (mm)	296	189
Number of lamellae	21	15
Lamellar frequency	7	8

Mean values given. The length reduction in *M. exilis* is accompanied by a reduction in the number of lamellae from 21 to 15, so that lamellar frequency (number of lamellae per 10 cm length of tooth) rises only slightly, from seven to eight. Data from Madden (1981). Had *M. exilis* retained 21 lamellae, it would have had a lamellar frequency of approximately $21/189 \times 100 = 11.1$.

If we had taken a molar of *M. columbi* and retained its 21 lamellae while shrinking it from 296 to 189 mm as in the island dwarf (Table 3), we can calculate that it would have had an average lamellar frequency of 11.1, interestingly similar to that seen in the Wrangel dwarfs, and perhaps above the functional optimum. With a longer period of evolution than the Wrangel dwarfs, natural selection may have reduced the number of lamellae in the Californian island mammoths, with adaptive function thus preserved.

Elephants from Malta, Sicily, and Cyprus

The most extreme examples of island dwarfing are seen among Mediterranean elephants, where the mainland species *Elephas (Palaeoloxodon) antiquus* produced dwarfed derivatives on several islands, including Cyprus and Sicily–Malta, the latter having been joined as one island during the Pleistocene.

The best-studied species is *Elephas falconeri* of Sicily and Malta (Ambrosetti 1968). Adult shoulder heights were around 1 m, compared to 4 m in the ancestral *Elephas antiquus*. The implied weight reduction, based on height–weight relations in living elephants (Laws, Parker, & Johnstone 1975) is around 100-fold, from around 10 t in *E. antiquus* to 100 kg in *E. falconeri*. The time available for the dwarfing is unfortunately not known, but seems to have been considerable. Several different size categories of elephant are recorded in the deposits, the earliest of which may be as much as 500 000 years old (Caloi, Kotsakis, Palombo, & Petronio in press). On Cyprus, *Elephas* remains at least as small as those from Malta have been found, but have not been studied in detail.

Ambrosetti (1968) indicated that there are no complete last molars in his sample from Sicily, but material from Malta and Cyprus at the Natural History Museum, London, includes several specimens on which total lamellar number can be counted. Two Maltese specimens have 14 lamellae, while three from Cyprus range from 12 to 14. This can be compared to a mean of 15 (range 12–18, $n = 17$) in a sample of mainland *E. antiquus* (Maglio 1973). (All of these figures exclude the 'talons', small incomplete lamellae at the front and back of the tooth.) There has been some reduction in lamellar number, but not as significant as in the Californian mammoths. This may be because lamellar frequency was already low in *E. antiquus* (between five and six: Maglio 1973), so the compression due to dwarfing did not so seriously threaten their function.

Further features of dwarf elephantids

In their skeletal morphology, the dwarf elephants show features which can be interpreted as paedomorphic (retaining in the adult features which in their ancestors appear in juveniles: Gould 1977). *Elephas falconeri* had a simple, low-domed skull which contrasts with the high, pneumatized parietal region of its ancestor *E. antiquus*, which was formed into two large bosses, giving the animal a 'double-domed' head in life (Ambrosetti 1968; Accordi & Palombo 1971; Sondaar 1977; Fig. 3). The difference would be sufficient to throw doubt on the phylogenetic connection between the two species, were their highly characteristic molar morphology not identical. Sondaar (1977) described the cranial difference as an allometric effect and pointed out its functional implication, the reduction in attachment area for muscles and tendons in the dwarf, which had a much lower weight of skull and tusks to support. Roth (1992) pointed out the likely paedomorphic mechanism of the change, and this is graphically illustrated by comparing the low skull of a modern juvenile elephant, which resembles that of *E. falconeri*, with the expanded, domed cranium of an adult (Fig. 4, based on Beden 1979). Unlike young elephants, however, the dwarfs grew large tusks in their paedomorphic skulls (Ambrosetti 1968), concordant with their adult status.

In their post-cranial skeleton, the dwarfed elephants also show interesting differences from their large ancestors (Ambrosetti 1968; Sondaar 1977; Roth 1992,

1993). Ambrosetti (1968) indicated that the limbs of *E. falconeri* were markedly shortened compared to those of their mainland ancestors. In addition, he described fusion between the radius and ulna, and between the tibia and fibula, which would appear to have allowed less lateral movement than in mainland elephants (Sondaar 1977).

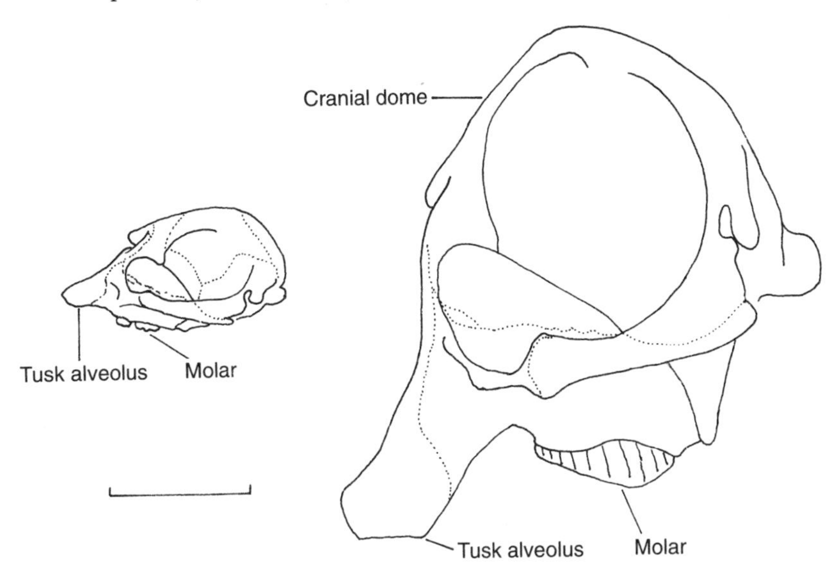

Fig. 3. Skulls of the dwarf *E. falconeri* (left) from Spinagallo Cave, Sicily, and its mainland ancestor *E. antiquus* (right) in left lateral view. The endocranial cavity is shaded. Note the relatively large brain and low cranium of the dwarf. From Accordi & Palombo (1971), reversed. Scale bar 20 cm.

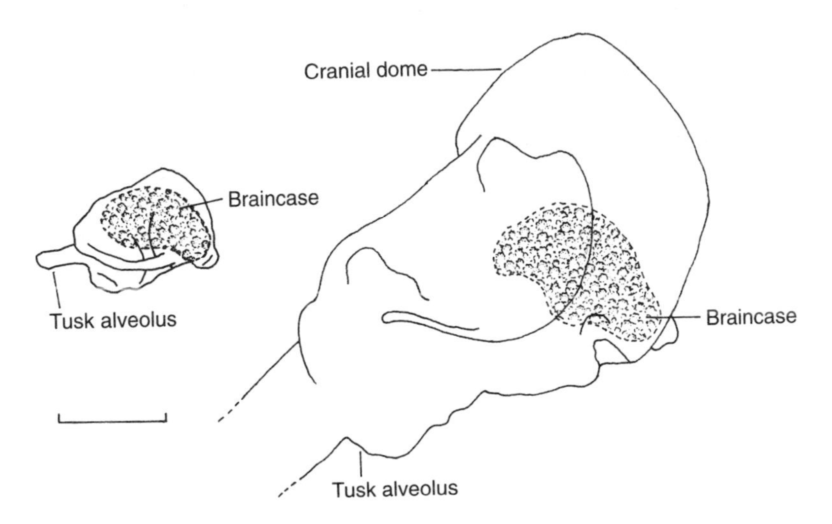

Fig. 4. Skulls of neonate (left) and adult female (right) African elephant, *Loxodonta africana*, in left lateral view, to show the increase in cranial height. Modified after Beden (1979). Scale bar 20 cm.

In large elephants, a columnar limb structure supports the great mass of the body. According to Ambrosetti (1968), the femur is twisted in such a way that the feet are positioned centrally beneath the weight of the body they support, but in the dwarf *E. falconeri* this torsion was reduced. Roth (1992) indicated that, for both *E. falconeri* and *M. exilis*, the distal morphology of the femur and humerus is modified from that of full-sized adult elephants in a way that shows the 'knee' and 'elbow' joints to have been more flexed, and the limb as a whole more arched, giving the animals a 'friskier' locomotion. In their postcranial morphology, the dwarfs resemble similar-sized (younger) individuals among modern elephants, so the change probably occurred by ontogenetic truncation (Roth 1992, 1993).

It is expected that the recovery and study of skulls and limb bones of the Wrangel mammoths, dwarfed for a relatively short period of time, will provide some interesting comparisons with the modifications seen in the more 'entrenched' dwarf elephants from California and the Mediterranean.

Discussion

Island dwarfs provide a wealth of interest in many areas, including the selective forces producing the size reduction, and the adaptive responses to it. In this paper, some attempt has been made to deduce the stages by which dwarfing occurs. It is suggested that there was a first stage in which there was a primary, strong selective pressure for reduction in body size. This reduction was probably initiated by non-genetic, ecophenotypic stunting. However, in all the cases described above, it is unlikely that phenotypic stunting alone could account for the final degree of size reduction. In the Jersey *C. elaphus*, for example, the estimated adult male body weight of 36 kg is well below the modern range of the species. The red deer is a variable and ecophenotypically labile species, with populations today ranging from the largest at 200–300 kg in eastern Europe (even larger in North America, although these are genetically distinct) to the smallest at approximately 90 kg in poor habitats such as North Africa or the Scottish moors. This range probably indicates the ecophenotypic limits of the species. To reduce in size so much further, down to 36 kg, must have entailed genetic change.

It is likely that the selected, genetic component of size reduction came about in part by genetic assimilation of the ecophenotypic effect. In other words, there was selection in favour of those individuals which showed the greatest stunting effect (Waddington 1942), a mechanism also implicated in island dwarfing by Bonner (1968) and Roth (1992, 1993).

This initial stage of dwarfing, dominated by pressure for size reduction, probably took no more than a few thousand years, and of the order of hundreds of generations or less. It was accompanied by allometric and other developmental effects, such as simplified display organs and an elevated tooth to body-size ratio. It is probable that the population did not become truly speciated from its mainland ancestor. Likely examples of populations which proceeded to this stage are the Jersey red deer and Wrangel mammoths.

In the second stage, if the dwarf survived long enough, natural selection produced further changes, both adjusting the development and structure of the animal to retain 'internal' integrity, and readapting it in terms of function in relation to the outside world. These processes would have begun in the first stage but, because they took longer, would continue much longer to form the second. Examples include the Cretan deer and the Californian and Mediterranean elephantids. Here, the longer-term processes resulted in a return to isometry of tooth and bone size, adaptive modification of limb morphology, and reduction of tooth plates in the elephants. At this stage, also, speciation may have occurred, illustrated by the deer antlers.

References

Accordi, F. S. & Palombo, M. R. (1971). Morfologia endocranica degli elefanti nani pleistocenici di Spinagallo (Siracusa) e comparazione con l'endocranio di *Elephas antiquus*. *Atti Accad. naz. Lincei, Rc.* **51**: 111–124.

Ambrosetti, P. (1968). The Pleistocene dwarf elephants of Spinagallo (Siracusa, south-eastern Sicily). *Geologica romana* 7: 277–398.

Beden, M. (1979). *Les éléphants (*Loxodonta *et* Elephas*) d'Afrique orientale: systématique, phylogénie, intérêt biochronologique.* Thesis no. 294: Faculty of Sciences, University of Poitiers.

Bonner, J. T. (1968). Size change in development and evolution. *J. Paleont.* **42** (5) (Suppl.): 1–15.

Bouchud, J. (1978). Les grands herbivores rissiens de l'Aven II des Ambîmes de la Fage à Noailles (Corrèze). *Nouv. Arch. Mus. Hist. nat. Lyon* No. 16: 9–39.

Caloi, L., Kotsakis, T., Palombo, M. R., & Petronio, C. (In press). The Pleistocene dwarf elephants of Mediterranean islands. In *The Proboscidea: trends in evolution and paleoecology*. (Eds Shoshani, J. & Tassy, P.). Oxford University Press, Oxford.

Case, T. J. (1978). A general explanation for insular body size trends in terrestrial vertebrates. *Ecology* **59**: 1–18.

de Vos, J. (1979). The endemic Pleistocene deer of Crete. *Proc. K. Ned. Akad. Wet. Ser. B* **82**: 59–90.

de Vos, J. (1984). The endemic Pleistocene deer of Crete. *Verh. K. Ned. Akad. Wet. Afd. Natuurk., Eerste Reeks* **31**: 1–100.

Fortelius, M. (1985). Ungulate cheek teeth: developmental, functional, and evolutionary interrelations. *Acta zool. fenn.* No. 180: 1–76.

Foster, J. B. (1964). Evolution of mammals on islands. *Nature, Lond.* **202**: 234–235.

Gould, S. J. (1975). On the scaling of tooth size in mammals. *Am. Zool.* **15**: 351–362.

Gould, S. J. (1977). *Ontogeny and phylogeny*. Belknap Press, Harvard.

Heaney, L. R. (1978). Island area and body size of insular mammals: evidence from the tri-colored squirrel (*Callosciurus prevosti*) of Southeast Asia. *Evolution* **32**: 29–44.

Kahlke, H.-D. (1965). Die Cerviden-Reste aus den Tonen von Voigstedt in Thüringen. *Paläont. Abh. Berl.* (A) **2** (2–3): 380–426.

Laws, R. M., Parker, I. S. C., & Johnstone, R. C. B. (1975). *Elephants and their habitats. The ecology of elephants in North Bunyoro, Uganda*. Clarendon Press, Oxford.

Lister, A. M. (1989). Rapid dwarfing of red deer on Jersey in the Last Interglacial. *Nature, Lond.* **342**: 539–542.

Lister, A. M. (1993a). Patterns of evolution in Quaternary mammal lineages. *Linn. Soc. Symp. Ser.* **14**: 71–93.

Lister, A. M. (1993b). Mammoths in miniature. *Nature, Lond.* **362**: 288–289.

Lister, A. M. (1993c). 'Gradualistic' evolution: its interpretation in Quaternary large mammal species. *Quaternary Int.* **19**: 77–84.

Lister, A. M. (1995). Sea levels and the evolution of island endemics: the dwarf red deer of Jersey. *Geol. Soc. Spec. Publ.* **96**: 151–172.

Lister, A. M. & Joysey, K. A. (1992). Scaling effects in elephant dental evolution—the example of Eurasian *Mammuthus*. In *Structure, function and evolution of teeth*: 185–213. (Eds Smith, P. & Tchernov, E.). Freund Publishing House Ltd., London & Tel Aviv.

Lomolino, M. V. (1985). Body size of mammals on islands: the island rule re-examined. *Am. Nat.* **125**: 310–316.

Madden, C. T. (1981). *Mammoths of North America.* PhD dissertation: University of Colorado at Boulder.

Maglio, V. J. (1973). Origin and evolution of the Elephantidae. *Trans. Am. Phil. Soc.* **63**: 1–149.

Mattioli, S. (1993). Antler conformation in red deer of the Mesola Wood, northern Italy. *Acta theriol.* **38**: 443–450.

Orr, P. C. (1967). Geochronology of Santa Rosa island, California. In *Proceedings of the symposium on the biology of the California Islands*: 317–325. (Ed. Philbrick, R.). Santa Barbara Botanic Garden, Santa Barbara, CA.

Roth, V. L. (1982). *Dwarf mammoths from the Santa Barbara, California Channel Islands: size, shape, development, and evolution.* PhD dissertation: Yale University.

Roth, V. L. (1990). Insular dwarf elephants: a case study in body mass estimation and ecological inference. In *Body size in mammalian paleobiology: estimation and biological implications*: 151–179. (Eds Damuth, J. & MacFadden, B.J.). Cambridge University Press, Cambridge.

Roth, V. L. (1992). Inferences from allometry and fossils: dwarfing of elephants on islands. *Oxford Surv. Evol. Biol.* **8**: 259–288.

Roth, V. L. (1993). Dwarfism and variability in the Santa Rosa Island mammoth (*Mammuthus exilis*): an interspecific comparison of limb-bone sizes and shapes in elephants. In *Third California Islands symposium*: 433–442. (Ed. Hochberg, F. G.). Santa Barbara Museum of Natural History, Santa Barbara, CA.

Scott, K. M. (1987). Allometry and habitat-related adaptations in the postcranial skeleton of Cervidae. In *Biology and management of the Cervidae*: 69–80. (Ed. Wemmer, C. M.). Smithsonian Institution Press, Washington, DC.

Sondaar, P. Y. (1977). Insularity and its effect on mammalian evolution. In *Major patterns in vertebrate evolution*: 671–707. (Eds Hecht, M. K., Goody, P. C., & Hecht, B. M.). Plenum Press, New York.

Thaler, L. (1973). Nanisme et gigantisme insulaires. *Recherche* **4**: 741–750.

Vartanyan, S. L., Garutt, V. E. & Sher, A. V. (1993). Holocene dwarf mammoths from Wrangel Island in the Siberian Arctic. *Nature, Lond.* **362**: 337–340.

Waddington, C. H. (1942). Canalization of development and the inheritance of acquired characters. *Nature, Lond.* **150**: 563–565.

Zeuner, F. E. (1946). *Cervus elaphus jerseyensis*, and other fauna in the 25-ft. beach of Belle Hougue Cave, Jersey, C.I. *Bull. a. Soc. Jerseiaise* **14**: 238–254.

Symp. zool. Soc. Lond. (1996) No. 69: 293–308

The psychology of human–small vertebrate interactions

ANNE McBRIDE

Anthrozoology Institute
Southampton University
Bassett Crescent East
Southampton SO16 7PX, UK

Synopsis

Humans develop a variety of relationships with animals. At one end of the range, where the animal serves a special psychological function for the human, strong attachment bonds may develop. Contrastingly, the human may simply interact with the animal, treating it as an object devoid of individuality *per se*. In this latter case the relationship is between the human and some distinct aspect of the species or breed which is divorced from the individual animal's character. The animal provides a means by which humans can make a statement about themselves or provides some form of intellectual stimulation and/or social facilitation.

It is generally accepted that people readily form attachment bonds with animals such as cats and dogs. Similar bonds can also be formed with small vertebrates by owners of all ages. Humans develop attachments with animals for a variety of reasons which reflect the owner's own personality and attitudes. In all cases the bond is formed because the animal satisfies some part of the human intraspecific attachment system, most commonly reflecting aspects of human–infant attachment.

Animals which form reciprocal attachments to humans tend to have substantial overlap of communication channels, such as visual signalling. The more similar the communication system, the greater the likelihood that the participants will perceive each other as a conspecific or symbiont. Small vertebrates have little overlap in communication channels with humans, relying largely on olfaction and ultrasound. In addition, they are primarily prey species. It follows that many of these animals will not naturally perceive humans as benign aspects of their environment. Such animals may never become completely habituated to the presence of humans in general, though they may recognize individuals. This may have important consequences for the behaviour they display, part, but not all, of which may be obvious as anti-predator tactics.

Introduction

Anthrozoology, the study of human–animal interactions, has been recognized as a formal interdisciplinary area of research for only just over a decade. Prior to this,

ZOOLOGICAL SYMPOSIUM No. 69
ISBN 0–19–857787–7

research was conducted in a piecemeal fashion by psychologists, ethologists, zoologists, and the medical fraternity, each working more or less in isolation within the framework of their own disciplines.

Anthrozoologists investigate both intra- (non-human) and interspecific (human/non-human) interactions. The species most intensely studied to date have been the cat (*Felis sylvestris catus*) and the dog (*Canis familiaris*), reflecting both the interest provided by these species to a variety of academic disciplines and the comparative ease with which they can be studied. Until recently these companion animals have been largely ignored by zoologists. Surprisingly little is known about their basic biology and ethology or about how their behaviour and, in particular, communication have been affected by artificial selection for breed characteristics. The study of cat and dog ethology also has implications for the welfare of these animals. Both cats and dogs are species whose population is increasing as ownership of companion animals becomes ever more popular (Gee & Veevers 1984a). This has implications for the management of our urban societies, both socially and from the zoonosis perspective (S. Robertson pers. comm.). A further area of research interest is concerned with the effects, on both human and non-human participants, of the various relationships people have with companion animals. Whilst, to date, the emphasis has been on the cat and dog, the interactions between humans and other species, both domestic and otherwise, are of interest to many anthrozoologists and no doubt will increasingly become a focus for research. Communications presented at the *Sixth International Conference on Human–Animal Interactions* at Montreal in 1992 by M. M. Baun, F. Cardiello & J. Jassen, C. D. Bryant, A. H. Fine, P. Isaac, A. Mead, S. Robertson, I. Slotkin, B. A. Smith, S. Swanson, J. Veevers, and M. Zemanek — some of which are cited elsewhere in this paper — demonstrate the growing interest in this area.

This paper will consider the interactions between humans and small vertebrates initially from the human perspective and then from that of the animal participant.

Human–animal relationships

Humans develop a variety of relationships with animals. At one end of the spectrum, humans can make substantial emotional investment in relationships with individual animals. At the other, the human treats the animal as an object devoid of any individuality *per se*. In the former instance, anthrozoologists talk of the establishment of a human–animal bond. There is a subtle but definite distinction between human–animal interactions and human–animal bonds. The term 'bond' is appropriate when some degree of social attachment (Estep & Hetts 1992) has been formed between animal and human. This may be either mutual or unidirectional. In the latter case the animal may be attached to an indifferent human or the human may have formed an unreciprocated attachment to the animal.

Human–animal bonds

Attachment has been defined as 'a special affectional relationship between two individuals that is specific in its focus and endures over time' (Gubernick 1981). It is assumed within this that the attachment object 'serves a special psychological function for which others cannot substitute and that he elicits affective and social responses that differ from those elicited by other figures' (Cohen 1974).

The function of attachment

Attachment serves to keep individuals together; its ultimate function (in the biological sense) is to enhance the fitness of one or both of the participants. The most frequently studied attachment is that between parent and offspring (Ainsworth 1973; Bowlby 1973; Klaus & Kennell 1976; Gubernick 1981). This bond has obvious benefits to the offspring in the exclusive provision of resources such as shelter, warmth, and food. In addition, the security of a strong attachment between parent and infant allows the infant to explore its immediate environs with as little risk as possible, as any distress signals will elicit a response by the attached parent.

The parent or primary caregiver (usually a close relative) benefits from a strong attachment to the infant in terms of inclusive fitness. Selection would favour the animal which cares for its own offspring rather than potentially reducing its fitness by caring for the offspring of others (Bertram 1976).

Parent–infant attachments are not universal, not even among a single group such as the mammals. Whether or not a species has evolved attachment as a part of the parenting system seems to depend on a combination of factors. Amongst others, these include the neonatal stage of maturation, the social organization of the species, and resource availability and distribution. This point will be reiterated later in this paper with regard to the relationships small vertebrates form with humans. Parent–infant attachments in humans are reciprocal and have obviously conferred evolutionary benefits on the species.

Attachments between other individuals, such as peers, are also functional, especially where the group relies on social cohesion in order to procure resources, for example by group hunting in the canids and indeed in early hominids. Attachment behaviour in *Homo sapiens* is a functionally adaptive strategy (Voith 1984). Thus, humans are biologically prepared to form attachments both to other adults and to their own parents and infants.

Assessing attachment

The term attachment, as with other constructs such as territoriality, needs to be used with caution. The existence of an attachment object, and thereby a bond, can be inferred from qualitative differences in behaviour directed towards it rather than towards other objects. In the case of humans, such differential behaviours do not necessarily need to be observed directly but can be appraised through techniques such as questionnaires or interviews. Whichever method is used, it remains true that the intensity of an attachment bond cannot be directly assessed. Rather, all that can be measured is the intensity of attachment behaviours elicited by the attachment object (Ainsworth 1973). Consequently, it is important that circularity is avoided and that the concept of attachment is not used to explain the behaviours from which it was initially inferred. The operational criteria from which an attachment can be inferred are given in Table 1. These apply to both human and non-human species and further discussion of these criteria is provided by Gubernick (1981).

Table 1. Operational criteria for indicating attachment (from Gubernick 1981)

1	Preference for one individual over another. Preferred individual is the presumed attachment figure
2	Seeking and maintaining proximity to presumed attachment figure
3	Response to brief separation from presumed attachment figure
4	Response to extended separation from presumed attachment figure
5	Response to reunion with presumed attachment figure
6	Use of the attachment figure as a secure base from which to explore the environment

In the case of humans, verbal statements can indicate an attachment. Comments such as 'I know it's just an animal, but I feel about it as though it were my child' are regularly expressed in situations where the owner feels they will be accepted, such as to the veterinary surgeon or when the owner is being counselled after the pet has died (Lee & Lee 1992). People also write extensively about their emotional investment in their pet, be it dog or rat. There are innumerable collections of such works, often triggered by the loss of the animal. An unusual example of such a work, unusual in that it produced substantial financial rewards, was the pop-song 'Ben' by Michael Jackson, 'Ben' being Michael's pet rat. Other behaviour, particularly associated with separation from the pet, indicates the strength of the attachment bond. Rituals surrounding the disposal of the corpse are based on those performed for humans. These can range from a simple ceremony of burying the pet in the garden, to the considerable expenditure and cultural significance surrounding individual cremation or burial in a pet cemetery.

The mechanisms of attachment

The formation and maintenance of human intraspecific attachment bonds are facilitated by a variety of factors. Prolonged physical proximity including tactile stimulation, social sign stimuli such as facial expressions and eye contact, the sharing of experiences especially if these are either happy or life-threatening: all serve to create and maintain bonds. Attachments can also be facilitated by investment, such as the effort expended by those responsible for the welfare of a dependant, as in caring for an infant.

Social activities which parents share with their children, such as feeding, playing, sleeping, all serve as attachment mechanisms which reinforce the bond. People share similar activities with their pets, even to the point of dressing up animals to protect them from the weather, for instance with bootees and coats manufactured for dogs.

In addition, pets naturally engage in behaviours which the owner interprets as being indicative of an attachment, of being 'loved by the animal'. Typically, such behaviours include the pet appearing 'sad' when the owner departs, being 'happy' to greet the owner, seeking to initiate play with the owner, displaying attention-seeking behaviours such as demands to be stroked, and maintaining physical proximity to the owner. These behaviours are particularly noticeable in the dog and cat and, combined with the animal's dependency on the human to be groomed and fed, serve to establish and maintain bonds.

Attachment is also enhanced by physical characteristics of the pet. This is most clearly demonstrated in the domestic dog. Much of the artificial selection of dog breeds has served to neotenize them, thus enhancing their suitability as attachment objects for people. This is particularly so in those basically non-utilitarian breeds such as Cavalier King Charles spaniels. This neotenization has affected both physical features and behavioural characteristics such as the retention of play. In terms of physical characteristics, artificial selection has produced breeds with rounded heads, large eyes, and small body size, features which are consistent with signals that normally indicate an immature individual. Such signals are also present in other species of companion animal such as the rabbit, hamster, cavy, and the small cage-birds. Such species are often perceived as 'cuddly', 'friendly', 'cheeky' — attributes also ascribed to human youngsters. Indeed many such animals are portrayed in literature as having not only the qualities of children but partaking in child-like adventures, Peter Rabbit being perhaps the best known of these *enfants terribles*.

Whilst anthropomorphism extends across a wide range of species, it seems to be especially prevalent in reference to dogs. It has been demonstrated (Voith 1984) that significantly more dog owners than cat owners refer to their pet as a person. This suggests that, consciously or unconsciously, people identify more directly with dogs than with cats. Probably this reflects the considerable overlap between the social organization and communication systems of man and dog. In both species social grouping is based on a hierarchical system maintained through extensive visual and vocal signalling. It is parsimonious to expect that the more similar the communication systems of two species, the more likely that each will find the signals from the other salient and be able to respond appropriately. This in turn would increase the likelihood that one or both of the participants will perceive the other as a conspecific.

Nevertheless, important attachment relationships can be formed with pets other than dogs, cats, and horses (Riddick 1985). This is especially so where the species is long lived. Longevity facilitates one of the requirements for attachment formation, namely prolonged physical proximity (Klaus & Kennell 1976). Some of the more obvious small vertebrate species in this category are tortoises and parrots; even rabbits can live for a decade or more. Less obvious are the bonds formed by owners of snakes, spiders, and even slugs (Platts 1991), species which are not perceived as being particularly social or interactive. It is likely that bonds formed with such animals are primarily unidirectional, in that the non-human participant interacts but does not attach.

The roles pets play

Companion animals play important social roles for their owners. This implies that the animal satisfies some pre-existing criteria for the owner. J. Veevers (pers. comm.) has suggested that a companion animal is one which satisfies all or some of the following criteria: (i) non-utilitarian, (ii) aesthetically pleasing, (iii) singular, (iv) sociable, (v) cohabitant, (vi) affectionate, (vii) submissive, and (viii) a possession. Naturally, not all species kept as companion animals, in particular the more exotic species, fulfil all these criteria. However, the more closely a species approximates to this model the more popular and satisfactory it is likely to be as a pet, that is, more likely to be able to fulfil a psychological need for the owner and become an attachment object.

The roles which companion animals play for humans can be categorized according to the major psychological function they perform. Perhaps the most commonly recognized is that of the *surrogate function*, where interaction with the pet supplements, or even substitutes, human–human interaction. The second function of roles played by pets is the *projective function*, where the pet represents a symbolic extension of the owner. Third is the *sociability function* in which the pet facilitates human intraspecific interaction (Veevers 1985).

The surrogate function

For a pet to have a surrogate function its presence must provide the owner with some of the benefits usually derived from roles played by other humans such as friend, child, parent, or partner. Where extreme attachments have been formed, the pet may not merely supplement a human relationship but actually replace it.

The extent to which different species are perceived culturally as surrogates can be loosely assessed by considering the following.

1. Is it acceptable to eat the animal? The use of a species for meat is only acceptable if the society does not anthropomorphize that animal. Once the animal is invested with a social role then eating it becomes akin to anthropophagy (Beck & Katcher 1983). In Britain it is not acceptable to eat horse or dog meat, though the Europeans enjoy the former and gourmands in the Far East the latter. Yet the eating of rabbit is acceptable to many, even if the myxomatosis episode means that shoppers prefer the 'safe' commercially bred rather than wild-caught variety.

2. What names does the animal have? Dogs, cats, and horses are often given human names such as 'Lucy' or 'Sam'. To the author's knowledge, no work has been conducted on the names given to other companion animal species. It may be hypothesized that these species do not have so much of a surrogate function and are not so anthropomorphized. If so, then they would be more likely to be given descriptive rather than human names, such as 'Snowy', 'Hammy', 'Thumper', and 'Squeak'. Similarly, the extent to which humans talk to animals with the expectation that there is some understanding can indicate different levels of anthropomorphism. Beck & Katcher (1983) reported that 30% of their pet-owning group confided in the pet. Likewise, preliminary findings by S. Rubi & J. L. Renck (pers. comm.) indicated that there were similarities between the dialogues humans have between themselves and with their pet cats. Work on rabbit–human interaction (S. Swanson pers. comm.) also suggests a link between patterns of communication and the degree of human bonding, or the degree to which the rabbit plays a surrogate role.

3. A final indicator of the degree of anthropomorphism, or surrogacy, is the extent to which the animal is made the subject of social ceremonies. These include the celebration of the animal's birthday, the purchase of a Christmas present for it, and having a portrait or photograph of the pet on display. Ceremonies to mark the pet's death are also enlightening and have been described. Shirley & Mercier (1983) report that three-quarters of their subjects held a funeral 'ceremony' for their dead pet. Indeed, suicides following the death of a pet are not unknown and indicate that the individual animal probably acted as a replacement for an important human relationship for that owner (Lee & Lee 1992).

Again, the majority of the studies investigating the surrogacy role of a pet have concentrated on the dog and cat. Animals can play the role of a friend, sharing many of the activities also carried out with human companions such as walking and relaxing.

The role of the pet as a companion is especially important to those people who do not have dense social networks, either because of their personality or because of external factors, such as lack of mobility, as is the case with many elderly persons. For many isolated and bereaved elderly the pet can take on a special significance, whether it provides a link with the deceased, as in the case of a dog that accompanied the departed husband on fishing trips, or has been acquired since the death. These animals may become as emotionally important to the surviving partner as was the deceased (Gee & Veevers 1984b).

There have been few systematic studies on the health benefits of pets (Beck & Katcher 1984) and those conducted with non-institutionalized elderly have produced conflicting results (Riddick 1985). Mugford & M'Comisky (1975) compared three groups of British pensioners. The experimental group were given parakeets whilst the control subjects were given either a begonia or no intervention. The results indicated that the experimental group showed improvements in both physical measures of health and in mood compared to the control groups. Riddick (1985) conducted a similar study where the experimental subjects were given an aquarium of goldfish (*Carassius auratus*) and biweekly visits from the researchers. Other subjects were assigned to one of two control groups, one which received biweekly visits from the experimenters and the other which received no visits over the 6 month experimental period. Measures of health included blood pressure, happiness, loneliness and anxiety. These were conducted as pre- and post-test measures and indicated that the aquarium group benefited significantly in terms of diastolic blood pressure, overall leisure satisfaction, and relaxation states. Comments such as 'The fish gave me a reason to get up in the morning', 'We watched television together' and 'They depend on me to feed them' indicate that these elderly people had formed attachment bonds with their new pets.

In the case of children, the role of pet as friend can be extended to involving the animal in role-play activities such as dressing-up and imaginary tea parties. For many children, the pet as friend may also incorporate an element of pet as teacher/parent. Indeed, pet as 'teacher' is an accepted, if not so blatantly stated reason for many parents buying their child a pet of its own. The most popular species are rabbits (*Oryctolagus cuniculus*), cavies (*Cavia porcellus*), and hamsters (*Mesocricetus auratus*). Conventional wisdom asserts that providing the child with such a pet will be of benefit to the child's emotional development (A. H. Fine pers. comm.). The pet can act as a surrogate for other humans and the child can practise a variety of interactions and gain a sense of responsibility, caring, and empathy with the feelings of another individual (Bossard 1953). There is some evidence to support this cultural belief. Paul & Serpell's (1993) study of British university students demonstrated that there was a positive relationship between pet keeping in childhood and humane attitudes in young adulthood. These attitudes extended to both non-pet species and other humans.

The role of a pet as surrogate parent is not commonly considered. However, the importance of the pet as a provider of unconditional love and, more importantly, being constantly available, may be of substantial emotional significance to children (Robinson & McBride 1995). This would be especially true in our current society where both parents are working or are separated. In earlier times this role of a constant care giver would have been played by a nanny employed for the purpose or a grandparent.

Perhaps the most common stereotype of pet as surrogate is of the pet being a surrogate child to childless couples. Yet in reality, the probability of pet ownership is greater in couples with children than without (Gee & Veevers 1984a). With respect to ownership of small vertebrates, usually regarded as children's pets, this probability is likely to be increased.

The projective function

People express their personality in a variety of ways. The clothes they choose, the car they drive, and the type of pet they own all act as public expression of character. For example, the person who chooses to own a breed of dog perceived by society as aggressive may be using that dog as a way of acting out their own hostile feelings towards society (Rosenbaum 1972). Likewise, owners of dangerous pets, such as poisonous snakes, may be making a statement about their status, along the lines that only someone powerful could own such an animal. Similarly, owners of rare exotic species intimate wealth, power, and exclusivity.

For pets to act as projective statements the owner does not need to form any attachment to the animal. Rather, any form of attachment is made with some distinct aspect of the species or breed which is divorced from the individual animal's character. Pets that are owned as projective statements are subject to the whim of both the owner and what society deems fashionable. Examples are the 'use' of chameleons in the 1940s in America as fashion adornments or the ownership of alligators by New Yorkers in the 1970s, and alleged subsequent disposal of the same into the sewage system!

Pets as social facilitators

Animals can serve to change the way people interact with each other. Animals, whatever the species, attract the attention of people. Hunt, Hart & Gomulkiewicz (in press) investigated the number and type of social approaches made to a woman sitting in a suburban park in California under four conditions: sitting with a pet rabbit or turtle, watching a television, or blowing bubbles. Social approaches made by adult passers-by were most frequent when the rabbit was present, though, if the adults were accompanied by children, approaches were also frequent when the woman was blowing bubbles. A substantial number of approaches were also made when the turtle was present and virtually no approaches were made when the woman was watching television.

Animals act as social lubricants by providing a neutral topic of conversation (Riddick 1985) and/or by providing a network of social connections to the owner (C. D. Bryant pers. comm.). This is especially so where the owner develops a specific interest, such as showing or scientific research, in a particular type of animal. Many companion animal species, from dogs to rats (*Rattus norvegicus*), have 'fancier' clubs which hold competitive shows. These provide the owners with considerable opportunities for social interaction. Such social networks can become pervasive where the person's business endeavours, such as commercial breeding or research interests, mean that their friendship networks and business contacts become merged.

For animals to serve as social facilitators does not require any bonding on the part of the human. While it may be that bonds are formed with a few individual animals, the high turnover of stock both in the laboratory and between owners of 'fancy' breeds, including the disposal of stock 'unsuitable for show' suggests that such bonding is uncommon, especially amongst larger breeders where exhibition is the chief reason for keeping the animals (McKay 1991).

Animals can also be kept to deter social interaction with other people. People who own an excessive number of animals, be they cats, rats, or canaries (*Carduelis canaria*), may become 'too busy' to be involved in activities with other humans (Veevers 1985; I. Slotkin pers. comm.). The deterrent factor may be even more blatant, as in the case of the man whose reason for keeping snakes was to prevent his relatives from visiting (Rosenbaum 1972; M. Zemanek pers. comm.). While in these cases the human may form an attachment, it is not necessary in order for the animal to fulfil the emotional need of the owner.

Animals as relaxants

Friedmann & Thomas (1985) have reviewed several studies on the beneficial effects on blood pressure of the ownership of cats and dogs and of the presence of familiar and unfamiliar dogs. In a series of studies undertaken by Katcher and his colleagues (Katcher, Friedmann, Beck, & Lynch 1983; Katcher, Segal, & Beck 1984) it was shown that fish also had the effect of lowering blood pressure. Decreases in blood pressure were found to be sustained longer in people watching fish swimming in an aquarium than in people looking at a poster or at identical tanks which did not contain fish. It was also found that watching fish prior to dental surgery was as effective as hypnosis in increasing patient compliance and decreasing perceptions of pain. Indeed, the beneficial effect of watching fish plus being hypnotized was no greater than that of just watching the fish.

To the author's knowledge there is no research on the relaxation effects of the presence of small mammals in stressful situations.

Small vertebrate–human interactions

The majority of small vertebrates kept by humans are rodents, kept as either companion or laboratory animals. The discussion of small vertebrate–human interactions will concentrate on this group.

Do rodents form attachments?

The primary intraspecific attachment within *H. sapiens* is between parent and infant. This seems to form the basis of most human attachments to animals, where the animal plays the dependent role. In order for such attachment bonds to be mutual the animal must be able to both form attachments and recognize the human as a suitable attachment figure.

The evidence as to whether or not rodents form intraspecific attachments is scanty and studies have been limited to only a few species. The rat, mouse (*Mus musculus*), and spiny mouse (*Acomys cahirinus*) are popular as both laboratory and companion animal species, though, of the three, the spiny mouse is the least common as a pet. Studies of these species (Leon & Moltz 1971, 1972; Sayler & Salmon 1971; Breen & Leshner 1977; Porter & Doane 1978; Gubernick 1981) suggest that intraspecific parent–infant attachment bonds are absent, a maternal pheromone serving to keep infants close to the nest. The attractant properties of these pheromones (as detected by attraction to soiled bedding) disappear around the age of weaning.

Given that rats and house mice give birth to altricial young in nests which are located away from other conspecifics, there is little need for either parent or young to recognize individuals so long as they can recognize the nest. The parents are unlikely to encounter other young and the young are unlikely to meet other adults until they emerge from the nest around the time of weaning. This also explains why these animals tend to nurse and show maternal behaviour to the young of other litters and even of other species.

The mechanism to ensure that the young remain near the nest appears to be the same in the rat, mouse, and spiny mouse. This is the recognition of and attraction to a maternal pheromone, which appears to be produced in the caecum of lactating females and emitted with the faeces (Leon 1974). The spiny mouse is different from the mouse and rat in that the young are precocial, uniquely amongst the murid rodents. Whilst this has affected the timing of when the recognition of the odour develops, the mechanism remains the same. In the spiny mouse, imprinting occurs between 2 and 12 h post-partum (Porter & Etscorn 1974). For comparison, the altricial rat does not show a preference for soiled bedding associated with its dam until approximately 12 days post-partum (Leon & Moltz 1971).

Guinea-pigs (*Cavia*) are a precocial species which, under natural conditions, have early contact with other conspecifics as they feed together in groups (Rood 1972). Wild female guinea-pigs (*Cavia aperea*) are highly aggressive to each other and nurse their own young exclusively. Under these conditions it would be expected that mother–infant attachments should form. However, domestic female guinea-pigs (*Cavia porcellus*) show minimal intrasexual aggression and will nurse young other than their own. Studies of attachment in these animals suggest that, as with the murid species, the young recognize odours associated with their environment during the first 3 days post-partum, but there is no convincing evidence for attachment (Carter & Marr 1970; Fullerton, Berryman & Porter 1974). Female domestic guinea-pigs do seem to recognize and form a loose attachment to their young (Fullerton *et al.* 1974).

Thus, it would appear that the small mammals commonly kept in laboratories and as pets are not biologically prepared to form attachments. In addition, the differences in size, in modes of communication, and in reliance on sensory systems between humans and small mammals would indicate that these animals are unlikely to regard humans as conspecifics.

Small vertebrate perceptions of humans

Small vertebrates perceive humans probably in one of two ways, either as a part of the inanimate environment or as a predator (Hediger 1969). Perception of people as a component of the inanimate environment is most likely when the animal has become habituated to an individual handler and recognizes scent and other sensory cues from the latter (H. Davis & C. Norris pers. comm.). The size of humans compared to small vertebrates makes it likely that they do not perceive us as an entity, but rather as several different aspects of the environment. Some of these aspects may well represent the presence of a predator.

For prey species, vigilance and the evasion of potential predators are constant conditions which pervade their daily life and activities. It follows that many of these animals will not naturally perceive humans as benign aspects of their environment. For example, to a hamster, being picked up from above is reminiscent of the swooping of a predator and is likely to initiate the anti-predator tactic of turning on its back and biting. If such treatment is repeated it can lead to increased aggression, as the animal becomes more rather than less fearful. Likewise, picking up a mouse by its tail, by hand, or with the use of forceps, will elicit stress vocalizations. This method is almost certain to produce an animal which is wary of being handled, since it requires that the whole of the animal's body weight is supported by a part of the body not designed for load bearing. This will cause pain, even if only for an instant.

Biting, struggling, and vocalization are behaviour patterns which are easily recognized as anti-predator tactics. These can be reduced somewhat by habituating young animals to being gently handled and/or training the animal to perceive handling as a positive experience. Such training can be achieved by placing small bits of food on the palm of the hand and allowing the animal to climb on to the hand. Initially, this should be done within the safety of the cage, only later gently holding the animal when it is on the hand and lifting it out of the cage. However, the time involved for such a procedure is often not available in a laboratory situation.

The perception of the environment as hostile can also produce more subtle changes in behaviour such as increased vigilance or a reduction in time spent grooming or feeding (Caine 1992; Mason & Mendl 1993). Physiological changes are also associated with both acute and chronic stress (Mason & Mendl 1993). Even animals considered to be habituated to observation or experimental techniques may be displaying subtle defensive tactics which compromise the validity of data. Even tenuous evidence of a 'predator', such as olfactory cues remaining from unfamiliar individuals having handled the home cage or experimental apparatus, could be sufficient to elicit reactions in the subject (Barclay, Herbert, & Poole 1988; Caine 1992).

Chronic stressors such as ambient ultrasound levels in the laboratory, emissions from cage washers, ringing telephones, and pressure hoses, can have important implications for both welfare and data validity (Sales, Evans, Milligan, & Langridge 1988). Such stressors may also have effects on less studied vertebrates such as aquarium fish (Barrie 1990).

The majority of small vertebrates kept as laboratory subjects or as pets spend their entire lives in the same environment — the confines of their cage. In contrast, the natural environment, aquatic or terrestrial, is one of constant and varied stimulation — albeit dominated by the need to be vigilant for predators. In such an environment there are also consistent factors such as familiar olfactory and visual stimuli, a point to remember when animals are moved from one home cage to another. Relocating some dirty bedding as well provides a familiar element.

For all species, and especially the opportunistic rodents, the sterile conditions of cage life pose questions concerning the psychological welfare of these animals and the validity of behavioural data. Animals maintained in suboptimal environments may develop stereotypies. Stereotypies can indicate an environment which is restrictive, for example a lack of tunnelling opportunities. Alternatively, stereotypies may be indicative of a source of unavoidable stress or fear.

For pet animals a common means of providing stimulation is an exercise wheel. Wheel running has been perceived as a positive activity but evidence suggests that it can be a stereotyped behaviour resulting from social and/or environmental deprivation (de Kock & Rohn 1971; Roper & Polioudakis 1977). Significant differences in wheel running were found between solitary gerbils (*Meriones unguiculatus*) maintained in normal laboratory conditions and those kept in a semi-natural environment with conspecifics and the opportunity to burrow, forage, gnaw, and nest build as well as wheel run. The isolated animals tended to be nocturnal rather than to follow the more diurnal activity pattern shown by the social group, a pattern resembling that of their wild counterparts. In terms of wheel running, the solitary animals ran at speeds of up to 2000 revolutions per hour, exceeding 20 000 revolutions per day. They ran for hours at a time, stopping for only brief pauses. In contrast, animals in the social group rarely stayed in the wheel for more than a few revolutions at a time, never reaching speeds above 50 revolutions per hour. This suggests that, for this species, wheel running is a stereotyped activity indicating a suboptimal environment (Roper & Polioudakis 1977).

Wheel-running in bank voles (*Clethrionomys glareolus*) and the Norway lemming (*Lemmus lemmus*) has been investigated by de Kock & Rohn (1971). Again, they found that wheel running was part of a stereotyped behaviour pattern, with distances covered by a solitary lemming in a small cage exceeding distances normally covered in the wild. Perhaps of more interest was the relationship found between wheel running and social rank, age, and sexual status in animals kept in semi-natural environments. In both species, the wheel was not used purely for exercise. Other motivations suggested were the desire to escape, the desire to wander, and the desire to cover a distance. Wheel use appeared to be connected to hormonal events and group structure.

The effects of human interaction on experimental animals

Studies such as those described above have important implications for the study of small vertebrates. If a behaviour traditionally perceived as neutral or positive, such as wheel running, can be changed so radically by social group or hormonal status, for instance, then how much more likely is it that the presence of a potential predator, or trace thereof, will influence behaviour? Some of these changes may be subtle and unrecognized as anti-predator tactics, as in the case of behaviours relating to surveillance (Caine 1992).

As scientists, we attempt to be as objective as we can and to merely interact, rather than to bond, with our subjects (Lehman 1992). With regard to small vertebrates, it is likely that they perceive human observers as negative, or (if very familiar with the handler) at best neutral aspects of the environment. Such an interaction is likely to have consequences for data integrity. Even in the field, evidence of the researcher's presence, such as lingering olfactory cues, may be sufficient to elicit changes in behaviour of the study animals. Every effort should therefore be made to reduce potential sources of such bias, and it should be recognized that many of the species we study will perceive us as threatening and never fully habituate to our presence.

Acknowledgements

My thanks to my colleagues at the Anthrozoology Institute for all their support and, in particular, to Dr John Bradshaw whose comments and criticisms have greatly benefited the readers of this paper.

References

Ainsworth, M. D. S. (1973). The development of infant–mother attachment. In *Review of child development research* **3**: 1–94. (Eds Caldwell, B. M. & Ricciuti, H. N.). University of Chicago Press, Chicago.

Barclay, R. J., Herbert, W. J., & Poole, T. B. (1988). Effects of routine procedures on the behaviour of laboratory rats and mice. In *Laboratory animal welfare research: rodents*: 62–63. Universities Federation for Animal Welfare, Potters Bar.

Barrie, A. (1990). *Goldfish as a new pet*. TFH Publications Ltd, New Jersey.

Beck, A. & Katcher, A. (1983). *Between pets and people: the importance of animal companionship*. Putnam's Sons, New York.

Beck, A. & Katcher, A. (1984). A new look at pet-facilitated therapy. *J. Am. vet. med. Ass.* **184**: 414–421.

Bertram, B. C. R. (1976). Kin selection in lions and in evolution. In *Growing points in ethology*: 281–301. (Eds Bateson, P.P.G. & Hinde, R.A.). Cambridge University Press, Cambridge.

Bossard, J. H. (1953). Domestic animals: their role in family life and child development. In *Parent and child: studies in family behaviour*: 236–252. (Ed. Bossard, J.). University of Pennsylvania Press, Philadelphia.

Bowlby, J. (1973). *Separation, attachment and loss* **2**. Basic Books, New York.

Breen, M. F. & Leshner, A. I. (1977). Maternal pheromone: a demonstration of its existence in the mouse (*Mus musculus*). *Physiol. Behav.* **18**: 527–529.

Caine, N. G. (1992). Humans as predators: observational studies and the risk of pseudohabituation. In *The inevitable bond*: 357–364. (Eds Davis, H. & Balfour, D.). Cambridge University Press, Cambridge.

Carter, C. S. & Marr, J. N. (1970). Olfactory imprinting and age variables in the guinea-pig, *Cavia porcellus*. *Anim. Behav.* **18**: 238–244.

Cohen, L. J. (1974). The operational definition of human attachment. *Psychol. Bull.* **81**: 207–217.

de Kock, L. L. & Rohn, I. (1971). Observations on the use of the exercise-wheel in relation to the social rank and hormonal conditions in the bank vole (*Clethrionomys glareolus*) and the Norway lemming (*Lemmus lemmus*). *Z. Tierpsychol.* **29**: 180–195.

Estep, D. Q. & Hetts, S. (1992). Interactions, relationships and bonds: the conceptual basis for scientist–animal relations. In *The inevitable bond*: 6–26. (Eds Davis, H. & Balfour, D.). Cambridge University Press, Cambridge.

Friedmann, E. & Thomas, S. A. (1985). Health benefits of pets for families. In *Pets and the family*. (Ed. Sussman, M.). *Marriage Fam. Rev.* **8**: 191–203.

Fullerton, D., Berryman, J. C., & Porter, R. H. (1974). On the nature of mother–infant interactions in the guinea-pig (*Cavia porcellus*). *Behaviour* **48**: 145–156.

Gee, E. M. & Veevers, J. E. (1984a). *Everyman and his dog: the demography of pet ownership*. Department of Sociology, University of Victoria, British Columbia, Canada. (Mimeographed.)

Gee, E. M. & Veevers, J. E. (1984b). *The pet prescription: assessing the therapeutic value of pets for the elderly*. Department of Sociology, University of Victoria, British Columbia, Canada. (Mimeographed.)

Gubernick, D. J. (1981). Parent and infant attachment in mammals. In *Parental care in mammals*: 243–305. (Eds Gubernick, D. J. & Klopfer, P. H.). Plenum Press, New York.

Hediger, H. (1969). *Man and animal in the zoo: zoo biology*. Seymour Lawrence/Delacorte Press, New York.

Hunt, S. J., Hart, L. A., & Gomulkiewicz, R. (In press). The role of small animals in social interactions between strangers. *J. soc. Psychol.*

Katcher, A. H., Friedmann, E., Beck, A. M., & Lynch, J. J. (1983). Talking, looking and blood pressure: physiological consequences of interaction with the environment. In *New perspectives on our lives with animal companions*: 351–359. (Eds Katcher, A. H. & Beck, A. M.). University of Pennsylvania Press, Philadelphia.

Katcher, A. H., Segal, H., & Beck, A. M. (1984). Contemplation of an aquarium for the reduction of anxiety. In *The pet connection: its influence on our health and quality of life*: 171–178. (Eds Anderson, R. K., Hart, B. L. & Hart, L. A.). Globe Publishing Co., South St Paul, Minneapolis.

Klaus, M. H. & Kennell, J. H. (Eds) (1976). *Maternal–infant bonding*. Mosby, St Louis, Mo.

Lee, L. & Lee, M. (1992). *Absent friend*. Henston Press, London.

Lehman, H. (1992). Scientist–animal bonding: some philosophical reflections. In *The inevitable bond*: 383–396. (Eds Davis, H. & Balfour, D.). Cambridge University Press, Cambridge.

Leon, M. (1974). Maternal pheromone. *Physiol. Behav.* **13**: 441–453.

Leon, M. & Moltz, H. (1971). Maternal pheromone: discrimination by pre-weanling albino rats. *Physiol. Behav.* 7: 265–267.

Leon, M. & Moltz, H. (1972). The development of the pheromonal bond in the albino rat. *Physiol. Behav.* **8**: 683–686.

McKay, J. (1991). *The new hamster handbook*. Blandford Press, London.

Mason, G. & Mendl, M. (1993). Why is there no simple way of measuring animal welfare? *Anim. Welfare* **2**: 301–319.

Mugford, R. & M'Comisky, J. (1975). Some recent work on the psychotherapeutic value of cage birds with old people. In *Pet animals and society*: 54–65. (Ed. Anderson, R.). Ballière Tindall, London.

Paul, E. S. & Serpell, J. A. (1993). Childhood pet-keeping and human attitudes in young adulthood. *Anim. Welfare* **2**: 321–337.

Platts, E. (1991). *Out of the slime*. *BBC Wildl.* **9**: 420–423.

Porter, R. H. & Doane, H. M. (1978). Studies of maternal behaviour in spiny mice (*Acomys cahirinus*). *Z. Tierpsychol.* **47**: 225–235.

Porter, R. H. & Etscorn, F. (1974). Olfactory imprinting resulting from brief exposure in *Acomys cahirinus. Nature, Lond.* **250**: 732–733.

Riddick, C. C. (1985). Health, aquariums and the non-institutionalised elderly. In *Pets and the family*. (Ed. Sussman, M.). *Marriage Fam. Rev.* **8**: 163–173.

Robinson, I. H. & McBride, E. A. (1995). Relationships with other pets. In *The Waltham book of human–animal interactions*. (Ed. Robinson, I. H.). Pergamon Press.

Rood, J. P. (1972). Ecological and behavioural comparisons of three genera of Argentine cavies. *Anim. Behav. Monogr.* **5**: 1–83.

Roper, T. J. & Polioudakis, E. (1977). The behaviour of Mongolian gerbils in a semi-natural environment, with special reference to ventral marking, dominance and sociability. *Behaviour* **61**: 207–237.

Rosenbaum, J. (1972). *Is your Volkswagen a sex symbol?* Bantam, New York.

Sales, G., Evans, J., Milligan, S., & Langridge, A. (1988). Effects of environmental ultrasound on the behaviour of laboratory rates. In *Laboratory animal welfare research: rodents*: 7–16. Universities Federation for Animal Welfare, Potters Bar.

Sayler, A. & Salmon, M. (1971). An ethological analysis of communal nursing by the house mouse (*Mus musculus*). *Behaviour* **40**: 60–85.

Shirley, V. & Mercier, J. (1983). Bereavement of older persons: death of a pet. *Gerontologist* **23**: 276.

Veevers, J. E. (1985). The social meaning of pets. In *Pets and the family*. (Ed. Sussman, M.). *Marriage Fam. Rev.* **8**: 11–30.

Voith, V. L. (1984). Human/animal relationships. In *The pet connection: its influence on our health and quality of life*: 147–156. (Eds Anderson, R. K., Hart, B. L., & Hart, L. A.). Globe Publishing Co., South St Paul, Minneapolis.

Index

Note: page numbers in *italics* refer to figures and tables